AF567074

HANDBOOK OF GREEN MATERIALS

❷ Bionanocomposites: processing, characterization and properties

World Scientific Series in Materials and Energy* **ISSN: 2335-6596**

Series Editor: Leonard C. Feldman *(Rutgers University)*

Published

Vol. 1 Handbook of Instrumentation and Techniques for Semiconductor Nanostructure Characterization
edited by Richard Haight (IBM TJ Watson Research Center, USA), Frances M. Ross (IBM TJ Watson Research Center, USA), James B. Hannon (IBM TJ Watson Research Center, USA) and Leonard C. Feldman (Rutgers University, USA)

Vol. 2 Handbook of Instrumentation and Techniques for Semiconductor Nanostructure Characterization
edited by Richard Haight (IBM TJ Watson Research Center, USA), Frances M. Ross (IBM TJ Watson Research Center, USA), James B. Hannon (IBM TJ Watson Research Center, USA) and Leonard C. Feldman (Rutgers University, USA)

Vol. 3 The World Scientific Handbook of Energy
edited by Gerard M. Crawley (University of South Carolina, USA)

Vol. 4 Materials for Biofuels
edited by Arthur J. Ragauskas (Georgia Institute of Technology, USA)

Vol. 5 Handbook of Green Materials: Processing Technologies, Properties and Applications (In 4 Volumes)
edited by Kristiina Oksman (Luleå University of Technology, Sweden), Aji P. Mathew (Luleå University of Technology, Sweden), Alexander Bismarck (Vienna University of Technology, Austria), Oriando Rojas (North Carolina State University, USA) and Mohini Sain (University of Toronto, Canada)

*The complete list of titles in the series can be found at
http://www.worldscientific.com/series/mae

MATERIALS AND ENERGY – Vol. 5

HANDBOOK OF GREEN MATERIALS

❷ Bionanocomposites: processing, characterization and properties

Editors

Kristiina Oksman and **Aji P Mathew**
Luleå University of Technology, Sweden

Alexander Bismarck
Vienna University of Technology, Austria

Orlando Rojas
North Carolina State University, USA

Mohini Sain
University of Toronto, Canada

NEW JERSEY • LONDON • SINGAPORE • BEIJING • SHANGHAI • HONG KONG • TAIPEI • CHENNAI

Published by

World Scientific Publishing Co. Pte. Ltd.
5 Toh Tuck Link, Singapore 596224
USA office: 27 Warren Street, Suite 401-402, Hackensack, NJ 07601
UK office: 57 Shelton Street, Covent Garden, London WC2H 9HE

British Library Cataloguing-in-Publication Data
A catalogue record for this book is available from the British Library.

World Scientific Series in Materials and Energy
HANDBOOK OF GREEN MATERIALS
Processing Technologies, Properties and Applications
(In 4 Volumes)

ISBN 978-981-4566-45-2 (Set)
ISBN 978-981-4566-48-3 (Vol. 1)
ISBN 978-981-4566-49-0 (Vol. 2)
ISBN 978-981-4566-50-6 (Vol. 3)
ISBN 978-981-4566-51-3 (Vol. 4)

Typeset by Stallion Press
Email: enquiries@stallionpress.com

Printed by FuIsland Offset Printing (S) Pte Ltd Singapore

Contents

List of Corresponding Authors

Volume 2

Chapter 1 Alexander Bismarck[1] and Kristiina Oksman[2]
[1] *Polymer and Composite Engineering (PaCE) group, Institute of Materials Chemistry and Research, Faculty of Chemistry, University of Vienna, Waehringerstr. 42, 1090 Vienna & Polymer and Composite Engineering (PaCE) group, Department of Chemical Engineering, Imperial College London, SW7 2AZ London, UK*
Email: a.bismarck@imperial.ac.uk
[2] *Composite Center Sweden, Division of Materials Science, Luleå University of Technology, Luleå, SE-97187 Sweden*
Email: kristiina.oksman@ltu.se

Chapter 2 Alexander Bismarck
Polymer and Composite Engineering (PaCE) group, Institute of Materials Chemistry and Research, Faculty of Chemistry, University of Vienna, Waehringerstr. 42, 1090 Vienna and Polymer and Composite Engineering (PaCE) group, Department of Chemical Engineering, Imperial College London, SW7 2AZ London, UK
Email: a.bismarck@imperial.ac.uk

Chapter 3 Christoph Weder
Adolphe Merkle Institute and Fribourg Center for Nanomaterials, University of Fribourg, Rte de l'Ancienne Papeterie, CH-1723 Marly, Switzerland
Phone: +41 26300 94 65
Email: christoph.weder@unifr.ch
www.am-institute.ch

Chapter 4 Aji P Mathew
Composite Center Sweden, Division of Materials Science, Luleå University of Technology, Luleå, SE-97187 Sweden
Phone: +46 920 493336
Email: aji.mathew@ltu.se

Chapter 5 Kristiina Oksman
Composite Center Sweden, Division of Materials Science, Luleå University of Technology, Luleå, SE-97187 Sweden
Phone: +46 920 493371
Email: kristiina.oksman@ltu.se

Chapter 6 Youssef Habibi
Center of Innovation and Research in Materials and Polymers (CIRMAP) University of Mons - UMONS, Place du Parc 23, 7000 Mons, Belgium
Email: Youssef.Habibi@umons.ac.be

Chapter 7 Kristiina Oksman
Composite Center Sweden, Division of Materials Science, Luleå University of Technology, Luleå, SE-97187 Sweden
Phone: +46 920 493371
Email: kristiina.oksman@ltu.se

Chapter 8 Stephen J. Eichhorn
College of Engineering, Maths & Physical Sciences, University of Exeter, Exter, EX4 4QL, UK
Email: S.J.Eichhorn@exeter.ac.uk

Chapter 9 Lars Berglund
Wallenberg Wood Science Center, and Dept of Fiber and Polymer Technology Royal Institute of Technology
SE-100 44 Stockholm, Sweden
www.wwsc.se
Phone: +46 8 790 8118, mobile +46 70 608 4505
Email: blund@kth.se

Chapter 10 Yvonne Aitomäki
Composite Center Sweden, Division of Materials Science, Luleå University of Technology, Luleå, SE-97187 Sweden
Email: yvonne.aitomaki@ltu.se

Chapter 11 Koon-Yang Lee
Polymer and Composite Engineering (PaCE) group, Institute of Materials Chemistry and Research, Faculty of Chemistry, University of Vienna, Waehringerstr. 42, 1090 Vienna and Polymer and Composite Engineering (PaCE) group, Department of Chemical Engineering, Imperial College London, SW7 2AZ London, UK
Email: koonyang.lee04@imperial.ac.uk

Chapter 12 Antonio Norio Nakagaito
Dept. of Mechanical Engineering, The University of Tokushima Minami Josanjima-cho 2-1, 770-8506 Tokushima, Japan
Phone: +81-88-656-7364
Email: nakagaito@tokushima-u.ac.jp

Chapter 13 Mirta I. Aranguren
INTEMA, Facultad de Ingeniería,
Universidad Nacional de Mar del Plata, Argentina
National Scientific and Technical Research Council
(CONICET), Argentina
Email: marangur@fi.mdp.edu.ar

Chapter 14 Takashi Nishino
Depaprtment of Chemical Science & Engineering, Kobe University
Rokko, Nada, Kobe 657-8501, Japan
Phone: +81-78-803-6164
Email: tnishino@kobe-u.ac.jp

Chapter 15 Dieter Klemm
Polymet Jena Association,
Wildenbruchstraße 15, 07745 Jena, Germany
Phone.: +49 3641 548281
Email: Dieter.Kemm@uni-jena.de

Chapter 16 Mohini Sain
Faculty of Forestry, Centre for Biocomposites & Biomaterials
Processing University of Toronto, 33 Willcocks Street,
Toronto, ON M5S 3B3, Canada
Phone: 416-978-5480
Email: m.sain@utoronto.ca

Chapter 17 Wolfgang Gindl-Altmutter
Institute of Wood Technology and Renewable Materials,
Department of Materials Science and Process Engineering,
BOKU-University of Natural Resources and Life Sciences
Vienna, Austria
Email: wolfgang.gindl-altmutter@boku.ac.at

Chapter 1

Bionanocomposites: Processing Methods, Characterization, and Properties

Alexander Bismarck[1] and Kristiina Oksman[2]
[1] *Polymer and Composite Engineering (PaCE) Group, University of Vienna, Vienna, Austria and Department of Chemical Engineering, Imperial College London, London, UK*
[2] *Composite Centre Sweden, Division of Materials Science, Luleå University of Technology, Luleå, Sweden*

The second volume of the Handbook of Green Materials focuses on bionanocomposites or renewable nanocomposites, their processing methods, characterization, and most important properties. The volume contains 17 chapters covering not only processing technologies for thermoset and thermoplastic bionanocomposites but also green chemistry of nanocellulose modification for composite applications. Specific composites materials such as advanced bacterial cellulose nanocomposites, all cellulose composites, and responsive nanocomposites are discussed. Properties of cellulose nanocomposites such as toughness, strength, stiffness, and optical transparency are described in detail. Moreover, the reinforcing efficiency of cellulose in nanocomposites will be assessed in one of the chapters. The last chapters cover potential applications of bionanocomposites in medical devices, such as blood vessel replacements, as substrates for printed electronics, and additives in wood adhesives.

1.1. Introduction

The reinvention of colloid science as nanotechnology has gained huge interest over the past 15 years, initially in academic research and in the past few years has also seen increasing interest from industry. This volume of the Handbook of Green Materials is meant to be a reference work for biobased nanomaterials, their nanocomposites, self-assembled structures, and green (nano)composites. In this second volume of the handbook, the different cellulose nanofibers (CNFs) or nanocrystals containing bionanocomposites will be discussed. The authors of the chapters are all experts in the research areas they are describing and are well known and respected in field.

Chapter 2 summarizes chemical modifications of nanoscale cellulose and its impact on the nanofiber–matrix interface, as well as the mechanical performance of the resulting polymer nanocomposites. It starts with the discussion of conventional methods used for (nano)cellulose modification. Methods for direct and indirect quantification of the nanofiber–matrix interface are reported. Green(er) gas phase esterification reactions and the utilization of ionic liquids as solvent and catalyst for the modification of nanocellulose are also discussed. In addition to this, this chapter also discusses recent work on the carboxymethylation of nanocellulose and the deformation mechanism of a nanocellulose fibril network within a composite.

Chapter 3 summarizes recent efforts to create renewable nanocomposites using sol-gel processes. One particularly useful method, dubbed the "template approach", relies on the formation of a three-dimensional cellulose network through self-assembly of originally well-individualized cellulose nanocrystals (CNCs), which is subsequently filled with a polymer of choice. The template is made by first forming a homogeneous aqueous CNC dispersion, followed by gelation through solvent exchange with a water-miscible solvent. The resulting CNC organogel is subsequently impregnated with a matrix polymer by immersion into a polymer solution. The process is widely applicable and has allowed for the fabrication of numerous CNC-based nanocomposites. It is particularly useful for the fabrication of otherwise inaccessible nanocomposites of immiscible components, such as the polar CNCs and hydrophobic polymers.

Chapter 4 describes solution casting as a route to renewable cellulose nanocomposites, which is an easy and versatile method to produce thin nanocomposite films/sheets in laboratory scale. In order to produce polymer nanocomposites by solution casting, the polymer phase is initially dissolved in water or a volatile organic solvent and mixed with nanosized reinforcements dispersed in the same solvent prior to be casted onto a flat substrate. The solvent phase is subsequently removed by evaporation and thereafter the dried film is released from the substrate. The effect of the processing route on the composite morphology and mechanical properties is discussed.

Chapter 5 describes the compounding of nanocomposites using twin screw extrusion. This processing method is expected to yield nanocomposites that can be easily injected or compression molded to produce parts, which eventually will enable large-scale production. The chapter is dicussing two specific processing routes, liquid feeding of the nanomaterials into the extruder and the utilization of a masterbatch, which is diluted to the required concentration during compounding. Examples of prepared nanocomposites and their mechanical and optical properties will be discussed.

Chapter 6 focuses on *in situ* polymerization as route to produce cellulose nanocomposites. This is one of the most viable and promising approaches because the inherently good dispersion of the nanomaterial in the matrix and excellent interfacial adhesion between the nanoreinforcement and matrix. This chapter collates key advances realized in the field of *in situ* polymerization of bionanocomposites, particularly based on polylactide and polycaprolactone matrices and nanocellulose fillers.

Chapter 7 discusses characterization methods of cellulose nanocomposites; methods for the quantification of cellulose nanofibril dispersion, distribution, orientation, and size within polymer matrixes. The microscopy techniques ultilized to determine those quantities are optical, scanning electron, transmission electron, and atomic force microscopy. Also, the use of X-ray diffraction to determine the orientation of cellulose nanomaterials within a matrix is discussed. The sample preparation and the characterization of bionanocomposites are challenging because these materials are non-conductive and are often soft and moisture sensitive and usually contain only low atomic number elements. Furthermore, different sample preparation techniques for these materials are discussed and some examples of the characterization of the nanocomposite structure by each of the microscopy techniques are shown.

Chapter 8 addresses the challenges and approaches in obtaining detailed and accurate information on the nature and mechanics of interfaces in CNF-based nanocomposites. Covering the nanofiber types of nanocrystals, nanofibrils, and bacterial cellulose (BC), the chapter directs the reader to issues governing the mechanics of nanocomposite materials in general. The chapter also introduces a number of techniques that have been successfully used to characterize the micromechanics of cellulose nanocomposites, namely, Raman spectroscopy, X-ray diffraction, and Förster resonance energy transfer imaging.

Chapter 9 highlights the fact that cellulose nanopapers or nanofiber networks possess superior mechanical properties and allow for new functional characteristics compared with paper and fiberboard materials and even thermoplastic composites containing a reviewable filler, such as wood polymer composites (WPCs), which are commercially available. The chapter analyzes the potential to combine toughness and strength in polymer matrix nanocomposites based on CNF networks.

Chapter 10 investigates the reinforcing efficiency of CNFs. The chapter summarizes the current status of the mechanical properties of CNF-reinforced polymer nanocomposites. A reinforcing efficiency parameter, back-calculated using established micromechanical models from composite data, was used to compare the reinforcing efficiency of CNF in composites. Included is a brief review of the factors affecting the reinforcing efficiency, such as the sources and production methods of CNF, as well as the matrices and methods used to manufacture nanocomposites. Some comparisons are made to current reinforced polymer composites, such as glass fiber composites. Highlighted are some of the issues that need to be addressed if these materials are to be used in structural applications.

Chapter 11 highlights the potential of BC, which is one of the strongest materials produced by nature, possessing a high modulus and strength, estimated to be 114 GPa and 1500 exceeding MPa, respectively. BC is produced by fermentation from low molecular weight sugars by non-pathogenic bacteria, such as the *Gluconacetobacter* species, and is a dimensionally stable hydrogel. When extracted from the gel and dried, without hornifying the CNF, it has been shown to be an effective nanoreinforcement for polymers to produce lightweight and mechanically strong nanocomposites when high loading fractions of BC were used. This chapter discusses the application of BC in advanced polymer materials. These materials include optically transparent nanocomposites and BC–reinforced, natural fiber-reinforced hierarchical composites. The use of BC in "all-cellulose nanocomposites" and biomimetic

BC cellulose-reinforced nanocomposites is described. The application of BC simultaneously as stabilizer and nanoreinforcement to produce macroporous polymers by mechanical frothing of acrylated epoxidized soybean oil is also discussed in this chapter.

Chapter 12 discusses the fabrication of flexible and optically transparent nanocomposites and flexible circuit boards. Circuit boards are ubiquitous in electronic products but flexible circuit boards might eventually be used as substrate for flexible image displays. Plastics are prospective candidates for flexible electronic devices due to their inherent flexibility and optical qualities but unfortunately they also have large thermal expansion coefficients. The expansion of the substrate has to be compatible with those of the active layers to be deposited on it, to avoid damage during the thermal cycles involved in the display manufacture. One way to reduce the thermal expansion of polymers is to reinforce them with nanofibers, without losing transparency.

Chapter 13 describes how shape memory behavior has been utilized in the production of nanocellulose-reinforced responsive composites. Additionally, the hygroscopic nature of cellulose has led to the production of nanocomposites with water-responsive behavior, displaying a dramatic reduction of the material rigidity in the presence of the water or switchable swelling behavior. More recent developments in the area of responsive nanocellulose materials are the grafting of active polymers or functional nanoparticles onto the surface of cellulose crystals or fibrillar structures. These developments are opening new roads for activating different material responses by indirect or non-contact methods.

Chapter 14 covers the area of self-reinforced polymer composites, which is one of the fastest growing areas in engineering polymers, but until now, these materials have been mainly developed on the basis of thermoplastic fibers of moderate performance. This chapter reviews a new type of self-reinforced polymer composites based on cellulosic fibers to produce *all*-cellulose composites or *all*-cellulose nanocomposites, in which both the fibers and matrix are cellulose. Natural cellulose boasts a high elastic modulus and high tensile strength, implying that cellulose possesses the potential to replace glass fibers. The concept of *all*-cellulose composites allows for the production of composites with higher fiber contents than traditional fiber-reinforced plastics. Moreover, since the matrix and reinforcement phases of these biocomposites are completely compatible with each other, *all*-cellulose composites allow for efficient stress transfer and adhesion at their interface. Fabrication, structure, and properties of such *all*-cellulose composites are reviewed.

Chapter 15 covers the application of BC in medical devices. The unique valuable properties of BC are mainly based both on the collagen-like hierarchical three-dimensional nanofiber network structure and its exciting controllability and design directly during biosynthesis. The potential medical applications of BC ranges from wound dressings up to innovative implants for visceral and cardiovascular surgery. This chapter reviews initially the current knowledge in biocompatibility, bioactivity, and mechanical properties of BC. Examples for *in situ* design of channeled

BC membranes and bioactive tubes implantable implants for small-diameter blood vessels are described.

Chapter 16 describes the application of nanocellulose in printed electronics. Reports on printed electronic devices are increasing exponentially. The emergence of this concept is strongly evident in many applications such as diode, transistor, as well as for radio-frequency identification. This chapter provides an overview of techniques to produce printed electronic devices and some useful applications for nanocellulose are shown.

Chapter 17 deals with nanocellulose augmented adhesives. A brief overview of the state of the art of wood adhesives is given in this chapter demonstrating the feasibility of wood adhesive modification through the addition of nanocellulose. Very significant improvements of the mechanical bond stability in both wet and dry conditions upon the incorporation of nanocellulose into wood adhesives have been demonstrated and will be discussed. The effects are consistent for brittle thermosets such as urea–formaldehyde as well as comparably soft thermoplastics, such as polyvinylacetate. It will be shown that both adhesives for solid wood and wood composite production do benefit from the addition of only a few percent nanocellulose.

Chapter 2

Green Chemical Modifications of Nanocellulose for Use in Composites

Koon-Yang Lee and Alexander Bismarck
Polymer and Composite Engineering (PaCE) Group, Institute of Materials Chemistry and Research, Faculty of Chemistry, University of Vienna, Vienna, Austria

Department of Chemical Engineering, Imperial College London, London, UK

The hydrophilicity of cellulose, arising from the presence of large number of hydroxyl groups, reduces its potential in nanocomposites containing nanocellulose (typically <10 wt%). In order to manufacture nanocomposites containing nanocellulose with improved properties, chemical modification is often needed. This chapter summarizes the chemical modification of nanoscale cellulose and its impact on the fiber–matrix interface, as well as the mechanical performance of the resulting polymer nanocomposites. We start with the discussion of conventional esterification of nanocellulose via acetylations reaction with acetic anhydrides and carboxylic acids. Methods for direct and indirect quantification of the fiber–matrix interface are reported. Green(er) esterification reactions in the gas phase and utilizing ionic liquids as the solvent and catalyst for the modification of nanocellulose are also discussed. In addition to this, this chapter also discusses the recent work on the carboxymethylation of nanocellulose and the deformation mechanism of bacterial cellulose networks within a composite.

2.1. Introduction

Cellulose is a linear macromolecule consisting of two D-anhydroglucose ring repeating units linked by β $(1 \rightarrow 4)$ glycosidic bonds. It is a renewable material that has superior mechanical properties and is rich in functional hydroxyl groups. Currently, much research activity focus on the isolation and production of nanoscale cellulose fibers. For comprehensive reviews on the production, isolation, and applications of nanocellulose, the readers are referred to Klemm *et al.*[1–3] and Siró *et al.*[4] Interests in nanocellulose come from the fact that nanoscale cellulose combines the physical and chemical properties of cellulose, such as hydrophilicity and the ability to be chemically modified by a broad range of reactions, with other features such as high specific surface area, aspect ratio, and good mechanical properties, estimated to possess a Young's modulus and tensile strength of up to 160 GPa[5] and 3 GPa,[6] respectively.

Nanocellulose can be obtained by two approaches: (i) top-down and (ii) bottom-up. The top-down approach involves the disintegration of (ligno)cellulosic biomass, such as wood fibers into nanoscale cellulose. This can be achieved by using grinders,[7] whereby wood pulp is passed through the slit between a static and rotating grindstone. The high shear fibrillation process converts a micrometer-scale cellulose pulp into nanoscale cellulose (herein termed as cellulose nanofibers or CNFs). In the bottom-up approach, the fermentation of low molecular weight sugars by cellulose-producing bacteria, such as from the *Acetobacter* species, produces nanoscale cellulose.[2,8–10] This nanoscale bacterial cellulose (BC) is pure cellulose without being associated with hemicellulose, pectin, or lignin.[11] BC is excreted by bacteria into the aqueous culture medium directly as nanofibers, with a diameter ranging from 25 to 100 nm.[9,11] These nanofibers make up the pellicles in the culture medium.[9]

Utilizing nanoscale cellulose as reinforcement in polymer matrices was first reported by Favier *et al.*[12] The authors reinforced latex (styrene-butyl acrylate copolymer) with cellulose nanowhiskers derived from tunicin to produce cellulose polymer nanocomposites. One major driving force for the utilization of nanoscale cellulose, as reinforcement in polymers, is the possible exploitation of the high stiffness of cellulose crystals. Although the true crystal modulus of cellulose is not clear,[13] measurements using X-ray diffraction, Raman spectroscopy, and numerical simulations estimated the stiffness of cellulose crystals to be approximately 100–160 GPa.[14–16] However, the hydrophilicity of cellulose due to the presence of large amount of hydroxyl groups reduces its potential in nanocomposites applications.[17] Therefore, chemical modification of cellulose for the development of biobased materials with improved mechanical performance received significant attention. This chapter summarizes the progress to date on the chemical modification of nanoscale cellulose and its impact on the mechanical performance of the resulting polymer nanocomposites.

2.2. Esterification of nanocellulose

2.2.1 *Acetylation with acetic anhydride*

One of the simplest methods of rendering the hydrophilic surface of nanocellulose hydrophobic is to react the hydroxyl groups of nanocellulose with an anhydride, thereby creating esters (Fig. 2.1).

This approach has been taken by numerous authors to produce green nanocomposites with enhanced nanocellulose–matrix interface.[18–22] A typical acetylation reaction includes a solvent exchange step, whereby suspension of nanocellulose in water is solvent exchanged from water through a second solvent before re-dispersing

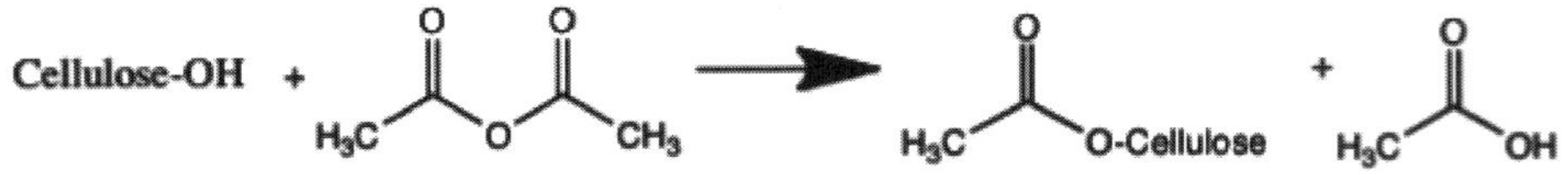

Cellulose-OH + H_3C O CH_3 → H_3C O-Cellulose + H_3C OH

Fig. 2.1. An example of the acetylation of nanocellulose with acetic anhydride.

the nanocellulose in the reaction medium, such as dimethyl formamide (DMF) or toluene. Solvent exchange is used because (vacuum) drying of nanocellulose to remove water does result in the hornification of the cellulose network.[23] Re-dispersing a hornified cellulose network in a second solvent is no longer possible. Although freeze-dried cellulose has the ability to be re-dispersed in a second solvent, a recent study showed evidence that severe bulk modification does occur on freeze-dried BC compared to BC that was solvent exchanged from water through methanol into the reaction medium.[24] The acetylation of cellulose is initially kinetically limited but in the later stages, becomes a diffusion limited reaction.[18,20] At the beginning of the reaction, the acetylation proceeds with the easily accessible hydroxyl groups on the nanocellulose. These include the exposed hydroxyl groups on the surface and in the amorphous region of the cellulose. Once all the easily accessible hydroxyl groups have been reacted, the modification proceeds with hydroxyl groups located inside the cellulose crystals. These hydroxyl groups are hindered by the cellulose structure itself and the recently grafted bulkier acetyl groups. Therefore, the rate-limiting step becomes the diffusion of reactive anhydride species through less-accessible regions of cellulose.

In order to quantify the quality of fiber–matrix interface before and after the chemical modification, dynamic mechanical thermal analysis (DMTA) can be used. Tan δ, a measure of the damping properties of a visco-elastic material, can be used to characterize the quality of the fiber–matrix interface in a composite. It can be expressed as the fiber–matrix interfacial strength indicator b as described in the relationship[25]:

$$b = \frac{\left(1 - \frac{\tan\delta_{\mathrm{c}}}{\tan\delta_{\mathrm{m}}}\right)}{V_{\mathrm{f}}} \tag{2.1}$$

where $\tan\delta_{\mathrm{c}}$, $\tan\delta_{\mathrm{m}}$, and v_{f} are the tan δ of the composite, the neat polymer, and reinforcement volume fraction, respectively. This equation assumes that the reinforcement is stiff and does not possess any energy dissipating properties. Therefore, the relative difference in the height of tan δ is only governed by the stress transfer between the nanofiber and the matrix. The stiffer the resulting composites (characterized by lower tan δ), the better the stress transfer between the fibers and the matrix (characterized by larger b value). It has been shown that the incorporation of acetylated BC into poly(lactic acid) (PLA) resulted in an increase in the storage moduli of the nanocomposites over neat BC-reinforced PLA nanocomposites,[21] which is due to the hydrophobization of cellulose as a result of the acetylation reaction. Table 2.1 summarizes b as a function of BC loading and it can be seen that the

Table 2.1. The fiber–matrix interfacial strength indicator, b, between neat BC/acetylated BC and PLA. Adapted from literature.[21]

BC loading in PLA	Neat BC	Acetylated BC
1 vol.%	20	31.4
4 vol.%	2.1	3.6

acetylation of BC did indeed result in an improved fiber–matrix interface, indicated by a larger b value associated with acetylated BC compared to neat BC.

CNFs were also acetylated with acetic anhydride and used as reinforcement for PLA.[18] However, no significant improvement in the storage modulus of the (modified) CNF-reinforced PLA nanocomposites for various CNF loadings and degrees of acetylation were reported. Moreover, no reasons have been given by the authors on the lack of improvement in the mechanical performance of the nanocomposites. Similar results were also reported by Bulota *et al.*[19] The tensile properties decreased with increasing CNF loading and degree of acetylation. This observation was postulated to be due to the presence of hemicellulose. Unlike BC, CNFs are not pure cellulose but cellulose containing a significant amount of amorphous hemicellulose. The acetylation of CNF does occur initially on the more *accessible* hydroxyl groups on hemicellulose. Even though the acetylation of CNF would result in higher measured θ_w (as a result of modified hemicellulose), the majority of the cellulose within CNF has not been modified enough to have an improved cellulose–polymer adhesion and hence, a lack of improvement in the mechanical performance of the nanocomposites. The acetylation of CNF, however, did have a significant impact on the water transmission rate of CNF films.[22] A water transmission rate of $234\,g\,m^{-2}\,d^{-1}$ and $167\,g\,m^{-2}\,d^{-1}$ was measured for neat CNF and acetylated CNF films, respectively. These results corroborate with the θ_w measurements of 41.2 ± 4.3 and 82.7 ± 5.8, respectively for neat CNF and acetylated CNF, respectively.

2.2.2 *Esterification with long chain carboxylic acids*

In addition to acetic anhydrides, carboxylic acids with various carbon chain lengths can also be used to modify the surface of nanocellulose. In this reaction, the carboxylic acids are postulated to be activated by the addition of *p*-toluenesulfonyl chloride, which generates anhydrides during modification reaction (Fig. 2.2).[26] By using various organic acids, namely, acetic acid, hexanoic acid, and dodecanoic acid, respectively, BC with different degrees of substitution (DS) with organic acids can be obtained.[27] θ_w measurements on cellulose films made from hydrophobized BC showed that by increasing the chain length of the organic acids used, the BC can be rendered hydrophobic, with θ_a varying from 19 to 133° (see Table 2.2).

In order to directly quantify the quality of the fiber–matrix interface, a method based on the measurements of the contact angle of polymer droplets on single nanofibers was developed.[17,28] This direct wetting method, also known as drop-on-fiber method, is generally used to quantify the contact angle formed between a single micrometer-sized fiber, such as carbon fibres and a single polymer droplet (see Fig. 2.3A). The contact angle was evaluated using generalized drop length–height method.[29] The advantage of the generalized drop length–height method (based on the principle of the Laplace–Young equation, neglecting the effect of gravity) is that the contact angle is calculated from the extracted drop profile of the entire droplet without the requirement to determine the maximum drop length. Figure 2.3B shows a PLLA droplet deposited on a single BC nanofiber. It can be seen from Table 2.2

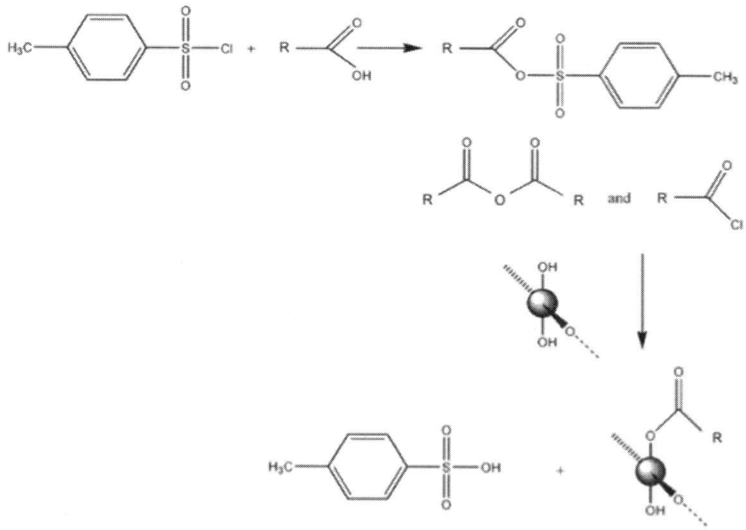

Fig. 2.2. Schematic reaction scheme of *in situ* activation of carboxylic acids for the esterification of cellulose.

Table 2.2. Advancing and receding contact angles, DS of BC modified with C_2 (acetic acid), C_6 (hexanoic acid), and C_{12} (dodecanoic acid), respectively. θ_a, θ_r, θ, DSS, E, and σ represent the advancing contact angle, receding contact angle, the contact angle formed between BC and *poly*(L-*lactic acid*) PLLA, degree of surface substitutions Young's modulus, and tensile strength of the resulting BC-reinforced PLLA nanocomposites, respectively. (Adapted from Ref. 17.)

Sample	θ_a	θ_r	θ	DSS (%)	E (GPa)	σ (MPa)
Neat BC	19 ± 3	12 ± 2	35.4 ± 0.8	—	1.89 ± 0.02	60.9 ± 0.5
C_2-BC	75 ± 1	35 ± 2	29.5 ± 2.9	98.9	1.70 ± 0.03	60.1 ± 0.9
C_6-BC	92 ± 1	45 ± 1	27.7 ± 2.0	58.5	1.79 ± 0.02	65.1 ± 0.9
C_{12}-BC	133 ± 4	80 ± 4	20.8 ± 2.5	51.9	1.98 ± 0.04	68.5 ± 1.5

[a]Neat PLLA possesses E and σ of 1.34 GPa and 60.7 MPa, respectively.
[b]The BC loading in the nanocomposites is 5 wt.%.

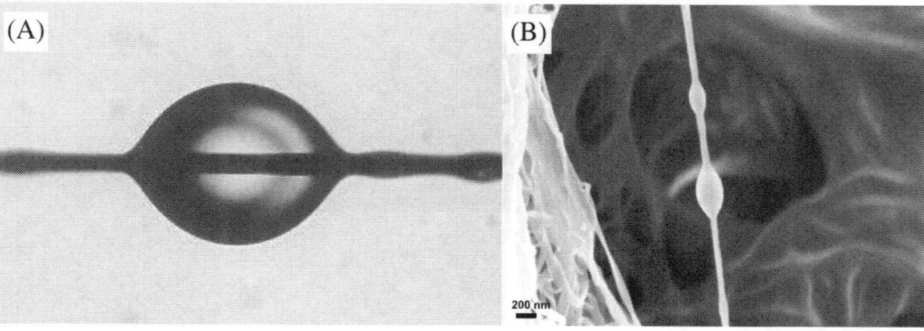

Fig. 2.3. Direct wetting of polymer on a single (A) carbon fiber and (B) BC nanofiber.[17] (Obtained with kind permission from Elsevier.)

that the contact angle between PLLA and modified BC decreases with increasing carbon chain length of the organic acids used. This implies that the quality of the fiber–matrix interface was improved by the hydrophobization of BC. This improvement is also reflected in the tensile properties of the resulting BC-reinforced PLLA nanocomposites. When PLLA is reinforced with C_{12}BC at 5 wt% BC loading, the tensile modulus and strength improved by as much as 46% and 13%, respectively.

2.2.3 *Green(er) esterification reaction of nanocellulose*

The aforementioned chemical modifications involve numerous solvent exchange steps, which are very laborious and use large quantities of potentially harmful organic solvents, such as DMF, toluene, and pyridine. In order to make the chemical modification process greener, Berlioz *et al.*[30] have developed a solvent-free gas-phase esterification of nanocellulose. This method utilizes the lowering in boiling points of reactants when the surrounding pressure is lowered. Freeze-dried cellulose (tunicate whiskers and BC) was placed on a grid above palmitoyl chloride to avoid contact between cellulose and the reactants. The pressure of the vessel was then reduced to 100 mbar and the temperature raised to evaporate palmitoyl chloride. During the diffusion process, the reactant reacts with the hydroxyl groups of nanocellulose forming ester bonds (see Fig. 2.4). By varying the reaction parameters, such as time and temperature, the authors obtained hydrophobized cellulose with DSs varying from 0.15 to 2. The DS obtained using this method is also comparable to the DS obtained via conventional wet-state esterification reactions.

Ionic liquid, were also explored as a greener solvent for the chemical modification of nanocellulose. BC can be rendered hydrophobic by modification with anhydrides of different aliphatic chain lengths (acetic, butyric, and hexanoic) in tetradecyltrihexylphosphonium bis(trifluoromethylsulfohnyl)imide [TDTHP] [NTf_2].[31] It has been shown that this particular ionic liquid does not dissolve or swell BC. Table 2.3 summarizes the reaction time and DS of BC modified by various anhydrides. However, the reaction time in ionic liquid was longer than in conventional solvents. For example, the modification of cellulose with hexanoyl chloride requires typically

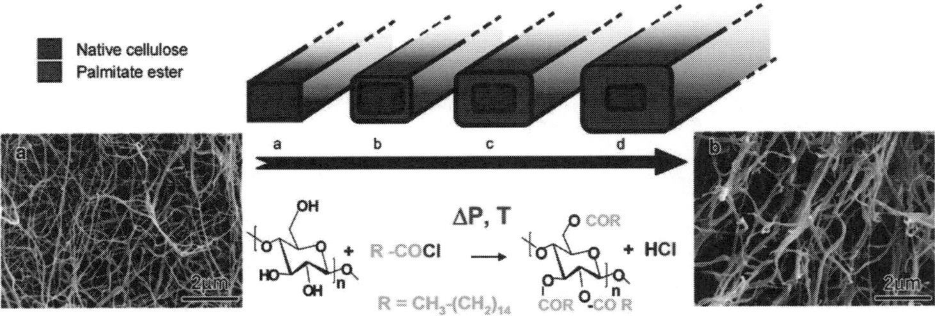

Fig. 2.4. Schematic diagram depicting the progress of gas-phase esterification of cellulose. (A) DS = 0, (B) DS = 0.25, (C) DS = 0.75, and (D) DS = 1.5, respectively. (Adapted from Ref. 30 with kind permission from ACS.)

Table 2.3. The modification of BC in [TDTHP][NTf_2] medium.[33]

Modification	Reaction time	DS
Acetic anhydride	6 h	0.17
Butyric anhydride	4 days	0.24
Hexanoic anhydride	11 days	0.13
Alkenyl succinic anhydride	15 days	0.04
Hexanoyl chloride	24 h	0.17

1–6 h in toluene[32] to obtain similar DS in this ionic liquid, it takes between 6 h and 15 days. CNF have also been modified with anhydrides of various aliphatic chain lengths (acetic, butyric, iso-butyric, and hexanoic) in 1-butyl-3-methylimidazolium hexafluorophosphate [bmim][PF_6].[33] A DS of 0.2–0.3 has been achieved with 2 h of reaction at 100°C. Ionic liquids with high purity were recovered from the reaction in both studies by repeated washing with water.[32,33]

2.3. Grafting of polymers onto nanocellulose

To fully utilize the potential of nanocellulose in composite applications, polymers that would eventually form the matrix of the nanocomposites can be grafted onto cellulose surface. The grafted polymer chains are postulated to act as long tails protruding into the matrix, creating a "co-continuous" phase between cellulose and the surrounding matrix.[34] The grafting of polymers onto cellulose can be accomplished via two approaches: (i) "grafting-from" and (ii) "grafting-to". In the "grafting-from" approach, monomers are polymerized directly from the surface of cellulose. On the other hand, the "grafting-to" approach covalently attaches pre-formed polymers onto the surface of cellulose. High molecular weight polymer and a high grafting density can be achieved by using the "grafting-from" approach.[35,36] Various types of polymers have been "grafted-from" and "grafted-to" cellulose. These include polylactide,[34] polycaprolactone,[37–39] polyhydroxyalkanoate,[39] polymethylmethacrylate,[40] maleated polypropylene,[41] and polyurethane.[42]

Grafting of polylactide and polycaprolactone can easily be achieved by the ring opening polymerization of L-lactide or ε-caprolactone using $Sn(Oct)_2$ as the catalyst. The grafted polymer chain length can also be varied by introducing a co-catalyst, such as benzyl alcohol.[38] The M_n of grafted PCL on CNF varies between 700 and 2200 g mol^{-1}, depending on the ratio between ε-caprolactone and benzyl alcohol (see Table 2.4). Slight improvements in the tensile strength of the resulting nanocomposites reinforced with 3 wt% (modified) CNF have also been observed. When neat CNF was used as reinforcement, the tensile strength decreased from 22 MPa for neat PCL to 17 MPa for neat CNF-reinforced PCL nanocomposites. This decrease can be attributed to the poor fiber–matrix interface, which is a direct result of the hydrophilic nature of CNF and hydrophobic PCL. When the surface of CNF is grafted with PCL of various M_n, the tensile strength increases back to ~22 MPa, with no significant changes of the tensile modulus of the nanocomposites.

Table 2.4. The molecular weight of PCL grafted onto CNF (M_n) and the tensile modulus (E) and strength (σ) of the resulting (modified) NFC-reinforced PCL nanocomposites at 3 wt% loading fraction.[38]

Materials	M_n (g mol^{-1})	E (MPa)	σ (MPa)
Neat NFC	—	256 ± 24	17 ± 1
NFC-g-PCL_short	700	229 ± 16	21 ± 1
NFC-g-PCL_medium	1100	257 ± 14	20 ± 2
NFC-g-PCL_long	2200	230 ± 34	22 ± 2

[a]Neat PCL possesses E and σ of 190 MPa and 22 MPa, respectively.

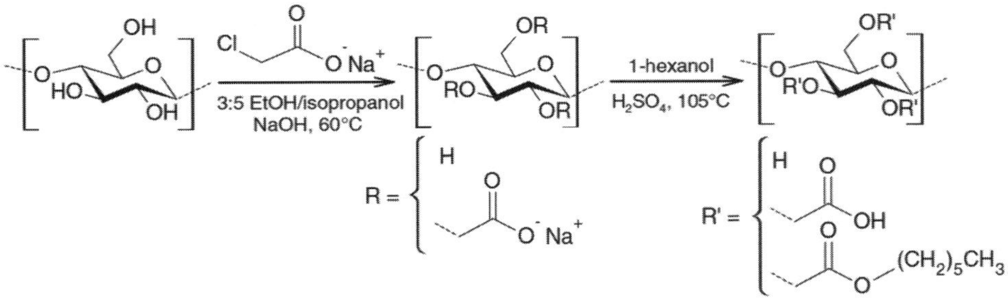

Fig. 2.5. Schematic diagram depicting the synthesis of carboxymethylated nanocellulose fibers (c-CNF) (center) and c-CNF-hex (right). The DS of COOH on c-CNF and hexanoate on c-CNF-hex was found to be 0.09 and 0.14, respectively. (Obtained from Ref. 44 with kind permission from Springer.)

Even though no positive impact can be observed on the tensile properties of the nanocomposites compared to neat PCL, the same authors showed that the peeling energy of the PCL-grafted CNF composites is significantly higher than that of neat CNF-reinforced PCL.[37] A maximum peeling energy of 64.9 J m^{-2} was observed for CNF grafted with very long PCL chains, as compared to 14.9 J m^{-2} for neat CNF. Free radical graft copolymerization can also be used to graft acrylic monomers, namely glycidyl methacrylate, ethyl acrylate, methyl methacrylate, butyl acrylate, and 2-hydroxyehyl methacrylate onto CNF using ammonium cerium (IV) nitrate as the initiator.[40]

2.4. Esterification of carboxymethylated nanocellulose

The surface functionalization of nanocellulose using conventional acetylation and polymer grafting is very well studied. Another method of nanocellulose modification is to modify the starting cellulose raw material, typically bleached pulp, followed by mechanical disintegration of this modified cellulose to nanocellulose. This method is thought to reduce the energy intensity and therefore, the overall cost of the cellulose nanofibrillation process. The starting pulp can be carboxymethylated prior to mechanical disintegration. This is typically done by swelling the pulp fibers in a mixture of ethanol/isopronaol, followed by the addition of aqueous NaOH to activate the fibers for reactions with monochloroacetic acid (see Fig. 2.5).[43]

Table 2.5. Tensile modulus and strength of c-CNF and c-CNF-hex-reinforced PLA nanocomposites.[44]

Samples	Tensile modulus (GPa)	Tensile strength (MPa)
Neat PLA	3.63 ± 0.13	65.8 ± 1.7
2.5% c-CNF/PLA	3.71 ± 0.09	64.2 ± 1.7
2.5% c-CNF-hex/PLA	3.67 ± 0.09	63.3 ± 0.8
5.0% c-CNF/PLA	3.75 ± 0.04	61.7 ± 0.8
5.0% c-CNF-hex/PLA	3.71 ± 0.08	60.0 ± 0.8
7.5% c-CNF/PLA	3.79 ± 0.04	61.3 ± 0.8
7.5% c-CNF-hex/PLA	3.79 ± 0.08	58.3 ± 1.7

The produced carboxymethylated cellulose can then be nanofibrillated to produce c-CNFs. The c-CNF can now be modified with alcohols to produce cellulose esters. This approach was taken by Eyholzer *et al.*[44] to modify bleached beech pulp with 1-hexanol, producing cellulose hexanoate (c-CNF-hex). However, the resulting PLA nanocomposites reinforced by c-CNF and c-CNF-hex did not show significant mechanical improvements over neat PLA (see Table 2.5). This is thought to be due to the poor dispersion of c-CNF and c-CNF-hex in PLA. In addition to this, the low DS of c-CNF-hex could also result in the lack of tensile strength improvements in the resulting nanocellulose-reinforced PLA composites. The same group of authors also compared the reinforcing ability of neat CNF and c-CNF in hydroxypropyl cellulose matrix. However, the study showed that the tensile properties of the resulting CNF- and c-CNF-reinforced composites were similar, indicating that carboxymethylation might not be the way forward to enhance the CNF–polymer matrix interface.

2.5. Glyoxalization of nanocellulose

Nanocellulose can be used as reinforcements in two different forms: (i) loose (freeze-dried) form or (ii) as paper. Sections 2.2 to 2.4 discuss the mechanical performance of cellulose-reinforced polymer nanocomposites when the nanocellulose is dispersed (as loose fibers) within a polymer matrix. These composites are typically manufactured using commercially available polymer processing technologies, such as extrusion or solution casting. However, the major drawback of utilizing nanocellulose in the loose fibrous form is the low nanocellulose loading achieved (typically less than 10 vol.%). The use of nanocellulose papers (otherwise known as "nanopapers") as reinforcements is well documented. This approach was first described by Yano *et al.*,[45] whereby the authors impregnated a BC network obtained from a BC pellicle with epoxy, acrylic, and phenol-formaldehyde resins using resin transfer molding processes. In all these resin systems, fiber volume fractions of between 60 and 70 wt% had been achieved. The resulting BC nanopaper-reinforced epoxy composites possess a tensile modulus and strength of 21 GPa and 325 MPa, respectively. These values are much higher than any of the nanocellulose-reinforced polymer composites discussed in Secs. 2.2 to 2.4.

Table 2.6. Properties of (glyoxalized) BC nanopaper and its (maleic anhydride grafted) PLLA nanocomposites. E, σ, and ε indicate tensile modulus, tensile strength, and engineering strain-to-failure, respectively.[49,50]

Sample	Loading fraction (vol.%)	E (GPa)	σ (MPa)	ε (%)
Neat BC nanopaper	—	10.1 ± 1.5	165.1 ± 33.9	2.6 ± 0.6
Glyoxalized BC nanopaper	—	10.7 ± 1.4	71.1 ± 36.1	0.6 ± 0.3
Neat PLA	—	2.0 ± 0.2	27.7 ± 2.5	23.5 ± 17.0
BC/PLA	13.8 ± 3.2	2.7 ± 0.3	65.9 ± 2.3	3.0 ± 0.5
$BC_{glyoxoal}$/PLA	12.3 ± 0.1	2.5 ± 0.1	55.9 ± 4.5	2.4 ± 0.2
BC/MAPLA	13.9 ± 1.2	2.7 ± 0.2	32.2 ± 5.4	1.2 ± 0.2
$BC_{glyoxal}$/MAPLA	15.5 ± 2.1	3.2 ± 0.2	39.0 ± 9.3	1.4 ± 0.2

MAPLA denotes maleic anhydride-grafted PLLA.

However, when such BC nanopapers are used to reinforce PLLA, it was observed that delamination occurred between the network of BC nanopapers instead of the interface between BC and PLLA.[46] This suggests prior to the nanocellulose modification to enhance the fiber–matrix interface, the fiber–fiber interaction should first be considered and optimized to enhance the stress transfer within the fibrous network. One approach to enhance the fiber–fiber interaction is to crosslink the nanocellulose fibrous network within this nanopaper. Numerous crosslinking agents have been reported. These include glyoxal and its derivatives, formaldehydes, epichlorohydrines, epoxies, diisocynates, and dichloroethanes.[47] Out of all these chemical substances, glyoxal is an ideal candidate due to its low toxicity and the fact that it can be derived from renewable resources.[48] Table 2.6 shows the mechanical properties of glyoxal-modified BC and its composites.

The crosslinking of a BC network did not result in a decrease of the tensile modulus of the nanopaper but both the tensile strength and the engineering strain-to-failure decreased significantly (see Table 2.6). This decrease is a result of the lack of hydrogen bonding between the nanofibers after glyoxalization reaction.[49] The decrease in the strain-to-failure of glyoxalized BC nanopaper is indicative of less delamination within the BC network. The mechanical properties of (modified) BC nanopaper-reinforced PLLA and maleic anhydride-grafted PLLA (MAPLA) nanocomposites are also tabulated in Table 2.6. It can be seen from this table that the glyoxalization of BC nanopaper did improve the tensile modulus of the resulting composites when MAPLA was used as the matrix. The tensile strength of the composites, however, was much worse than that of neat BC nanopaper-reinforced PLLA composites. Nonetheless, the fracture surface of the resulting composites showed that the delamination of BC network within MAPLA composites was significantly decreased when the BC nanopaper was glyoxalized. This led to the proposed deformation mechanism, presented in Fig. 2.6.

This proposed deformation mechanism assumed to possess two main interfaces: (i) the fibrillar network of BC and (ii) the network of BC and resin. These two interfaces are thought to be crucial to the mechanical performance of BC nanopaper–polymer composites. Due to the dense network of BC nanopaper, the nanopaper is assumed to be impenetrable to resin.[46] By crosslinking both the BC network

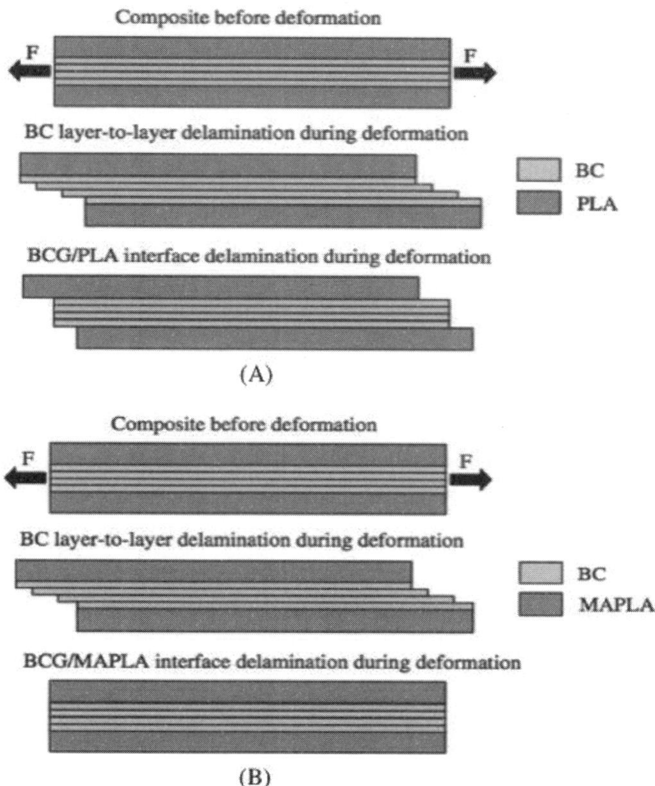

Fig. 2.6. (A) The proposed deformation mechanism of BC nanopaper-reinforced PLA (B) MAPLA composites. (Reprint from Ref. 50 with kind permission from Springer.)

within the nanopaper as well as using MAPLA as the resin, both the interfaces are thought to be enhanced, resulting in an improvement in the mechanical performance of the nanocomposites. However, it should be noted that the maleic anhydride grafting onto PLLA led to severe chain scission of PLLA. The weight averaged molecular weights of PLA and MAPLA are measured to be 212 kDa and 122 kDa, respectively.[50] This resulted in a significantly reduced tensile strength of $BC_{gloxyal}$/MAPLA.

2.6. Conclusions

This chapter has covered the progress to date on the chemical modification of nanocellulose with the aim to improve interactions with a polymer matrix. The esterification of nanocellulose is well studied and most commonly used to render the hydrophilic surface of nanocellulose hydrophobic. It has also been proven that this method could result in the improvement of the mechanical performance of BC-reinforced polymer nanocomposites. Mixed results were obtained for acetylated CNF-reinforced polymer nanocomposites, postulated to be due to the presence of amorphous hemicellulose, which could be preferentially modified by the chemical reaction or the presence of hemicellulose embrittles/softens the surrounding polymer matrices. The "grafting-from" approach can also be utilized to produce nanocomposites with improved mechanical performance. In addition to this, pulp can also be modified with carboxymethyl groups prior to fibrillation. Not only will this aid the fibrillation process, the carboxymethylated nanocellulose can be modified with alcohol to produce cellulose esters.

We have also discussed the use of (modified) BC nanopaper as reinforcements for composites. In this context, the delamination of the BC network played an important role in the overall performance of the nanopaper-reinforced polymer composites. Although some of these chemical modifications did not show significant improvements in the mechanical performance of the resulting nanocomposites, it can be anticipated that large volume of nanocellulose with various surface functionality can be produced in a more environmental friendly and cost-effective manner, leading to truly green materials with various functionalities.

Acknowledgments

The authors would like to thank University of Vienna and the UK Engineering and Physical Science Research Council (EPSRC) Follow-on Fund (EP/J013390/1) for funding.

References

1. D. Klemm, B. Heublein, H.P. Fink and A. Bohn, Cellulose: fascinating biopolymer and sustainable raw material, *Angew. Chem. Int. Ed.* **44**(22) (2005) 3358–3393.
2. D. Klemm, F. Kramer, S. Moritz, T. Lindstrom, M. Ankerfors, D. Gray and A. Dorris, Nanocelluloses: a new family of nature-based materials, *Angew. Chem. Int. Ed.* **50**(24) (2011) 5438–5466.
3. D. Klemm, D. Schumann, F. Kramer, N. Hessler, D. Koth and B. Sultanova, Nanocellulose materials — different cellulose, different functionality, *Macromol. Symp.* **280** (2009) 60–71.
4. I. Siro and D. Plackett, Microfibrillated cellulose and new nanocomposite materials: a review, *Cellulose* **17**(3) (2010) 459–494.
5. S.J. Eichhorn, Cellulose nanowhiskers: promising materials for advanced applications, *Soft Matter* **7**(2) (2011) 303–315.
6. T. Saito, R. Kuramae, J. Wohlert, L.A. Berglund and A. Isogai, An ultrastrong nanofibrillar biomaterial: the strength of single cellulose nanofibrils revealed via sonication-induced fragmentation, *Biomacromolecules* **14**(1) (2013) 248–253.
7. S. Iwamoto, A.N. Nakagaito, H. Yano and M. Nogi, Optically transparent composites reinforced with plant fiber-based nanofibers, *Appl. Phys. A-Mater. Sci. Process.* **81**(6) (2005) 1109–1112.
8. A.J. Brown, The chemical action of pure cultivations of bacterium aceti, *J. Chem. Soc. Trans.* **49** (1886) 172–187.
9. R. Jonas and L.F. Farah, Production and application of microbial cellulose, *Polym. Degrad. Stabil.* **59**(1–3) (1998) 101–106.
10. P. Ross, R. Mayer and M. Benziman, Cellulose biosynthesis and function in bacteria, *Microbiol. Rev.* **55**(1) (1991) 35–58.

11. M. Iguchi, S. Yamanaka and A. Budhiono, Bacterial cellulose — a masterpiece of nature's arts, *J. Mater. Sci.* **35**(2) (2000) 261–270.
12. V. Favier, G.R. Canova, J.Y. Cavaille, H. Chanzy, A. Dufresne and C. Gauthier, Nanocomposite materials from latex and cellulose whiskers, *Polym. Adv. Technol.* **6**(5) (1995) 351–355.
13. S.J. Eichhorn, A. Dufresne, M. Aranguren, N.E. Marcovich, J.R. Capadona, S.J. Rowan, C. Weder, W. Thielemans, M. Roman, S. Renneckar, W. Gindl, S. Veigel, J. Keckes, H. Yano, K. Abe, M. Nogi, A.N. Nakagaito, A. Mangalam, J. Simonsen, A.S. Benight, A. Bismarck, L.A. Berglund and T. Peijs, Review: current international research into cellulose nanofibres and nanocomposites, *J. Mat. Sci.* **45**(1) (2010) 1–33.
14. S.J. Eichhorn and G.R. Davies, Modelling the crystalline deformation of native and regenerated cellulose, *Cellulose* **13**(3) (2006) 291–307.
15. M. Matsuo, C. Sawatari, Y. Iwai and F. Ozaki, Effect of orientation distribution and crystallinity on the measurement by X-ray-diffraction of the crystal-lattice moduli of cellulose-I and cellulose-II, *Macromolecules* **23**(13) (1990) 3266–3275.
16. Y.C. Hsieh, H. Yano, M. Nogi and S.J. Eichhorn, An estimation of the Young's modulus of bacterial cellulose filaments, *Cellulose* **15**(4) (2008) 507–513.
17. K.-Y. Lee, J.J. Blaker and A. Bismarck, Surface functionalisation of bacterial cellulose as the route to produce green polylactide nanocomposites with improved properties, *Compos. Sci. Technol.* **69**(15–16) (2009) 2724–2733.
18. P. Tingaut, T. Zimmermann and F. Lopez-Suevos, Synthesis and characterization of bionanocomposites with tunable properties from poly(lactic acid) and acetylated microfibrillated cellulose, *Biomacromolecules* **11**(2) (2010) 454–464.
19. M. Bulota, K. Kreitsmann, M. Hughes and J. Paltakari, Acetylated microfibrillated cellulose as a toughening agent in poly(lactic acid), *J. Appl. Polym. Sci.* **126** (2012) E448–E457.
20. S. Ifuku, M. Nogi, A. Kentaro, H. Keishin, F. Nakatsubo and H. Yano, Surface modification of bacterial cellulose nanofibers for property enhancement of optical transparent composites: dependence on acetyl-group DS, *Biomacromolecules* **8**(6) (2007) 1937–1978.
21. L.C. Tome, R.J.B. Pinto, E. Trovatti, C.S.R. Freire, A.J.D. Silvestre, C.P. Neto and A. Gandini, Transparent bionanocomposites with improved properties prepared from acetylated bacterial cellulose and poly(lactic acid) through a simple approach, *Green Chem.* **13**(2) (2011) 419–427.
22. G. Rodionova, M. Lenes, O. Eriksen and O. Gregersen, Surface chemical modification of microfibrillated cellulose: improvement of barrier properties for packaging applications, *Cellulose* **18**(1) (2011) 127–134.
23. J. Diniz, M.H. Gil and J. Castro, Hornification — its origin and interpretation in wood pulps, *Wood Sci. Technol.* **37**(6) (2004) 489–494.
24. K.Y. Lee and A. Bismarck, Susceptibility of never-dried and freeze-dried bacterial cellulose towards esterification with organic acid, *Cellulose* **19**(3) (2012) 891–900.
25. A. Afaghi-Khatibi and Y.W. Mai, Characterisation of fibre/matrix interfacial degradation under cyclic fatigue loading using dynamic mechanical analysis, *Compos. Pt. A-Appl. Sci. Manuf.* **33**(11) (2002) 1585–1592.
26. T. Heinze, T. Liebert and A. Koschella, *Esterification of Polysaccharides*, Springer (2006).
27. K.-Y. Lee, F. Quero, J.J. Blaker, C.A.S. Hill, S.J. Eichhorn and A. Bismarck, Surface only modification of bacterial cellulose nanofibres with organic acids, *Cellulose* **18**(3) (2011) 595–605.
28. M.Q. Tran, J.T. Cabral, M.S.P. Shaffer and A. Bismarck, Direct measurement of the wetting behavior of individual carbon nanotubes by polymer melts: The key to carbon nanotube-polymer composites, *Nano Lett.* **8**(9) (2008) 2744–2750.

29. B.H. Song, A. Bismarck, R. Tahhan and J. Springer, A generalized drop length-height method for determination of contact angle in drop-on-fiber systems, *J. Colloid Interface Sci.* **197**(1) (1998) 68–77.
30. S. Berlioz, S. Molina-Boisseau, Y. Nishiyama and L. Heux, Gas-phase surface esterification of cellulose microfibrils and whiskers, *Biomacromolecules* **10**(8) (2009) 2144–2151.
31. L.C. Tome, M.G. Freire, I.P.N. Rebelo, A.J.D. Silvestre, C.P. Neto, I.M. Marrucho and C.S.R. Freire, Surface hydrophobization of bacterial and vegetable cellulose fibers using ionic liquids as solvent media and catalysts, *Green Chem.* **13**(9) (2011) 2464–2470.
32. C.S.R. Freire, A.J.D. Silvestre, C.P. Neto, M.N. Belgacem and A. Gandini, Controlled heterogeneous modification of cellulose fibers with fatty acids: effect of reaction conditions on the extent of esterification and fiber properties, *J. Appl. Polym. Sci.* **100**(2) (2006) 1093–1102.
33. K. Missoum, M.N. Belgacem, J.P. Barnes, M.C. Brochier-Salon and J. Bras, Nanofibrillated cellulose surface grafting in ionic liquid, *Soft Matter* **8**(32) (2012) 8338–8349.
34. A.L. Goffin, J.M. Raquez, E. Duquesne, G. Siqueira, Y. Habibi, A. Dufresne and P. Dubois, From interfacial ring-opening polymerization to melt processing of cellulose nanowhisker-filled polylactide-based nanocomposites, *Biomacromolecules* **12**(7) (2011) 2456–2465.
35. S.T. Milner, Polymer brushes, *Science* **251**(4996) (1991) 905–914.
36. B. Zhao and W.J. Brittain, Polymer brushes: surface-immobilized macromolecules, *Prog. Polym. Sci.* **25**(5) (2000) 677–710.
37. H. Lonnberg, L. Fogelstrom, Q. Zhou, A. Hult, L. Berglund and E. Malmstrom, Investigation of the graft length impact on the interfacial toughness in a cellulose/poly(epsilon-caprolactone) bilayer laminate, *Compos. Sci. Technol.* **71**(1) (2011) 9–12.
38. H. Lonnberg, K. Larsson, T. Lindstrom, A. Hult and E. Malmstrom, Synthesis of polycaprolactone-grafted microfibrillated cellulose for use in novel bionanocomposites — influence of the graft length on the mechanical properties, *ACS Appl. Mater. Interfaces* **3**(5) (2011) 1426–1433.
39. X. Samain, V. Langlois, E. Renard and G. Lorang, Grafting biodegradable polyesters onto cellulose, *J. Appl. Polym. Sci.* **121**(2) (2011) 1183–1192.
40. K. Littunen, U. Hippi, L.S. Johansson, M. Osterberg, T. Tammelin, J. Laine and J. Seppala, Free radical graft copolymerization of nanofibrillated cellulose with acrylic monomers, *Carbohydr. Polym.* **84**(3) (2011) 1039–1047.
41. N. Ljungberg, C. Bonini, F. Bortolussi, C. Boisson, L. Heux and J.Y. Cavaille, New nanocomposite materials reinforced with cellulose whiskers in atactic polypropylene: effect of surface and dispersion characteristics, *Biomacromolecules* **6**(5) (2005) 2732–2739.
42. X.D. Cao, Y. Habibi and L.A. Lucia, One-pot polymerization, surface grafting, and processing of waterborne polyurethane-cellulose nanocrystal nanocomposites, *J. Mater. Chem.* **19**(38) (2009) 7137–7145.
43. C. Eyholzer, F. Lopez-Suevos, P. Tingaut, T. Zimmermann and K. Oksman, Reinforcing effect of carboxymethylated nanofibrillated cellulose powder on hydroxypropyl cellulose, *Cellulose* **17**(4) (2010) 793–802.
44. C. Eyholzer, P. Tingaut, T. Zimmermann and K. Oksman, Dispersion and reinforcing potential of carboxymethylated nanofibrillated cellulose powders modified with 1-hexanol in extruded poly(lactic acid) (PLA) composites, *J. Polym. Environ.* **20**(4) (2012) 1052–1062.
45. H. Yano, J. Sugiyama, A.N. Nakagaito, M. Nogi, T. Matsuura, M. Hikita and K. Handa, Optically transparent composites reinforced with networks of bacterial nanofibers, *Adv. Mater.* **17**(2) (2005) 153–155.

46. F. Quero, M. Nogi, H. Yano, K. Abdulsalami, S.M. Holmes, B.H. Sakakini and S.J. Eichhorn, Optimization of the mechanical performance of bacterial cellulose/poly (L-lactic) acid composites, *ACS Appl. Mater. Interfaces* **2**(1) (2010) 321–330.
47. D. Klemm, B. Phillipp, T. Heinze, U. Heinze and W. Wagenknecht, *Comprehensive Cellulose Chemistry: Fundamentals and Analytical Methods*, Germany: Wiley-VCH Verlag (1998).
48. E.C. Ramires, J.D. Megiatto, Jr., C. Gardrat, A. Castellan and E. Frollini, Biobased composites from glyoxal-phenol matrices reinforced with microcrystalline cellulose, *Polimeros* **20**(2) (2010) 126–133.
49. F. Quero, M. Nogi, K.-Y. Lee, G. Vanden Poel, A. Bismarck, A. Mantalaris, H. Yano and S.J. Eichhorn, Cross-linked bacterial cellulose networks using glyoxalization, *ACS Appl. Mater. Interfaces* **3**(2) (2011) 490–499.
50. F. Quero, S.J. Eichhorn, M. Nogi, H. Yano, K.Y. Lee and A. Bismarck, Interfaces in cross-linked and grafted bacterial cellulose/poly(lactic acid) resin composites, *J. Polym. Environ.* **20**(4) (2012) 916–925.

Chapter 3

Preparation of Cellulose Nanocrystal/Polymer Nanocomposites *via* Sol-Gel Processes

Mehdi Jorfi, Pratheep K. Annamalai and Christoph Weder
Adolphe Merkle Institute and Fribourg Center for Nanomaterials, University of Fribourg, Rte de l'Ancienne Papeterie, CH-1723 Marly, Switzerland

As a result of high specific surface area and energy, cellulose nanocrystals (CNCs) usually have a strong tendency for aggregation. This makes it difficult to produce nanocomposites with polymers in which these filler particles are homogeneously dispersed and form percolating networks. This chapter summarizes recent efforts to create such nanomaterials by applying sol-gel processes. One particularly useful method, dubbed the "template approach", relies on the formation of a three-dimensional network through self-assembly of originally well-individualized nanofibers, and subsequently filling the template with a polymer of choice. The template is made by first forming a homogeneous aqueous CNC dispersion, followed by gelation through solvent exchange with a water-miscible solvent. The resulting CNC organogel is subsequently imbibed with a matrix polymer by immersion into a polymer solution. The process is broadly applicable and has allowed for the fabrication of several CNC-based nanocomposites. It is particularly useful for the fabrication of otherwise inaccessible nanocomposites of immiscible components, such as the polar CNCs and hydrophobic matrix polymers.

3.1. Introduction

The general approach of incorporating the nanoscale particles in polymer is highly attractive because it allows one to create materials with new or improved properties by exploiting the synergistic effects of both polymer and nanoparticles.[1–3] The increasing interest in polymer nanocomposites with cellulose nanocrystals (CNCs) is primarily associated with their outstanding mechanical reinforcing capability, but they offer several other advantages over other fillers, including biosustainability, biorenewability, low production cost, and possibly lower cytotoxic and (pro-)inflammatory effects when inhaled.[4–8] CNCs have also been instrumental for the design of mechanically adaptive nanocomposites, whose mechanical properties can be controlled by modulating the CNC–CNC interactions within the polymer by an external stimulus.[9,10] However, while many attractive properties and functions have been demonstrated, the broad technological exploitation of CNCs has so far been stifled by the difficulty to process these materials. Indeed, it has proved challenging to create simple protocols that allow one to appropriately disperse the nanocrystals in polymer hosts. Typically, CNCs are isolated from the biosource by acid hydrolysis and ultrasonic treatment. The CNCs obtained by these processes exhibit a strong tendency to aggregate owing to their high specific surface area and energy. The traditional approach to solve this problem has been the combination of aqueous polymer solutions or emulsions and CNC suspensions and subsequent drying.[11] Unfortunately, this process can only be applied for selected polymers. Further, several solubilizing schemes were explored to improve CNC dispersibility in organic media, e.g., the use of surfactants,[12] silylation,[13] grafting of PEO,[14] and acylation.[15] However, surface modifications reduce the interactions between the CNCs in a matrix and thereby the macroscopic mechanical properties of the corresponding nanocomposites, unless pronounced interactions with the matrix are at play. Several groups demonstrated that stable suspensions of CNCs with negatively charged sulfate or carboxylate groups, commonly produced by hydrolysis of the native cellulose with sulfuric acid,[16] or by 2,2,6,6-tetramethylpiperidine-1-oxyl radical (TEMPO)-mediated oxidation, respectively, can also be produced in certain polar organic solvents, for example by lyophilization of aqueous CNC dispersions and re-dispersion of the resulting aerogel in the organic solvent.[17,18] Nanocomposites are then accessible by mixing organic CNC dispersions with an organo-soluble polymer of choice. As an alternative, a new and simple process was recently introduced, which allows the formation of homogeneous, percolating CNC composites with *virtually any* host polymer (Fig. 3.1).[19–21] This “template approach” is based on the formation of a three-dimensional CNC network, which is assembled through a sol-gel process, i.e., the gelation of a dispersion of well-individualized CNCs through a solvent exchange. The gelled nanofiber template is then imbibed with a solution of a polymer of choice, before the nanocomposite is dried and shaped. The available data suggest that this approach is broadly applicable to a diverse range of nanofibers (not only cellulose) and that it allows for the fabrication of nanocomposites that may otherwise be inaccessible, such as poly(butadiene)/nanocellulose nanocomposites. This chapter reviews the template approach as it pertains to the processing of CNC nanocomposites, presents examples of “conventionally reinforced” materials prepared by this method, and discusses examples of advanced CNC-based materials made with this method. Related uses of solvent-exchange-based sol-gel schemes in the context of CNC processing are also discussed, which include the re-dispersion of CNC organogels and the impregnation of such gels with monomers and subsequent polymerization (Fig. 3.1).

3.2. Preparation of CNC organogels *via* a sol-gel process

The sol-gel process departs from a very good dispersion of CNCs in water or a suitable organic solvent such as N,N-dimethylformamide (DMF),[17,22] N-methylpyrrolidine (NMP),[17] formic acid,[17] or m-cresol.[17] The required concentration for subsequent gel formation depends on the aspect ratio of the CNC type used, and varies between *ca.* 8.0 mg/mL for high-aspect-ratio CNCs such as those isolated from tunicates (aspect ratio ~70)[19,21,23] and *ca.* 20 mg/mL for low-aspect-ratio

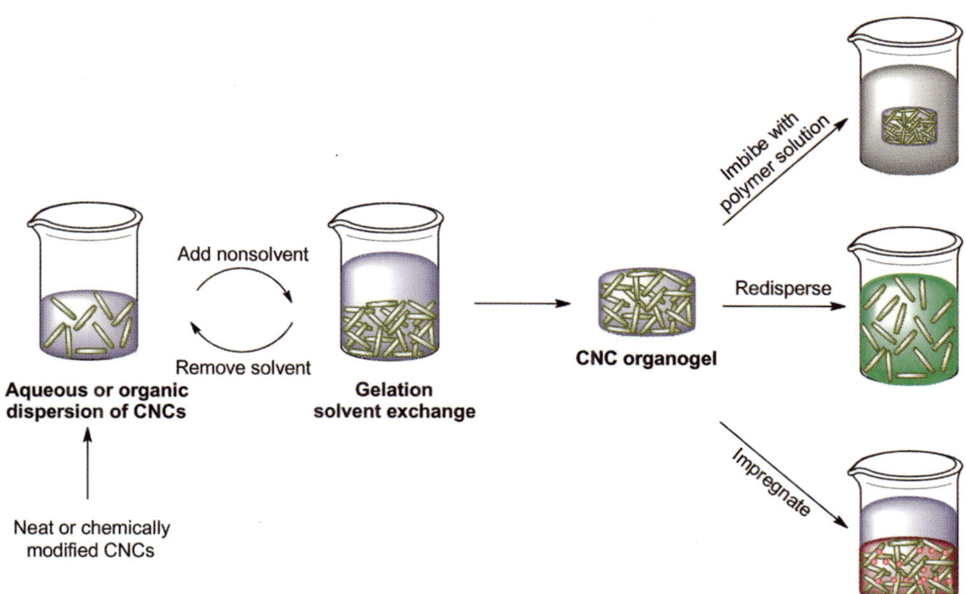

Fig. 3.1. Schematic representation of sol-gel methods used for the processing of CNCs. After forming a dispersion of CNCs in water or a suitable organic solvent, gelation is caused by a solvent exchange with an organic solvent. In the resulting organogel, the CNCs assemble into a percolating network. Possible subsequent processing steps include (i) imbibing the organogel with a polymer solution, (ii) re-dispersion of the organogel in a suitable solvent, and (iii) impregnation of the organogel with monomers/precursors and subsequent *in situ* polymerization.

CNCs such as those isolated from cotton (aspect ratio ~10) or soft-pulp wood.[9]

Gelation is subsequently induced through solvent exchange with a solvent that is miscible with water or the organic solvent employed. Acetone, acetonitrile, ethanol, isopropanol, methanol, and tetrahydrofuran are examples of solvents that have been used.[19] In the laboratory, the solvent exchange is achieved by gently adding the organic non-solvent on top of the CNC dispersion, and exchanging the supernatant organic layer frequently, until the bottom portion had assembled into a mechanically coherent organogel (Fig. 3.2). The CNC content of the gels can be controlled by the concentration of the initial CNC dispersion, the ratio of organic solvent to CNC dispersion (more solvent lowers the CNC content in the gel), and the choice of the organic solvent.[19,21,23] Gels with a CNC content of 1.0–3.0% w/w have been reported.[18,19,24–26] The subsequent exchange with another solvent, for example one that is not miscible with the solvent used to prepare the original CNC dispersion is also possible; an example includes the exchange from an aqueous CNC dispersion to an acetone gel to a toluene gel.[19]

To study the structure of the CNCs in the organogel, CNC organogels were converted into aerogels by supercritical fluid extraction with CO_2, a process that preserves the structure of the original assembly. Scanning electron microscopy (SEM) images of aerogels prepared from organogels of CNCs isolated from tunicates in

Fig. 3.2. Top: Picture of a typical CNC organogel isolated from cotton in acetone.[20] Bottom: SEM image of a CNC aerogel. This material was prepared by supercritical extraction of a CNC organogel isolated from tunicates with CO_2 (scale bar = 400 nm).[19] (The SEM image is reproduced from Ref. 19 with permission from Nature Publishing Group.)

acetone confirm the formation of a robust porous network of individualized CNCs (Fig. 3.2). It is evident that such a porous, three-dimensional CNC network is a perfect template, which in principle can be filled with any polymer of choice to create CNC-based nanocomposites.

3.3. Preparation of polymer/CNC nanocomposites

3.3.1 *Imbibing CNC organogels with polymers*

Arguably, the most important application of sol-gel processing of CNCs is the preparation of polymer/CNC nanocomposites, which is readily achieved by filling the CNC template with a matrix polymer (Fig. 3.1). This can be accomplished by immersing the organogel into a solution of the polymer of choice. The main requirements for this template process are that the solvent used to dissolve the polymer is miscible with the solvent used to produce the CNC organogel and that it does not re-disperse the CNCs.

In the laboratory, the imbibing procedure simply involves placing a CNC organogel for several hours into a solution of the polymer of choice, removing the imbibed gel and drying it, and re-shaping the resulting nanocomposite, for example by compression molding. The CNC content of the nanocomposites can be controlled by the content of the CNCs in the organogel (which can be adjusted as discussed above), and the concentration of the polymer solution. A broad compositional range is therefore readily accessible. While the processing of polymer nanocomposites with high contents of high-aspect-ratio fillers is traditionally limited by the significant viscosity increase caused by the introduction of the filler, the here-discussed template approach readily affords materials with a CNC content of up to 100%, if very dilute polymer solutions (or no polymer at all) are used. The lower limit of the CNC content that can be achieved by the template process is to a large extent dictated by the viscosity of the polymer solution, which above a certain concentration simply becomes too high to allow for an efficient exchange. It depends thus on the molecular weight of the polymer used; typical values are of the order of *ca.* 2% v/v and *ca.* 12% v/v for polystyrene (PS) with an average molecular weight, M_w of ~230,000 and polybutadiene (PBD) with a M_w of ~2,000,000–3,000,000, respectively.[19]

A first systematic study that utilized the template process for the preparation of polymer/CNC nanocomposites involved an ethylene oxide-epichlorohydrin copolymer (EO-EPI) as a soft, low-glass-transition temperature (T_g) matrix and CNCs isolated from tunicates and cotton, respectively.[19] These systems were chosen because these nanocomposites can also be prepared by casting from DMF as a common solvent for polymer and CNCs,[19] and this allowed a direct comparison of the properties of materials prepared by the two methods. Figure 3.3, which shows the shear moduli G' of the nanocomposites thus prepared as a function of CNC content, illustrates that G' increased from 1.3 MPa for the neat EO-EPI to 300 MPa for a nanocomposite comprising 23% v/v CNCs isolated from tunicates and to 43 MPa for a nanocomposite comprising 26% v/v CNCs isolated from cotton. The mechanical reinforcement observed for samples made by the template approach (filled squares) by casting from DMF follows the same trends and is virtually indistinguishable for similar compositions. The data are well described by the percolation model that is described in detail elsewhere,[10,18,19,27,28] which serves to illustrate that the CNCs are indeed well dispersed in the matrix and form a three-dimensional percolating network of CNCs, even after compression molding. Similar comparative studies were published for other compositions, including nanocomposites of PS and CNCs isolated from cotton,[19] and nanocomposites of EO-EPI and CNCs isolated from microcrystalline cellulose.[20]

Probably, the most important feature of the template approach, with respect to the processing of nanocomposites, is its capability to make percolating nanocomposites of otherwise immiscible components accessible, such as polar CNCs and hydrophobic matrix polymers. This was first demonstrated by imbibing acetone gels of CNCs isolated from tunicates and cotton[29] with solutions of PBD[19,26] and styrene-butadiene rubber (SBR).[26] In one of the studies,[26] it was shown that the use of toluene as a solvent for the SBR led to some shrinkage of the gels, which could be largely suppressed when anhydrous tetrahydrofuran was used. The reason for this effect is so far unknown. Whereas reference experiments based on conventional

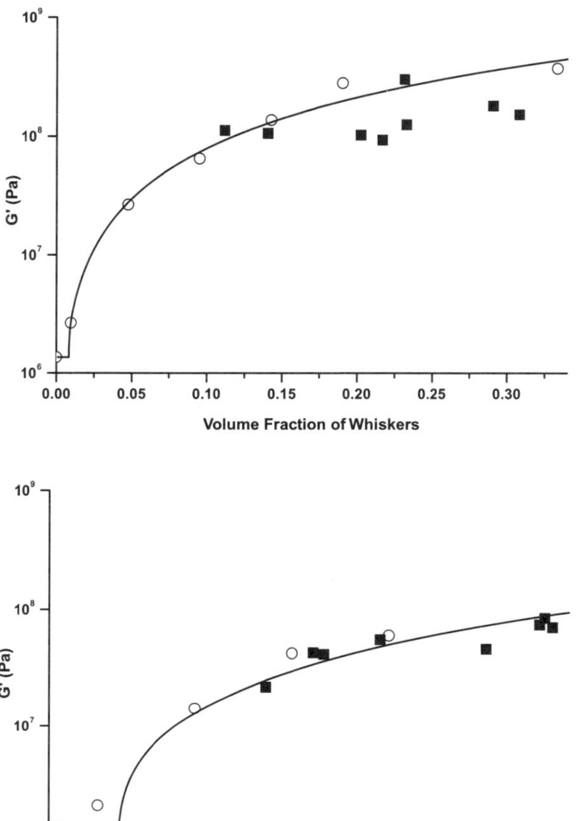

Fig. 3.3. Shear moduli G' of nanocomposites based on an EO-EPI matrix and CNCs isolated from tunicates (top) or cotton (bottom).[19] The nanocomposites were prepared by either solution-casting (open circles) or the template approach (filled squares) and the mechanical data were determined by dynamic mechanical analysis at 25°C. Solid lines represent the theoretical predictions calculated on the basis of a percolation model. (Reproduced from Ref. 19 with permission from Nature Publishing Group.)

mixing of the CNC dispersions in DMF with PBD solutions in toluene resulted in macroscopically phase-separated films, the template process afforded homogeneous materials that displayed the mechanical properties expected from percolating network structures. The incorporation of the CNCs did not affect the main relaxation temperature of the matrix polymers which is indicative of weak CNC–polymer interactions and points to the fact that the reinforcement is largely due to CNC–CNC interactions, which involve hydrogen bonding. It was previously shown that CNC-based polymer nanocomposites can be viewed as it simply mimics the architecture and function of the skin of sea cucumbers, which can change its stiffness when needed.[10,27] It was shown for several systems that the hydrogen-bonding interactions between the CNCs within the polymer matrix can be mediated by exposure to water.[30] Such stimuli-responsive, mechanically adaptive behavior was also observed for PBD/CNC and SBR/CNC nanocomposites made by the template approach. Exposure to water resulted in modest aqueous swelling and dramatic softening, consistent with disengagement of the CNC network as a consequence of competitive hydrogen bonding with water. Since the matrix polymers employed are hydrophobic, it appears that the CNCs create a percolating network of hydrophilic channels within the hydrophobic matrices.

3.3.2 *Re-dispersion of CNC organogels in organic solvents*

Suspensions of CNCs in organic solvents are routinely produced *via* lyophilization of aqueous dispersions and subsequent re-dispersion,[17] but re-dispersion is sometimes incomplete, especially in the case of CNCs with low aspect ratio. An alternative to lyophilization are solvent exchange schemes from water to organic solvents by repeated centrifugation and ultrasonication steps. These schemes can involve several transfers, for example from water to acetone to dichloromethane to toluene, and may lead to slight modifications of the surface energy of the nanocrystals and thus facilitate the dispersion of CNCs in common rather organic solvents.[31] As an alternative, it was shown that a simple sol-gel-sol exchange process may be used instead. Starting from aqueous dispersions comprising CNCs isolated from tunicates or cotton, at a concentration of ~5–8 mg/mL, Tang *et al.* prepared gels of both cellulose types by solvent exchange with acetone as described in Sec. 3.2.[25] Rather than imbibing the CNC gels with polymer solutions, the gels were placed into DMF, in which they disintegrated within minutes and without ultrasonication to form homogeneous CNC dispersions, from which the acetone could readily be eliminated by evaporation. The DMF dispersions were subsequently combined with an oligomeric, difunctional diglycidyl ether of bisphenol and a diethyl toluenediamine-based curing agent, the solvent was evaporated, and the mixture was cured into epoxy resin/CNC nanocomposites.

It is evident that this process is broadly applicable and can be used instead of lyophilization and re-dispersion or solvent exchange schemes. For example, the procedure was adapted to fabricate stimuli-responsive mechanically adaptive polyurethane/CNC nanocomposites, which change their mechanical properties upon exposure to water and display a water-activated shape-memory effect.[32] The method also served to produce environmentally benign cellulose acetate butyrate (CAB)/CNC nanocomposites.[33] This was achieved by fabricating ethanol gels of CNCs isolated from microcrystalline cellulose and re-dispersing these gels in ethanol solutions of the CAB; in this case, simply increasing the amount of solvent and adding the polymer (which may also interact with and solubilize the CNCs), rather than changing the solvent led to re-dispersion of the CNCs (Fig. 3.4). Casting and solvent evaporation afforded nanocomposite films, which in addition to the pronounced reinforcement in mechanical properties showed a high level of optical transparency (Fig. 3.4), which is indicative of a high level of dispersion.

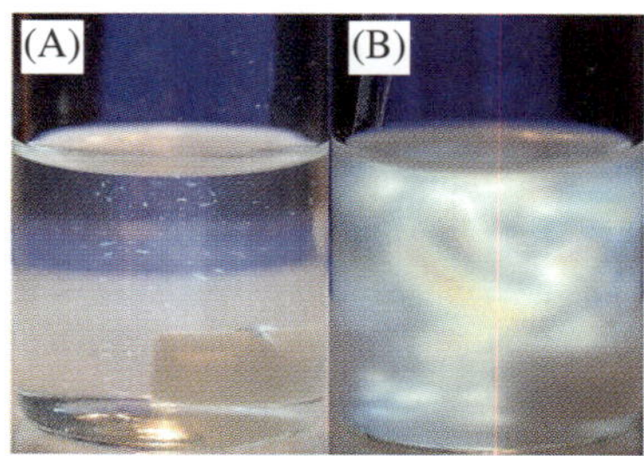

Fig. 3.4. Top: Appearance of (A) a solution of CAB with a piece of a CNC gel in ethanol and (B) a homogeneous solution of CAB with 9% w/w CNCs in ethanol, both viewed through crossed polarizers. Bottom: Visual examination of films of neat CAB and CAB/CNC nanocomposite containing 3, 6, and 12% w/w of CNCs, respectively.[33] (Reproduced from Ref. 33 with permission from Elsevier.)

3.4. Advanced soft materials made *via* sol-gel processes

3.4.1 *Electrically conductive gels and CNC aerogels*

The preparation of electrically conductive gels and aerogels has also been realized using a sol-gel process.[34] CNCs isolated from tunicates were first coated with the electrically conducting polymer poly(3,4-ethylenedioxythiophene): poly(styrene sulfonic acid) (PEDOT:PSS) by *in situ* polymerization of ethylenedioxythiophene (EDOT). Gels made by gelating an aqueous dispersion of the PEDOT: PSS-decorated CNCs thus produced by solvent exchange with acetone were dimensionally stable (Fig. 3.5) and electrically conductive. In the wet state, i.e., when swollen with acetone, the gels have a surface resistivity of *ca.* $200\,\mathrm{k}\Omega/\square$ and a bulk conductivity of *ca.* $6.6\,\mathrm{e}^{-6}\,\mathrm{S/cm}$. After drying into films, of rather inhomogeneous dimensions, the surface resistivity decreased to a level of *ca.* $100\,\Omega/\square$, and matched that of films made by casting PEDOT:PSS-coated CNCs from DMF.

3.5. CNC-based hydrogels

The versatility of the template approach was further exploited for the development of CNC-reinforced hydrogels,[35] with the intent of creating materials that have potential for use in biomedical applications such as articular cartilage. Here, the incorporation of CNCs served again the purpose of modifying the mechanical properties of these hydrogels, which are relevant for use in tissue engineering and drug delivery. In this study, sol-gel process was used to create CNC organogel (two different sources: aspen wood and cotton) from an aqueous CNC dispersion

Fig. 3.5. Picture of a PEDOT:PSS-coated CNC gel produced from an 8 mg/mL aqueous dispersion of the CNCs *via* solvent-exchange with acetone.[34] (Reproduced from Ref. 34 with permission from the Royal Society of Chemistry.)

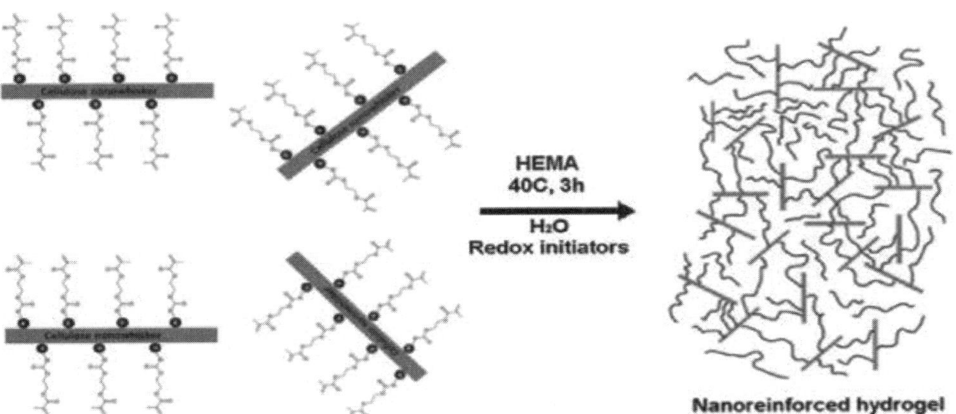

Fig. 3.6. Proposed mechanism for the formation of CNC-reinforced hydrogels.[35] (Reproduced from Ref. 35 with permission from Elsevier.)

and subsequent solvent exchange with acetone. Then, the hydrogels were created using "grafting from" approach to functionalize hemicellulose by *in situ* free radical polymerization of methacrylic groups immobilized on the CNC surfaces to develop multifunctional crosslinking sites. A proposed mechanism for the formation of the hydrogels is shown in Fig. 3.6. The characteristics of the produced hydrogels, such as mechanical, viscoelastic, and swelling properties, were found to be significantly influenced by the CNC content and the network structure, which was perceptibly created with the help of the sol-gel process.

3.6. Concluding remarks

Control over the nanoparticle dispersion remains a key issue for the design of homogenous polymer/nanofiber nanocomposites when ultimate transport properties (for example, thermal, electrical, or mechanical) are the target. Therefore, processes which allow one to produce polymer nanocomposites, which comprise a percolating, three-dimensional network of well-individualized CNCs, are important to the creation of a broad range of materials. The sol-gel processes presented here appear to be uniquely suited for this purpose, at least for quantities required for fundamental research studies. The available data seem to suggest that the methods permit rather straightforward access to nanocomposites of virtually any CNC type and polymer matrix, where the formation of percolating filler networks leads to outstanding and in some cases unusual transport properties. The scale-up of the key steps appears to be feasible, albeit the formation of gels would in an industrial setting have to be conducted in a more rapid manner (and under less-well-controlled conditions) than it is currently done on the laboratory scale. In addition, the minimization of volatile organic compounds required appears to be desirable. So far, only a limited number of examples for the sol-gel processing of CNC-based materials have been explored, but the versatility of the approaches makes a broader exploitation in the near future very likely.

Acknowledgments

The authors gratefully acknowledge financial support received from the Swiss National Science Foundation (NRP 62: Smart Materials, Nr. 406240_126046, NRP 64: Chances and Risks of Nanomaterials, Nr. 406440_131264/1, NRP 66: Resource Wood, Nr. 406640_136911/1), and the Adolphe Merkle Foundation.

References

1. G.M. Whitesides, Nanoscience, nanotechnology, and chemistry, *Small* **1** (2005) 172–179.
2. J.H. Fendler, Self-assembled nanostructured materials, *Chem. Mater.* **8** (1996) 1616–1624.
3. H.L. Frisch and J.E. Mark, Nanocomposites prepared by threading polymer chains through zeolites, mesoporous silica, or silica nanotubes, *Chem. Mater.* **8** (1996) 1735–1738.
4. M.J.D. Clift, E.J. Foster, D. Vanhecke, D. Studer, P. Wick, P. Gehr, B. Rothen-Rutishauser and C. Weder, Investigating the interaction of cellulose nanofibers derived from cotton with a sophisticated 3D human lung cell coculture, *Biomacromolecules* **12** (2011) 3666–3673.
5. D. Klemm, F. Kramer, S. Moritz, T. Lindström, M. Ankerfors, D. Gray and A. Dorris, Nanocelluloses: a new family of nature-based materials, *Angew. Chem. Int. Ed.* **50** (2011) 5438–5466.
6. S.J. Eichhorn, A. Dufresne, M. Aranguren, N.E. Marcovich, J.R. Capadona, S.J. Rowan, C. Weder, W. Thielemans, M. Roman, S. Renneckar, W. Gindl, S. Veigel, J. Keckes, H. Yano, K. Abe, M. Nogi, A.N. Nakagaito, A. Mangalam, J. Simonsen, A.S. Benight, A. Bismarck, L.A. Berglund and T. Peijs, Review: current international research into cellulose nanofibres and nanocomposites, *J. Mater. Sci.* **45** (2010) 1–33.

7. M.A.S.A. Samir, F. Alloin and A. Dufresne, Review of recent research into cellulosic whiskers, their properties and their application in nanocomposite field, *Biomacromolecules* **6** (2005) 612–626.
8. Y. Habibi and C. Weder, Cellulose nanocrystals: chemistry, self-assembly, and applications, *Chem. Rev.* **110** (2010) 3479–3500.
9. L. Hsu, C. Weder and S.J. Rowan, Stimuli-responsive, mechanically-adaptive polymer nanocomposites, *J. Mater. Chem.* **21** (2011) 2812–2822.
10. K. Shanmuganathan, J.R. Capadona, S.J. Rowan and C. Weder, Biomimetic mechanically adaptive nanocomposites, *Prog. Polym. Sci.* **35** (2010) 212–222.
11. V. Favier, H. Chanzy and J.Y. Cavaille, Polymer nanocomposites reinforced by cellulose whiskers, *Macromolecules* **28** (1995) 6365–6367.
12. C. Bonini, L. Heux, J.Y. Cavaille, P. Lindner, C. Dewhurst and P. Terech, Rodlike cellulose whiskers coated with surfactant: a small-angle neutron scattering characterization, *Langmuir* **18** (2002) 3311–3314.
13. C. Gousse, H. Chanzy, G. Excoffier, L. Soubeyrand and E. Fleury, Stable suspensions of partially silylated cellulose whiskers dispersed in organic solvents, *Polymer* **43** (2002) 2645–2651.
14. J. Araki, M. Wada and S. Kuga, Steric stabilization of a cellulose microcrystal suspension by poly(ethylene glycol) grafting, *Langmuir* **17** (2001) 21–27.
15. H.H. Yuan, Y. Nishiyama, M. Wada and S. Kuga, Surface acylation of cellulose whiskers by drying aqueous emulsion, *Biomacromolecules* **7** (2006) 696–700.
16. R.H. Marchessault, F.F. Morehead and N.M. Walter, Liquid crystal systems from fibrillar polysaccharides, *Nature* **184** (1959) 632–633.
17. O. van den Berg, J.R. Capadona and C. Weder, Preparation of homogeneous dispersions of tunicate cellulose whiskers in organic solvents, *Biomacromolecules* **8** (2007) 1353–1357.
18. J.R. Capadona, K. Shanmuganathan, S. Triftschuh, S. Seidel, S.J. Rowan and C. Weder, Polymer nanocomposites with nanowhiskers isolated from microcrystalline cellulose, *Biomacromolecules* **10** (2009) 712–716.
19. J.R. Capadona, O. Van Den Berg, L.A. Capadona, M. Schroeter, S.J. Rowan, D.J. Tyler and C. Weder, A versatile approach for the processing of polymer nanocomposites with self-assembled nanofibre templates, *Nat. Nanotechnol.* **2** (2007) 765–769.
20. J.R. Capadona, K. Shanmuganathan, S. Trittschuh, S. Seidel, S.J. Rowan and C. Weder, Polymer nanocomposites with nanowhiskers isolated from microcrystalline cellulose, *Biomacromolecules* **10** (2009) 712–716.
21. C. Weder, J.R. Capadona and O. Van Den Berg, Self-assembled nanofiber templates; versatile approaches for polymer nanocomposites. US 7935745 (2011).
22. M. Samir, F. Alloin, J.Y. Sanchez, N. El Kissi and A. Dufresne, Preparation of cellulose whiskers reinforced nanocomposites from an organic medium suspension, *Macromolecules* **37** (2004) 1386–1393.
23. C. Weder, J.R. Capadona and O. Van Den Berg, Self-assembled nanofiber templates; Versatile approaches for polymer nanocomposites. US 20080242765A1 (2008).
24. G. Siqueira, A.P. Mathew and K. Oksman, Processing of cellulose nanowhiskers/ cellulose acetate butyrate nanocomposites using sol–gel process to facilitate dispersion, *Compos. Sci. Technol.* **71** (2011) 1886–1892.
25. L.M. Tang and C. Weder, Cellulose whisker/epoxy resin nanocomposites, *ACS Appl. Mater. Interfaces* **2** (2010) 1073–1080.
26. K.L. Dagnon, K. Shanmuganathan, C. Weder and S.J. Rowan, Water-triggered modulus changes of cellulose nanofiber nanocomposites with hydrophobic polymer matrices, *Macromolecules* **45** (2012) 4707–4715.

27. J.R. Capadona, K. Shanmuganathan, D.J. Tyler, S.J. Rowan and C. Weder, Stimuli-responsive polymer nanocomposites inspired by the sea cucumber dermis, *Science* **319** (2008) 1370–1374.
28. K. Shanmuganathan, J.R. Capadona, S.J. Rowan and C. Weder, Bio-inspired mechanically-adaptive nanocomposites derived from cotton cellulose whiskers, *J. Mater. Chem.* **20** (2010) 180–186.
29. P.K. Annamalai, K.L. Dagnon, E.J. Foster, S.J. Rowan and C. Weder, Water-responsive mechanically adaptive nanocomposites based on styrene-butadiene rubber and cellulose nanocrystals – processing matters, submitted.
30. R. Rusli, K. Shanmuganathan, S.J. Rowan, C. Weder and S.J. Eichhorn, Stress-transfer in anisotropic and environmentally adaptive cellulose whisker nanocomposites, *Biomacromolecules* **11** (2010) 762–768.
31. G. Siqueira, J. Bras and A. Dufresne, New Process of Chemical Grafting of Cellulose Nanoparticles with a Long Chain Isocyanate, *Langmuir* **26** (2010) 402–411.
32. J. Mendez, P.K. Annamalai, S.J. Eichhorn, R. Rusli, S.J. Rowan, E.J. Foster and C. Weder, Bioinspired Mechanically Adaptive Polymer Nanocomposites with Water-Activated Shape-Memory Effect, *Macromolecules* **44** (2011) 6827–6835.
33. G. Siqueira, A.P. Mathew and K. Oksman, Processing of cellulose nanowhiskers/cellulose acetate butyrate nanocomposites using sol-gel process to facilitate dispersion, *Compos. Sci. Technol.* **71** (2011) 1886–1892.
34. J.D. Mendez and C. Weder, Synthesis, electrical properties, and nanocomposites of poly(3,4-ethylenedioxythiophene) nanorods, *Polym. Chem.* **1** (2010) 1237–1244.
35. M.A. Karaaslan, M.A. Tshabalala, D.J. Yelle and G. Buschle-Diller, Nanoreinforced biocompatible hydrogels from wood hemicelluloses and cellulose whiskers, *Carbohydr. Polym.* **86** (2011) 192–201.

Chapter 4

Processing of Bionanocomposites: Solution Casting

Aji P. Mathew and Kristiina Oksman
Composite Centre Sweden, Division of Materials Science,
Luleå University of Technology, Luleå, Sweden

Solution casting is the oldest technology to make plastic films and was developed in the 19th century to produce photographic films by Eastman Kodak. Solution casting is an easy and versatile method to produce nanocomposite thin films/sheets in laboratory scale. In the solution casting of polymer nanocomposites, the polymer phase is dissolved in water or a non-aqueous volatile solvent and mixed with nanosized reinforcements in the same solvent medium prior to casting on a flat surface. The solvent phase is removed by evaporation and thereafter the dried film is released from the substrate. This chapter discusses the solution casting process of nanocomposites where biobased nanofibers or crystals are used as the reinforcing phase. The effect of processing route on the composites morphology and mechanical properties is discussed.

4.1. Introduction

Solution casting is usually a low temperature process and provides films with uniform thickness, optical purity, low haze, and isotropy. The disadvantages of solution casting are slow processing speed, use of solvents, the cost for solvent recovery, and that the method is limited to laboratory-scale process in most of the cases.

Biobased nanosized reinforcements that are commonly used in bionanocomposites are cellulose nanofibers (CNFs), cellulose nanocrystals (CNCs), chitin nanofibers (ChNFs), chitin nanocrystals (ChNCs), and starch nanocrystals (SNCs).[2–6] Biobased nanoreinforcements are obtained and stored in the form of aqueous dispersions, and therefore, solution casting is the easiest and most versatile method to produce nanocomposites, especially when using polymers that are soluble or dispersive in water.[2,5,6]

In cases where polymers that are insoluble in water are used, the nanoreinforcement dispersion is usually a solvent exchanged with appropriate non-aqueous solvents before mixing it with the polymer solutions and casting into films.[3,7,8] The typical process of solution casting of a nanocomposite material is schematically shown in Fig. 4.1.

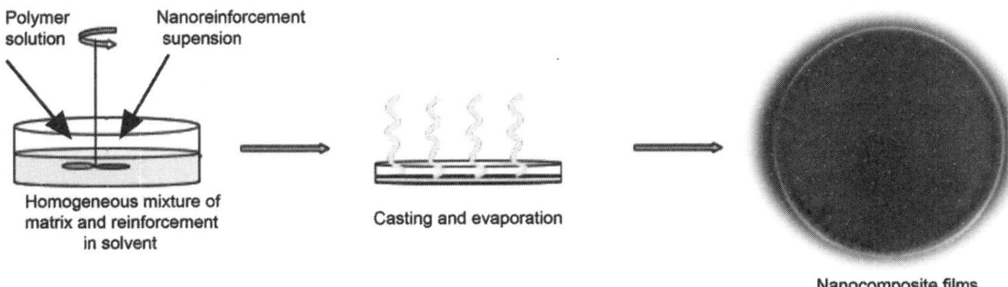

Fig. 4.1. Schematic representation of a typical solution casting technique.

The pioneering studies on cellulose nanocomposites were conducted at CERMAV-CNRS, Grenoble, France during 1990s in which majority of these studies used the solution casting technique as the processing technique of choice.[9–11]

The first studies on cellulose nanocomposites using solution casting used water-soluble or dispersible matrix materials such as starch or latex and the reinforcements were CNCs, also called nanowhiskers (CNWs). The process is simple and no specific processing equipment is needed to make nanocomposites films. Furthermore, good dispersion was usually obtained taking advantage of the common solvent effect (water) for the polymer phase and the reinforcing phase. It is also well known that the slow evaporation step allows the formation of a two-dimensional network of nanocrystals or nanofibers in the matrix phase which was seen to be very advantageous for mechanical properties of these materials. On the other hand, this slow evaporation step sometimes resulted in settling of the nanocrystals in the polymer solution and a concentration gradient in the resultant nanocomposites and non-homogeneous dispersion of the reinforcements in the matrix.[12]

This chapter discusses the processing of nanocomposites by solution casting by (i) casting from aqueous medium and (ii) casting from non-aqueous medium. The effect of the processing route on the dispersion and the mechanical properties of the resultant nanocomposite will also be discussed.

4.2. Solution casting of bionanocomposites in aqueous medium

Solution casting from aqueous medium is the easiest and the most straightforward route to produce biobased nanocomposites as the nanoreinforcements under consideration are most stable in aqueous medium. In this case, the matrix phase used should be water-soluble or dispersive polymers (such as latex).

4.2.1 *Nanocomposites of starch and CNCs*

4.2.1.1 *Processing*

Starch is an easy matrix to choose when considering processing of biobased nanocomposites by solution casting from aqueous medium. A homogeneous dispersion of the CNCs in starch can be obtained taking advantage of the solubility in aqueous medium.

Nanocomposites were prepared from waxy maize starch plasticized with sorbitol or glycerol as the matrix and a stable aqueous suspension of tunicin crystals — an animal source for cellulose — as the reinforcing phase.[13,14] Starch and plasticizer were first mixed and dispersed in water. The mixture contained 10 wt% of a waxy maize starch, 5 wt% sorbitol, and 85 wt% water which resulted in starch films with 33 wt% plasticizer after drying. For nanocomposite preparation, different percentages of starch and plasticizer were mixed in different ratios of nanocrystals to obtain composite films. The plasticizer content was fixed at 33 wt% (dry basis of starch matrix). The CNC content varied between 0 and 25 wt%. The gelatinization of starch was performed in a stirred autoclave reactor operating at 160°C for 5 min. After the mixing, the suspension was degassed under vacuum in order to remove air and then cast in a Teflon mold which was stored at 70°C under vacuum to allow water evaporation.

The determination of the disappearance of ghosts was carried out by optical microscopy, and the absence of starch degradation was controlled by visual inspection of film appearance, degradation leading to a tanning of resulting films.

4.2.1.2 *Morphology*

The scanning electron microscopy (SEM) images shows the fractured surfaces of sorbitol-plasticized starch and its nanocomposites with different CNCs contents (Fig. 4.2).

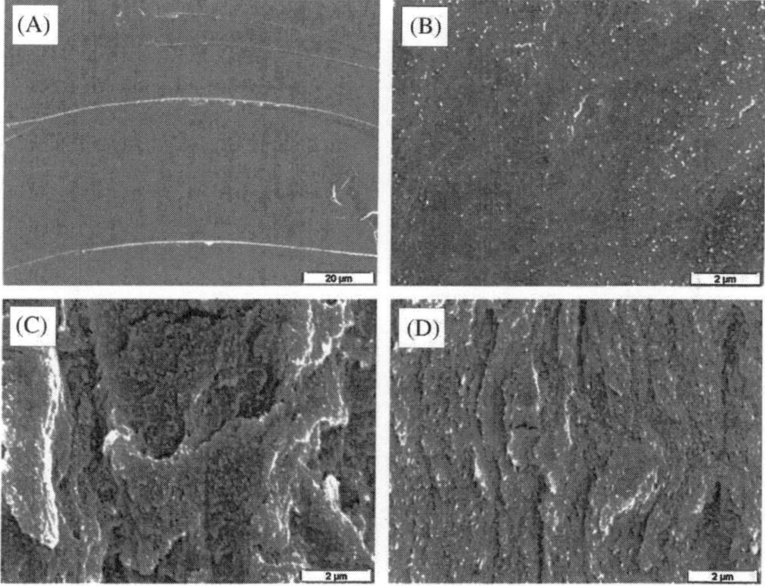

Fig. 4.2. Scanning electron micrographs of the fractured surfaces of (A) sorbitol-plasticized starch and its nanocomposites with (B) 5 wt%, (C) 15 wt%, and (D) 25 wt% tunicin crystals. (Reprinted with permission from Ref. 14, Copyright 2002 American Chemical Society.)

The tunicin CNCs appear as white dots, which are distributed evenly throughout the plasticized starch matrix. As the crystal content increases in the composites, an increase in the number of white dots can be seen. A uniform distribution of crystals in the matrix is clearly seen especially in composites with higher crystal content (15 and 25 wt%). Such an even and uniform distribution of the reinforcing phase in the matrix is essential for obtaining optimum mechanical performances. A very similar morphology was also presented by glycerol-plasticized starch nancomposites.[13]

4.2.1.3 *Properties*

Starch usually has poor moisture resistance and the incorporation of CNCs was shown to be an effective way to decrease its moisture sensitivity and to obtain better mechanical properties.

In Fig. 4.3 the equilibrium water uptake for glycerol- and sorbitol-plasticized starch–cellulose nanocomposites is plotted as a function of crystal content. In the glycerol nanocomposite system, the moisture uptake decreases with increasing nanocrystal content, which may be attributed to the uniform dispersion of CNC in starch matrix obtained during the solution casting process. Furthermore, the sorbitol-plasticized starch nanocomposite showed lower moisture sensitivity at all studied compositions. It was found that the unfilled sorbitol-plasticized matrix absorbs around 39.5% water compared to glycerol-plasticized waxy maize starch (62% water). An explanation can be proposed based on the chemical structure of both plasticizers. The chain length of sorbitol is about twice that of glycerol. Therefore, in glycerol the end hydroxyl groups, which are expected to be more accessible to water, are about twice those compared to sorbitol. At higher whiskers contents, it seems that the cellulose filler mainly influences the moisture sorption, since the

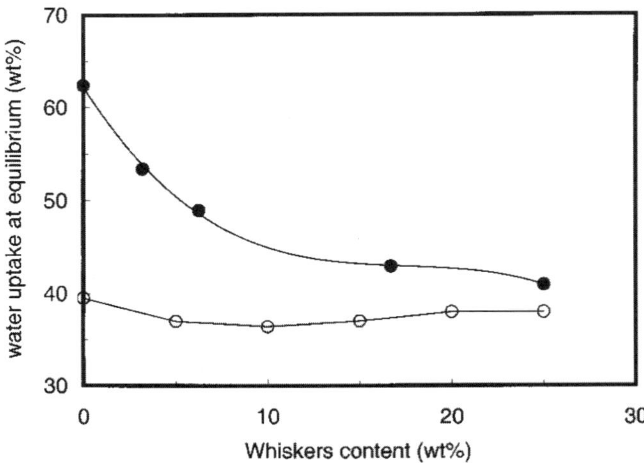

Fig. 4.3. Maximum relative water uptake, at 98% relative humidity (RH) for glycerol (solid circles) and (B) sorbitol (open circles) plasticized waxy maize starch with tunicin nanocrystals as a function of the crystal content. (Reprinted with permission from Ref. 14, Copyright 2002 American Chemical Society.)

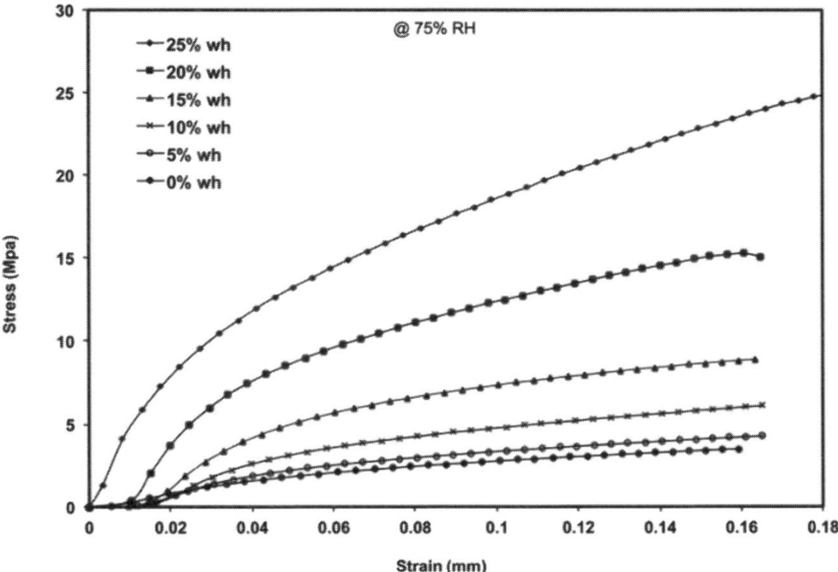

Fig. 4.4. Effect of nanocrystal content on the stress–strain curves of sorbitol-plasticized starch conditioned at 75% RH.

uptake becomes similar for both plasticizer-based materials. This can be ascribed to the formation of a cellulose network through hydrogen bonding between whiskers whereby the role of the matrix becomes negligible.

The plasticization by water appears to have a very significant effect on the composite performance irrespective of the plasticizer used. Generally, in these composites, the tensile strength and Young's modulus are high at lower RH levels, and elongation at break remains constant, irrespective of RH and filler content.

Figure 4.4 shows the mechanical properties of sorbitol-plasticized starch/tunicin nanocrystals nanocomposites, conditioned at 75% RH.[15] The elongation at break remains roughly constant around 16–18%. The large amount of water present in the system plasticizes the matrix, and the elongation at break is largely independent of CNC concentration. This shows that the plasticizer content governs the flexibility of the system and chain extensibility. The strength and the Young's modulus increased gradually with increased CNC content, showing that the crystals are the load-bearing entity in the composites.

4.2.2 *Nanocomposites of chitosan and ChNCs*

Chitosan is also a suitable matrix for solution casting of biobased nanocomposites as chitosan forms a homogeneous solution in 2 wt% acetic acid which can be directly mixed with aqueous suspensions of cellulose or ChNCs. The chitosan phase can also be easily crosslinked after solution casting to produce more chemically and mechanically stable nanocomposites.[16]

4.2.2.1 *Processing*

Chitosan with 81% deacetylation was used as matrix phase and crab shell chitin was used as starting material to produce ChNCs. Glutaraldehyde solution was used as crosslinker for chitosan. Aqueous dispersion of ChNCs (3 wt%) was mixed with 2 wt% chitosan solutions to obtain a final composition of 5 and 10 wt% of chitin crystals in a chitosan matrix.

The mixtures were homogenized by magnetic stirring for 12 h and sonicated for 5 min prior to casting. The formulations were left to evaporate in the vacuum oven at temperature of 45°C for 48 h. The prepared films had a thickness of 0.05–0.08 mm and a total dry weight of 0.5 g. These films were further neutralized by immersing in 1 M NaOH and then repeatedly washed with distilled water and allowed to dry at room temperature. To obtain crosslinking, the films were kept immersed in 0.02% (w/v) glutaraldehyde solution for 48 h at room temperature and then washed with distilled water and dried in air at room temperature.

4.2.2.2 *Morphology*

The chitosan and the nanocomposites prepared were in the form of thin films and had high optical clarity and the addition of 5 and 10 wt% of the chitin crystals did not affect the optical clarity of the films significantly.[16] In the case of the crosslinked chitosan and its nanocomposites, the films were optically clear but showed a light brown color. This coloration is typical for chitosan/glutaraldehyde gels and is attributed to the formation of a chromophoric imine group (–N=C–) during the crosslinking.

The SEM images in Fig. 4.5 show the fractured surface of the crosslinked and uncrosslinked chitosan and its nanocomposites with 10% chitin content. There is no visible difference in the morphology between the matrix and the nanocomposite and it is not possible to see any agglomeration on the fractured surface in this magnification, which indicates uniform dispersion of chitin crystals in the chitosan matrix. No concentration gradient of the chitin crystals was observed.

The atomic force microscopy (AFM) study to get more detailed information of the chitosan–chitin composites (uncrosslinked and crosslinked) is shown in Figs. 4.5E and F. The images show a two-phase system with a uniform dispersion. The lighter colored entities dispersed in the continuous phase are considered as chitin crystals embedded in the chitosan matrix phase. It is also possible to see that the crystals are randomly oriented in the matrix.

The results of the sorption experiments are presented as the percentage uptake of the water by the crosslinked and uncrosslinked chitosan matrix and chitosan–chitin as a function of time are given in Fig. 4.6. The average total uptake (wt%) for the uncrosslinked matrix and the nanocomposites was 180%, 150%, and 115% of water and about 80%, 70%, and 60% of the total water uptake occurred within the first 5 min. In the case of crosslinked system, the total water uptake was lowered to 105% for the matrix and 70% for the nanocomposites. In the crosslinked systems

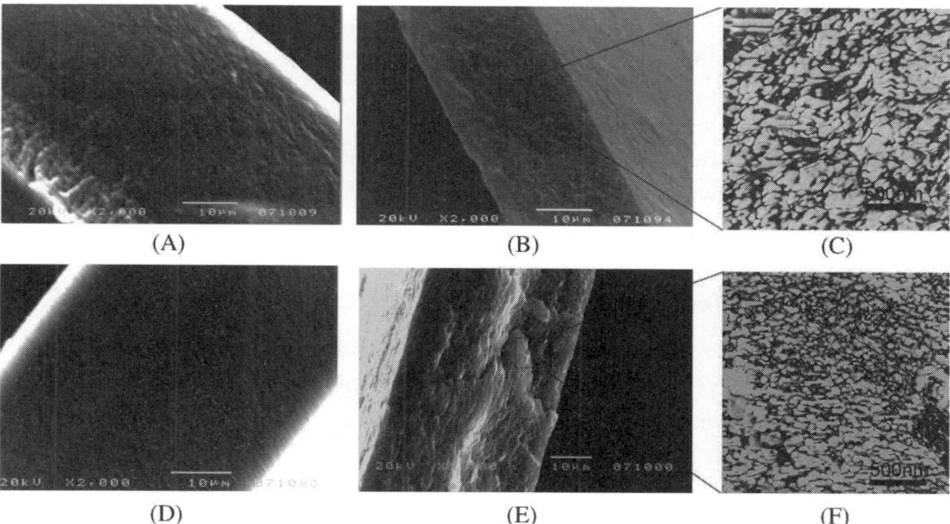

Fig. 4.5. Cross section of chitosan and chitosan–chitin nanocomposites films with and without crosslinking. (A) Overview of the chitosan film, (B) chitosan–chitin film, (C) crosslinked chitosan, (D) crosslinked chitosan–chitin film and a more detailed view of (E) chitosan–chitin, and (F) crosslinked chitosan–chitin films.

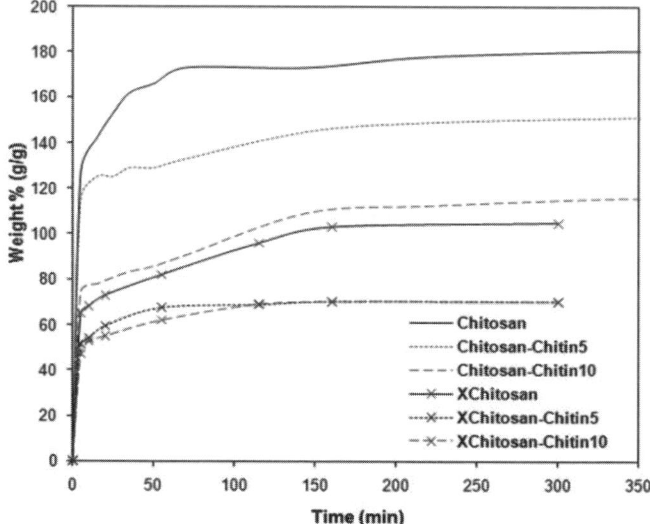

Fig. 4.6. Effect of crosslinking and chitin crystal concentration on the water uptake of chitosan–chitin nanocomposites (Reprinted with permission from Ref. 16, Copyright 2009 American Chemical Society.)

also, the uptake during the first 5 min accounted for 60–70% of the total water uptake by the system.

The sorption studies conducted using acidified water with a pH of 3 showed that crosslinked materials are very stable at this pH, whereas the uncrosslinked nanocomposites dissolved in the acidic water.[12] The crosslinked materials showed higher uptake of about 400% by weight of water in acidic medium. The uptake decreased with increase in ChNC content. It is, therefore, confirmed that crosslinking allows these chitosan nanocomposites to be used in acidic medium, while the uniformly distributed chitin crystals contribute to the perm-selectivity of the nanocomposite membranes.

4.2.3 *Hydrogel nanocomposites of poly(methyl vinyl ether-co-maleic acid)*

This study is about crosslinked nanocrystal composites of poly(methyl vinyl ether-co-maleic acid) (PMVEMA) and poly(ethylene glycol) (PEG) which showed to act as strong hydrogels with good mechanical properties.[17,18]

4.2.3.1 *Processing*

The PMVEMA–PEG matrix was mixed with different concentrations of CNCs to obtain CNCs contents of 0%, 25%, 50%, 75%, and 100%, by weight, in each film. The PMVEMA–PEG/CNC mixture was then solution cast onto Teflon Petri dishes, air dried overnight, and then cured at 135°C for 6.5 min. After curing, the films were allowed to cool to room temperature and stored in a desiccator at 54% RH for one week prior to further testing.

The crosslinking of CNCs with PMVEMA and PEG is anticipated to occur during the curing of the solution-casted films, via an esterification reaction between the C6 hydroxyl groups on the cellulose, terminal hydroxyl groups of PEG, and the carboxylic acid groups on the PMVEMA. The possibilities for intermolecular and interchain reactions exist as well.

4.2.3.2 *Morphology*

The AFM study of crosslinked nanocomposites with 50% CNCs is shown as a representative sample. Figure 4.7A is an overview of the surface showing light and dark phases, the light phase is expected to be the CNCs.

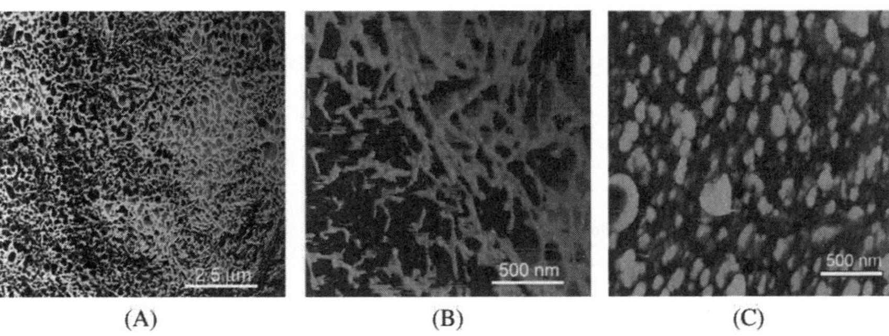

Fig. 4.7. AFM images of PMVEMA-CNC co-crosslinked composites: (A) overview, (B) detailed view of the film surface, and (C) detailed view of the film cross section.

The AFM images obtained from the film surface (Figs. 4.7A and B) show a two-phase system with a continuous network of reinforcing phase (light colored structures) in a soft phase (darker regions). In Fig. 4.7C, ultramicrotomed cross section of the film is shown. In this image, the oval-shaped structures are assumed to be the cellulose crystals embedded in the matrix phase. A broad range of size distribution is observed for the crystals in the nanocomposites images, which may be due to the fact that the crystals are cut at different angles during sample preparation. This is expected, as the crystals are randomly oriented in solution-casted films. The morphology study indicated that CNCs are very uniformly distributed in the matrix during solution casting and is further facilitated by the *in situ* co-crosslinking which prevents aggregation.

4.2.3.3 *Water sorption*

The water sorption studies show that the prepared nanocomposites can swell in water and form a strong and stable gel. The water sorption curves for the nanocomposite films are shown in Fig. 4.8. 100% CNC films and 100% PMVEMA-PEG films did not retain structural integrity throughout the entire water retention experiment. However, the nanocomposite films swelled in water and retained their film structure. This can be considered as an indication that the crosslinking reaction between the whiskers and the matrix yielded a networked gel. The time taken by the 25% CNC film to reach equilibrium was significantly longer than for both the 50% and 75% CNC films. In addition, at equilibrium, the 25% CNC film absorbed significantly higher amount of water (~900%) compared to both 50% CNC films (~100%) and 75% CNC films (~150%). In 50% CNC film, it appeared that there was a balance

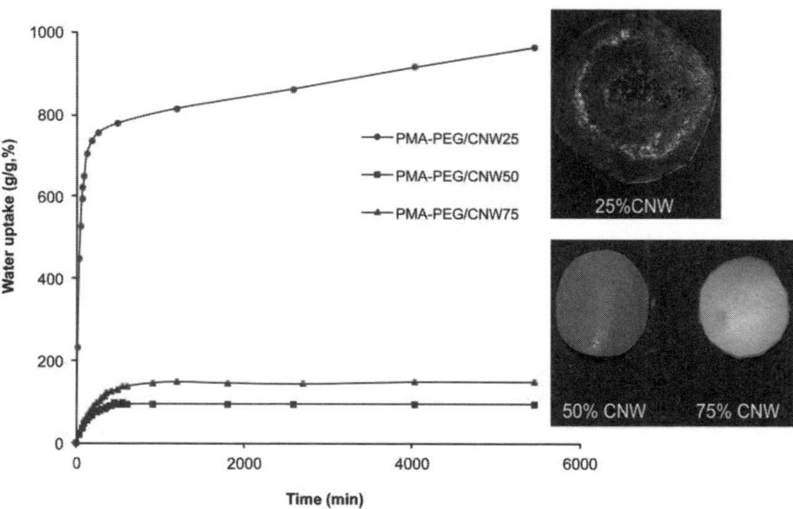

Fig. 4.8. Water uptake of co-crosslinked PMVEMA-PEG and CNC nanocomposites and appearance of the swollen materials with different CNC contents.

between the increased crosslinking and the presence of high amount of CNC, which added up to decrease the ability of the matrix to absorb water molecules.

4.2.4 *Nanocomposites of collagen–cellulose using in situ fibrillation and crosslinking*

Inspired by the hierarchical structure, biocompatibility, non-toxicity, and mechanical properties of the two well-known biopolymers, viz., collagen and cellulose, attempts are made to develop biomaterials with high mechanical properties and good biocompatibility using solution casting technique. In these materials, *in situ* fibrillation of collagen and crosslinking of collagen is carried out prior to solution casting.[19,20]

4.2.4.1 *Processing*

CNF suspensions (1 wt%) were thoroughly mixed with collagen suspension (0.5 wt%) in acetic acid and stirred overnight at room temperature to obtain a homogeneous mixture. The suspension weights were carefully adjusted during the mixing process to have 50 or 75 wt% CNFs in the final composite. Suspensions were cast on Petri dishes and dried at 35°C for 24 h to obtain flat films. These films were treated with saturated solution of NaCl for 3 h and washed with distilled water to remove all acid and residual NaCl. Finally, the films were dried completely by keeping in a vacuum oven at 35°C for 12 h.

The crosslinking of the nanocomposites were carried using 1 vol% glutaraldehyde or 0.04 mM genipin. After crosslinking, the films were washed and dried in vacuum oven at 35°C for 12 h.

4.2.4.2 *Morphology*

The morphology of the developed materials was studied using SEM. Figure 4.9A shows the microstructure of the collagen film after the pH-induced fibrillation. The film consists of regenerated collagen fibrils in a matrix phase of non-fibrous collagen, indicating that a partial fibrillation of collagen was achieved. In the crosslinked collagen, shown in Fig. 4.9B, the fractured surface was smoother and two phases (fibrous and non-fibrous) were visible. Figure 4.9C shows uncrosslinked collagen–cellulose nanocomposites and Figs. 4.9D and E show collagen composites, crosslinked using glutaraldehyde and genipin, respectively. In these images, CNFs were easily distinguishable, but the collagen fibrils were mostly invisible. It was also possible to see that in all cases, the uncrosslinked materials had rougher fracture surfaces compared to the corresponding crosslinked ones. It may be considered that the collagen fibrils and cellulose fibrils are embedded in a matrix of collagen though the two types of fibrils were not easily distinguishable from each other.

The nanocomposite materials were also studied using AFM and the images were obtained on XColl-Cell75 in tapping mode. Figure 4.10 shows the three-dimensional height image and it is possible to see the CNFs as well as collagen fibrils that were

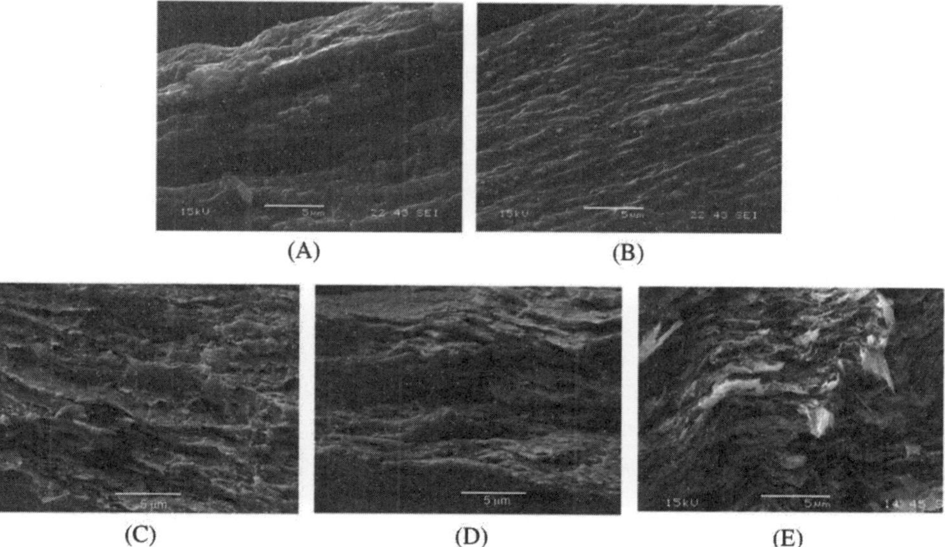

Fig. 4.9. Morphology of (A) collagen (Coll), (B) crosslinked collagen (Coll) compared with (C) Coll-Cell75 (D) glutaraldehyde crosslinked Coll-Cell75, and (E) genipin crosslinked Coll-Cell75.

Fig. 4.10. Three-dimensional AFM image of genipin crosslinked Coll-Cell75 nanocomposite with broader collagen fibrils embedded in finer CNFs.

regenerated (collagen fibrils and cellulose fibrils are marked in Fig. 4.10). It is also seen that the collagen fibrils as well as cellulose nanofibrils intermingle with each other, supporting the *in situ* fibrillation of collagen phase that has occurred during the processing step.

4.2.4.3 *Mechanical properties*

Table 4.1 shows the data for tensile testing, in room conditions. The increase in mechanical properties is remarkable in these composites and crosslinked as well

Table 4.1. Mechanical properties of fibrous nanocomposites based on collagen and CNFs.

Sample	Strength (MPa)	Strain at break (%)	E-modulus (GPa)
Collagen	56.2 ± 12.8	3.8 ± 1.0	3.5 ± 0.7
Cellulose	80.0 ± 5.0	3.2 ± 1.4	2.6 ± 0.1
Coll-Cell75	96.2 ± 12.2	$5.1 \pm 1.0*$	7.4 ± 0.8
XColl-Cell75 (glu)	132.6 ± 16.2	5.5 ± 1.2	5.8 ± 1.2
XColl-Cell75 (gen)	186.2 ± 34.2	25.4 ± 0.5	14.5 ± 0.7

as uncrosslinked nanocomposites showed better mechanical properties than both collagen and the CNF films.

The mechanical properties of the pure components (collagen fibers and CNFs), strength, strain at break, and stiffness increased when the materials were combined to a mix with 25% collagen. The crosslinking of the composite showed a further positive impact on the strength. The glutaraldehyde crosslinking of collagen phase increased the strength to 133 MPa for composites 75% CNF fibers, which is roughly 250% higher than pure collagen. This synergistic effect in nanocomposites (both uncrosslinked and crosslinked) indicates good chemical compatibility between the two phases, as well as good dispersion. Additionally, the *in situ* pH-induced fibrillation of collagen during solution casting seems to have positive impact on the mechanical characteristics of these materials. When the collagen phase gets partially fibrillated, collagen fibrils formed will develop intimate physical entanglement with CNFs while maintaining good compatibility and chemical interaction with the collagen matrix phase. In the crosslinked composites, the inter-crosslinking between collagen matrix and collagen fibrils is expected. It may be envisioned that collagen fibrils act as compatibilizer in the composites by providing physical interaction with CNFs and chemical interaction with collagen matrix.

The solution-casted films of genipin XColl-Cell75 composites showed good biocompatibility, whereas the glutaraldehyde crosslinked ones had poor biocompatibility. Cell adhesion and proliferation of ligament cells (HLC) cells showed good adhesion and proliferation on genipin crosslinked nanocomposites. These nanocomposites demonstrate the potential and flexibility of solution casting technique in the production of materials for high-end medical applications.

4.3. Nanocomposites of non-aqueous soluble polymers

4.3.1 *Polylactic acid nanocomposites*

4.3.1.1 *Processing*

Petersson *et al.*[21] prepared polylactic acid (PLA)-based CNC nanocomposites. The CNCs were pretreated to make them dispersible in PLA solution, using the following routes.

Route 1: Neutralized CNCs suspensions in aqueous medium were freeze-dried to remove water. The freeze-dried whiskers were then redispersed in chloroform, by sonication.

Route 2: Neutralized CNCs suspensions in aqueous medium were solvent exchanged to tert-butanol using a series of centrifugation steps and the aqueous phase is replaced with tert-butanol and was dispersed well. The suspensions were then freeze-dried to remove the solvent. The freeze-dried whiskers were then redispersed in chloroform, by sonication.

Route 3: The surfactant was added to the CNC suspension in the ratio 4:1 and the pH of the suspension was adjusted to 8.5 using 1 wt% NaOH solution. The suspension was then freeze-dried and subsequently redispersed in chloroform and used for nanocomposite processing.

The goal of the pretreatment processing was to restrict the aggregation on CNCs during freeze-drying and to obtain a good dispersion of CNCs in non-aqueous medium, i.e., chloroform which is the solvent for PLA. The nanocomposites were prepared by solution casting. The CNCs dispersed in chloroform prepared by routes 1, 2, and 3 were mixed with the PLA solution in chloroform and stirred well to obtain uniform dispersion. The homogeneous mixtures were then poured in Petri dishes and left to evaporate in room temperature for 1 day. The films were then placed in a vacuum oven at 40°C for 2 weeks to remove all remaining chloroform. The prepared films had a thickness of 0.25 mm.

4.3.1.2 *Morphology*

The morphology study of the nanocomposites showed interesting results (Fig. 4.11). The PLA–CNCs composites showed poor distribution of CNCs and bigger aggregates were visible on the fracture surface. The CNCs appeared to have better distributed in PLA–CNC prepared by solvent exchange to tert-butanol. In the case of composite with surfactant-treated CNCs, the dispersion was even better and it was not possible to see any agglomerations. However, some pores were visible in this case, which may be due to the foam generated in the mixtures during processing, due to the presence of surfactant. The nanostructure study using transmission electron microscopy (TEM) also showed that the dispersion was improved by the solvent exchange to tert-butanol and the use of surfactant compared to the control sample. PLA–CNC composites with tert-butanol showed regions of loose agglomerates with PLA chains penetrating between the CNCs. The PLA–CNC nanocomposites with surfactant showed more even distribution of CNCs throughout the sample.

4.3.1.3 *Dynamic mechanical thermal properties*

The DMTA study of the nanocomposites showed that the incorporated crystals were able to restrict the motion of the PLA chains and thereby increase the temperature stability of PLA.[21] The dispersion of the crystals inside the matrix was of great importance since it governed the available surface area of the nanocrystals. When a surfactant is used to improve the dispersion of the crystals, it is important to investigate the relationship between the surfactant and the matrix. The interaction between the surfactant and the matrix should be less pronounced and the amount

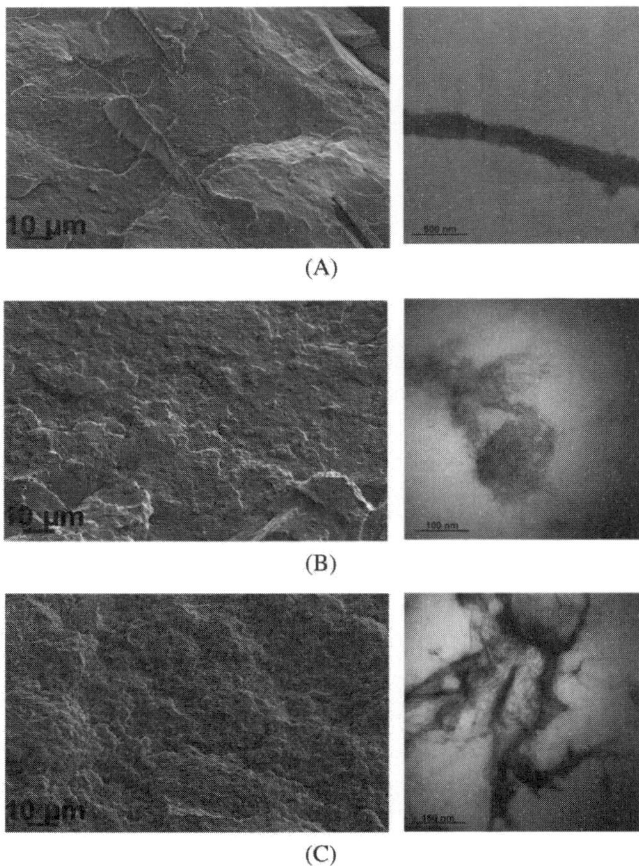

Fig. 4.11. SEM and TEM images of (A) PLA–CNC, (B) PLA–CNC with tert-butanol, and (C) PLA-CNC with surfactant.

of surfactant should be optimized in order for the whiskers to reinforce the matrix both in the elastic and plastic zones. It is possible that the surfactant was coating the crystals and preventing a direct interaction between the PLA and the cellulose crystals.

4.3.2 *Cellulose acetate butyrate nanocomposites*

Cellulose acetate butyrate (CAB-553-0.4) has a butyryl content of 46 wt%, acetyl content of 2 wt%, and hydroxyl content of 4.8%. CAB is soluble in low molecular weight alcohols (methanol, ethanol, isopropanol, and *n*-propanol) as well as other common organic solvents. The melting temperature is between 150 and 160°C and films of CAB are transparent and have good ultraviolet stability. CAB is supplied as a dry and free-flowing powder.[22]

CNCs suspensions in aqueous medium were solvent exchanged to acetone using a series of centrifugation steps to replace the aqueous phase with acetone and was dispersed well in acetone by sonication.

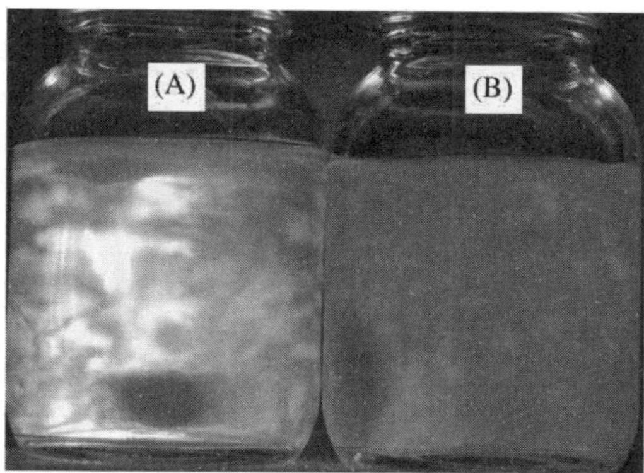

Fig. 4.12. Birefringence of suspensions of CNWs in (A) aqueous medium and in a (B) CAB acetone solution.

The CAB and CAB nanocomposites films were prepared by solution casting. A mixture of CAB, TEC (plasticizer), and different weight percentages of CNCs in acetone was prepared by stirring on a hot magnetic plate at 50°C for 3 h until a homogenous mixture was obtained. The prepared formulations were sonified for 5 min prior to casting on well-dried glass Petri dishes. The prepared casting suspensions showed birefringence indicating good dispersion of CNCs in CAB solution (see Fig. 4.12). The formulations were left to evaporate in the vacuum oven at 45°C for 48 h. The prepared films had a thickness of ≈0.15 mm and a total dry weight of 3 g.

4.3.2.1 *Morphology*

The AFM images of the unplasticized and plasticized CAB nanocomposites with 10 wt% CNCs (overview and detailed view) are shown in Fig. 4.13. All the nanocomposites showed a homogeneous two-phase system without any microscaled agglomerates. It is possible to see a second phase embedded in the continuous phase and having various sizes and shapes which may be CNCs embedded in the matrix. The sizes and shapes of the white colored spots were found to vary considerably which indicates that CNCs exist mostly as nanoscale aggregates and/or are oriented in different directions in the matrix, which is an expected behavior of nanocomposites prepared by solution casting. The dispersion and distribution of CNCs in the matrix were very uniform in the plasticized and unplasticized composites even at CNC concentration of 10 wt%.

The transparency study of these nanocomposites indicated very good nanoscale dispersion of CNCs in the CAB matrix. The plasticized CAB/CNC composites were more transparent than the corresponding unplasticized ones, which indicate that the plasticizer helps in the uniform dispersion of whiskers.

4.3.2.2 *Dynamic mechanical thermal properties*

The values for the storage modulus (E') and the tan δ peak of the nanocomposites are given in Table 4.2 for comparison. At 30°C, CAB/CNC5 nanocomposite showed

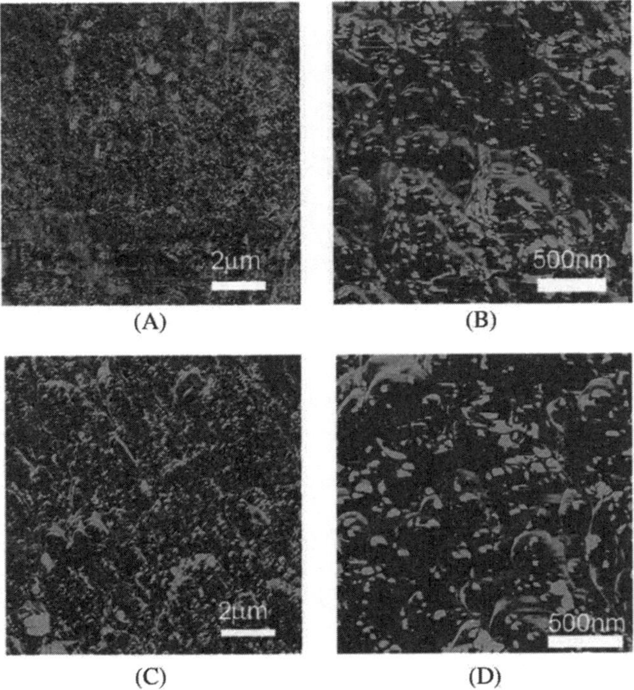

Fig. 4.13. The morphology of unplasticized CAB/CNC composites: (A) overview and (B) detailed view, and plasticized CAB/CNW composites: (C) overview and (D) detailed view.

Table 4.2. Storage modulus and tan δ of CAB and CAB nanocomposites.

	Storage modulus E' (GPa)			
Sample code	30°C	80°C	155°C	tan δ peak (°C)
CAB	1.3 ± 0.1	1.0 ± 0.1	2.9 ± 0.8	154 ± 0
CAB/CNC5	1.4 ± 0.4	1.2 ± 0.2	25.6 ± 0.4	157 ± 0
CAB/CNC10	1.8 ± 0.1	1.4 ± 0.6	85.4 ± 0.0	155 ± 0
CAB/TEC	1.2 ± 0.0	0.8 ± 0.0	0.4 ± 0.1	147 ± 1
CAB/TEC/CNC5	1.3 ± 0.0	0.8 ± 0.0	0.4 ± 0.2	142 ± 1
CAB/TEC/CNC10	1.4 ± 0.1	1.0 ± 0.1	2.1 ± 0.9	147 ± 1

an increase of 8%, while at same temperature, CAB/CNC10 nanocomposite showed an increase of 32% in the storage modulus. At 80°C, CAB/CNC5 nanocomposite showed an improvement of the storage modulus with 19%, which corresponded to 189 MPa. At a temperature of 155°C (see Table 4.2), CAB/CNC5 and CAB/CNC10 showed an increase of 788% and 2865%, respectively. The tan δ peak temperature showed a very slight shift toward higher temperature for both CAB/CNW nanocomposites compared to CAB.

The storage modulus increased at 30°C with 3% and 17% for CAB/TEC/CNC5 and CAB/TEC/CNC10 while at 80°C, the increase was 10% and 31%, respectively. The tan δ peak, for the plasticized nanocomposites did not show any tendency to shift to higher temperatures, compared to plasticized CAB. Comparing the unplasticized and plasticized nanocomposites, it can be seen that the improvement in modulus is more prominent in unplasticized nanocomposites. This was contrary to the expectation that the plasticizer would ultimately give rise to better performance of the nanoreinforcements by facilitating the dispersion of the nanowhiskers in the matrix. But we believe that in this case the plasticizer was functioning as a lubricant between CAB and the whiskers and thereby weakened the interface to some extent, even with better dispersion.

4.4. Conclusions

Solution casting has been extensively used as an easy and versatile method to produce green nanocomposites with a biopolymer as matrix phase and the biobased nanoreinforcement as reinforcing phase.

Solution casting process can involve casting and evaporation using aqueous phase or a non-aqueous phase depending on the type of the matrix polymer. In the case of water-soluble polymers like starch, chitosan, collagen, etc., the solution casting technique is quite straightforward and usually results in nanocomposites with homogeneous dispersion. The studies highlighted in this chapter demonstrate the potential to perform matrix crosslinking, co-crosslinking between matrix and nanoreinforcements, or *in situ* fibrillation of one phase during the processing by solution casting.

In the case of non-water-soluble polymers as matrix polymer, addition steps like solvent exchange, freeze-drying, use of surfactant are needed prior to casting and evaporation.

Improved mechanical properties and water resistance were observed in several biobased nanocomposites prepared by solution casting. Furthermore, crosslinking of matrix phase or co-crosslinking of matrix with the nanoparticles provided better mechanical and moisture stability for these nanocomposites. Some of these solution-casted nanocomposites have shown potential as biomedical scaffolds, hydrogels, barrier layers, optically clear films, etc.

References

1. D. Collins, *The Story of Kodak.* New York: HN Abrams Inc (1990).
2. A. Dufresne, D. Dupeyre and M.R. Vignon, Cellulose microfibrils from potato tuber cells: processing and characterization of starch–cellulose microfibril composites, *J. Appl. Polym. Sci.* **76** (2000) 2080–2092.
3. A.P. Mathew, A. Chakraborty, K. Oksman and M. Sain, The structure and mechanical properties of cellulose nanocomposites prepared by twin screw extrusion. In *Cellulose Nanocomposites Processing, Characterization and Properties*, K. Oksman and M. Sain (Eds.), ACS Symposium Series 938, Oxford University Press (2006), Ch 9, pp. 114–131.
4. K.-Y Lee, J.J. Blaker and A. Bismarck, Surface functionalisation of bacterial cellulose as the route to produce green polylactide nanocomposites with improved properties, *Compos. Sci. Technol.* **69** (2009) 2724–2733.

5. A. Morin and A. Dufresne, Nanocomposites of chitin whiskers from riftia tubes and poly (caprolactone), *Macromolecules* **35** (2002) 2190–2194.
6. H. Angellier, S.M. Boisseau, L. Lebrun and A. Dufresne, Processing and structural properties of waxy maize starch nanocrystals reinforced natural rubber, *Macromolecules* **38** (2005) 3783–3792.
7. M. Grunert and W.T. Winter, Nanocomposites of cellulose acetate butyrate reinforced with cellulose nanocrystals, *J. Polym. Environ.* **10**(2) (2002) 27–30.
8. I. Kvien, B.S. Tanem and K. Oksman, Characterization of cellulose whiskers and their nanocomposites by atomic force and electron microscopy, *Biomacromolecules* **6**(6) (2005) 3160–3165.
9. V. Favier, J.Y. Cavaille and H.Chanzy, Polymer nanocomposites reinforced by cellulose whiskers, *Macromolecules* **28** (1995) 6365–6367.
10. V. Favier, G. Canova, C.Y. Cavaille, H. Chanzy, A. Dufresne and C. Gauthier, Nanocomposite materials from latex and cellulose whiskers, *Polym. Adv. Technol.* **6** (1995) 351–355.
11. W. Helbert, J.Y. Cavaille and A. Dufresne, Thermoplastic nanocomposites filled with wheat straw cellulose whiskers. Part I: processing and mechanical behavior, *Polym. Comp.* **17** (1996) 604–611.
12. J. Bras, M.L. Hassan, C. Brizesse, E.A. Hassan, N.A. El-Wakil and A. Dufresne, Mechanical, barrier, and biodegradability properties of bagasse cellulose whiskers reinforced natural rubber nanocomposites, *Ind. Crops Prod.* **32**(3) (2010) 627–633.
13. M.N. Anglés and A. Dufresne, Plastisized starch/tunicin whiskers nanocomposites. 1. Structural analysis, *Macromolecules* **33** (2000) 8344–8353.
14. A.P. Mathew and A. Dufresne, Morphological investigation of nanocomposites from sorbitol plasticized starch and tunicin whiskers, *Biomacromolecules* **3** (2002) 609–617.
15. A.P. Mathew, W. Thielemans and A. Dufresne, Mechanical properties of nanocomposites from sorbitol plasticized starch and tunicin whiskers, *J. Appl. Polym. Sci.* **109** (2008) 4065–4074.
16. A.P. Mathew, M.-P., Laborie and K. Oksman, *Biomacromolecules* **10**(6) (2009) 1627–1632.
17. L. Goetz, A.P. Mathew, K. Oksman, P. Gatenholm and A.J. Ragauskas, A novel nanocomposite film prepared from crosslinked cellulosic whiskers, *Carbohydr. Polym.* **75**(1) (2009) 85–89.
18. L. Goetz, M. Foston, A.P. Mathew, K. Oksman and A.J. Ragauskas, Poly(methyl vinyl ether-co-maleic acid)-polyethylene glycol nanocomposites cross-linked in situ with cellulose nanowhiskers, *Biomacromolecules* **11**(10) (2010) 2660–2668.
19. A.P. Mathew, K. Oksman, D. Pierron and M.-F. Harmand, Fibrous nanocomposite scaffods prepared by partial dissolution for potential use as ligament or tendon substitutes, *Carbohydr. Polym.* **87**(3) (2012) 2291–2298.
20. A.P. Mathew, K. Oksman, D. Pierron and M.-F. Harmand, Crosslinked fibrous composites based on cellulose nanofibers and collagen with in situ PH induced fibrillation, *Cellulose* **19** (2012) 139–150.
21. L. Petersson, I. Kvien and K. Oksman, Structure and thermal properties of poly(lactic acid)/cellulose whiskers nanocomposite materials, *Compos. Sci. Technol.* **67** (2007) 2535–2544.
22. J.-E. Ayuk, A.P. Mathew and K. Oksman, The effect of plasticizer and cellulose nanowhiskers content on the dispersion and properties of cellulose acetate butyrate nanocomposites, *J. Appl. Polym. Sci.* **114**(5) (2009) 2723–2782.

Chapter 5

Melt Compounding Process of Cellulose Nanocomposites

Kristiina Oksman and Aji P. Mathew
Composite Center Sweden, Division of Materials Science,
Luleå University of Technology, Luleå, Sweden

Compounding of nanocomposites using twin-screw extrusion is a process whereby nanocellulosic materials are mixed with a polymer melt. The process can be scaled up to produce cellulose nanocomposites for industrial-scale use. This process method may also allow nanocomposites to be easily injection molded or compression molded. Our aim has been to develop a cellulose nanocomposite processing method using two specific processing routes: (i) liquid feeding of the nanomaterials into the extruder and (ii) preparation of a master batch which is diluted to the desired concentration during a compounding process. The prepared nanocomposites have usually been compression molded to sheets or injection molded to samples. The studies on nanocomposite structure and properties have enabled optimization of the process as well as improved performance. Depending on the extent of separation of cellulose nanocrystals or nanofibers in the liquid medium and the interaction of nanocelluloses with the polymer matrix, different types of nanocomposites have been obtained: composites with aggregated nanocelluloses, partially dispersed nanocellulose, or fully dispersed nanocelluloses. Ultimately, the objective of these studies has been to produce nanocomposites with good mechanical properties, thermal stability, and transparency and at the same time develop an energy-efficient and cost-effective processing methodology.

5.1. Introduction

Cellulose nanocomposites research began in the early 1990s, when a research group at CERMAV-CNRS in Grenoble, France, led by Prof. Cavaille R,[1] started to study different cellulose crystals and their reinforcing potential for different polymers. The first attempts were made with latex and tunicate nanocrystals (also called whiskers) and the second with polypropylene and wheat straw nanocrystals. The group also worked with starch polymers and showed that the addition of cellulose nanocrystals (CNCs) resulted in interesting properties.[1–3] A couple of years later W. Glasser *et al.*[4] and Winter and Grunert[5] made nanocomposites with cellulose acetate as matrix. All the first studies were made in a laboratory scale using the solvent casting method and the researchers did not focus on composite processing development but were mainly interested in studying the effects of additions of nanosized cellulose materials in different polymers.

Cellulose-based nanocomposites became a very popular research topic in the early 2000s and many new research groups entered this field.[1–8] From these earlier works, it was concluded that the separation of nanosized reinforcement from natural materials and the processing techniques were limited to laboratory scale.[1–7] Therefore, there was a need to develop new processing techniques which could be used in industrial scale. Early in 2002, K. Oksman's research group at the Norwegian University of Science and Technology began to study the compounding extrusion of CNCs with polylactic acid (PLA) as matrix polymer and published a first article on that topic.[8] The hypothesis in this work was to feed the CNCs into the extruder via a liquid medium instead of dry powder. The idea came from an earlier study in which a liquid plasticizer was pumped into the PLA melt during extrusion of natural fiber composites to be able to improve the composite's toughness.[9] Later, the same group presented a solution for cellulose nanofiber (CNF) composites. The group used different matrix materials and developed a master batch, which could be fed as dry power into the extruder.[10–22]

The main challenges in melt compounding of nanocomposites are to feed the nanomaterials to the extruder, disperse them in the melt polymer without causing degradation of the polymer or the used nanomaterials during the process, and also to remove liquid which is used as feeding medium. From an economic point of view, liquid feeding is more beneficial than dry feeding. Supplementary treatment of the suspension before extrusion, such as freeze-drying, will involve additional expenses like investment costs for equipment, longer processing times, and higher energy consumption. Also, from an environmental point of view, risk of free-flying extremely small nanofibers or crystals is minimized by keeping them in a liquid suspension. This chapter will give the reader a short introduction to the compounding technology and the nanocomposite materials and their properties.

5.2. Compounding extrusion

The compounding process is a way to incorporate different additives or reinforcements into a melt polymer. Compounding can be batchwise or continuous. This chapter will only discuss the continuous compounding process.

Continuous compounding is done using either co-rotating or counter-rotating twin-screw extruders. The co-rotating extruders have a more flexible screw design and different processing zones which are arranged in series, resulting in better dispersion and distribution of the nanomaterials in the polymer melt than can be achieved with the counter-rotating extruder.

Screw elements are important for nanomaterial dispersion and distribution (see Fig. 5.1). First, kneading elements help to melt the polymer and also contribute to the dispersion and distribution of the nanomaterials in the melt, mixing elements and backward-pumping elements also contribute to the mixing.[23]

This chapter will discuss the compounding process using co-rotating extruders. Figure 5.2 shows the screw configuration with marked processing zones.

Fig. 5.1. Different screw elements used in compounding. (1) Conveying elements, (2) kneading elements, (3) dispersing elements, and (4) distributing elements.

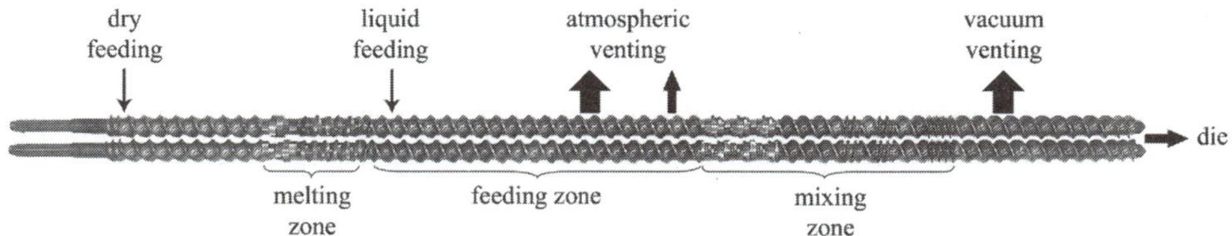

Fig. 5.2. Screw configuration showing the different processing zones.

Fig. 5.3. Compounding facility with Coperion Werner and Pfleiderer ZSK extruder suitable for preparation of cellulose nanocomposites.

In Fig. 5.3, a compounding facility with a co-rotating twin-screw extruder from Coperion Werner & Pfleiderer (ZSK 25 WLE) with a gravimetric feeding system for dry materials (K-Tron) and a peristaltic pump (Heidolph) for liquid feeding as well as vacuum venting for removal of excess liquid is shown. This is a laboratory-scale facility but can be easily upgraded to larger scale for manufacturing of different composite materials or polymer blends.

Fig. 5.4. DSM mini-extruder setup, showing the conical screws with conveying capability but no capability for dispersion or distribution.

There are also so-called mini-extruders on the market, which are used in the nanocomposite preparation. These are usually designed for batchwise processing but can be made continuous by allowing materials to pass the screws several times. These should not be confused with compounding extruders, because usually no effective dispersion or distribution of nanomaterials can be done and the screw configuration cannot be changed. Figure 5.4 shows a DSM mini-extruder; note that the screws are not modular as they are in the Coperion compounding extruder, and cannot be designed in the same way. Also, there are no venting zones to remove the liquid or air from the system and the screws are very short. But the system is ideal for preparation of small amounts of premixed materials or polymers and is easy to maintain and clean.

5.3. Processing strategies for compounding nanocomposites

One of the main challenges in compounding of cellulose nanocomposites is the feeding of the nanomaterials into the extruder. If the cellulose nanomaterials are dried, they are very light, in which case, the feeding will be difficult due to the low bulk density of the material. Dry nanosized materials in the form of free-flying nanoparticles might also cause health problems in the working environment. The drying process is not only expensive but will bond the CNFs or crystals very strongly together (hydrogen bonding and hornification) and it is very difficult to separate these after drying. Therefore, two processing strategies where this aggregation is avoided have been developed. The first is liquid feeding or wet feeding[8–13] of nanocellulose materials, and the second is master-batch preparation and its use in the extrusion of composites.[14–22]

5.3.1 *Liquid feeding of cellulose nanomaterials*

Liquid feeding, or liquid vehicle processing, means that CNFs or crystals are dispersed in a liquid medium, such as water, in which the nanocellulose materials are easily dispersed. Different solvents, which are evaporated during the extrusion process, and different plasticizers together with water or solvents have also been used as liquid medium.[8–13]

This type of process was reported for the first time by Oksman *et al.*[8] in 2006. PLA and CNC composites were prepared and N,N-dimethylacetamide (DMAc) with lithium chloride (LiCl) was used as a liquid vehicle for the pumping of the CNCs into the PLA melt. DMAc/LiCl was chosen because it was shown to act as an isolation agent for CNC and a medium in which CNC could be easily dispersed. In this study, polyethylene glycol (PEG1500) was also used as a processing aid and a maleic anhydride (MA)-grated PLA was used as a processing aid and coupling agent.[8]

The compounding was carried out in an extruder, shown in Fig. 5.5. The temperature ranged from 170°C at start to 185°C at the die. PLA polymer was fed into the main inlet and the dispersed nanocrystals were pumped into the melt polymer using a peristaltic pump. The liquid phase was removed by atmospheric venting and by vacuum venting.[8]

The speed of the main feeder and the pump was adjusted to final composition of CNC to be 5 wt%. The processing aid and coupling agent was premixed with the PLA polymer and fed into the main inlet. The final content of the additives were: coupling agent 10 wt% (PLA–MA), plasticizer 15 wt% (PEG), and nanoreinforcements 5 wt% (CNC). The liquid phase was 20 wt% and was removed during extrusion. The codes for the tested material combinations were PLA_{DMAC} as reference material, the composite with coupling agent $PLA_{DMAC}/PLA\text{–}MA_{10}/CNW_5$, and composite with plasticizer and coupling agent $PLA_{DMAC}/PEG_{15}/PLA\text{–}MA_{10}/CNW_5$. Transmission electron microscopy images of the material confirmed that the nanocrystals were partially dispersed in a PLA matrix but also some dark spots <500 nm were seen. Figure 5.6 shows the transmission electron microscopy image of the composite nanostructure with overview and detailed view.

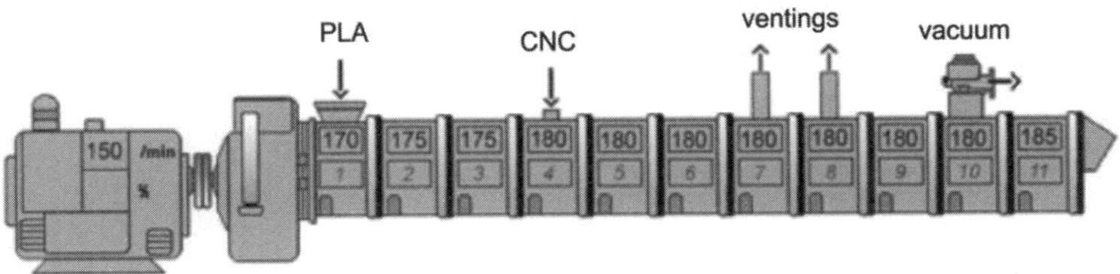

Fig. 5.5. Schematic view of the extrusion process showing the feeding of PLA polymer first into the extruder and liquid vehicle with CNC and PEG pumped downstream in the polymer melt. Also shown are the temperature profile and location of used venting ports.

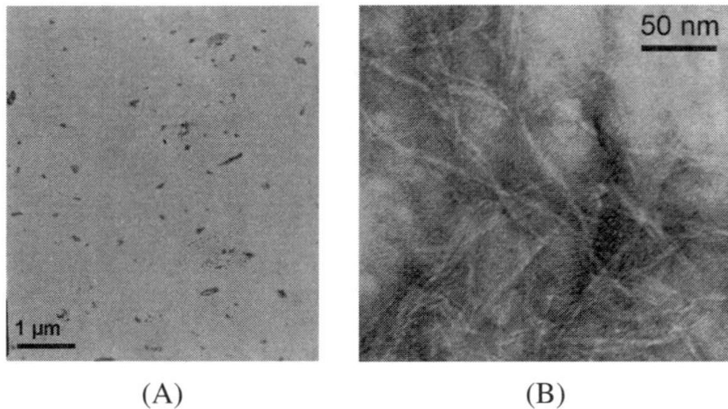

(A) (B)

Fig. 5.6. Transmission microscopy image of PLA/CNC composite: (A) an overview image with some small aggregates and (B) detailed view where separated nanocrystals are seen.

Table 5.1. Mechanical properties of the nanocomposites compared to PLA. (Reproduced from Ref. 8.)

Mechanical property	PLA	PLA-(PLA-MA_{10})-CNW_5	PLA-(PLA-MA_{10})-PEG_{15}-CNW_5
Tensile modulus (GPa)	2.9 ± 0.1	3.9 ± 0.3	2.6 ± 0.2
Strength (MPa)	41 ± 3	78 ± 7	48 ± 4
Elongation (%)	1.9 ± 0.2	2.7 ± 0.5	17.8 ± 9

The mechanical properties of materials were studied to get an idea of how the addition of CNCs affected the composites' modulus, strength, and elongation to break. The composites' mechanical properties are shown in Table 5.1.

The composites' properties are compared with reference PLA, which was subjected to same process as the studied nanocomposites. Both nanocomposites showed better mechanical properties and the composite with partially dispersed CNCs with the plasticizer showed 9 times better elongation. The ductility of this material was also much better than the others. The strength was improved in both nanocomposites and the E-modulus was slightly decreased when the plasticizer was used, as expected. This first study showed that even a very low content of CNCs can positively affect for the composite's mechanical properties and the effect was highest on the composite strength. Similar increase in the composite elongation was also seen in another cellulose nanocomposite, reported by Petersson and Oksman,[10] and an increase in all mechanical properties, including elongation to break, was observed when cellulose acetate butyrate (CAB) was used as matrix. This behavior is explained by the combination of nanosized crystals dispersed in a low molecular weight polymer and the brittle matrix. It is also possible that PEG interacts with cellulose crystals, covers them, and therefore improves the dispersion of the CNC, which improves toughness but will lower the reinforcing effect of CNC. Generally, the addition of CNC had a positive effect on the mechanical properties of the composites.[8]

Another example of liquid feeding is all-cellulose nanocomposites, where CAB was used as a matrix.[11] This CAB matrix differs from PLA. Although it is also

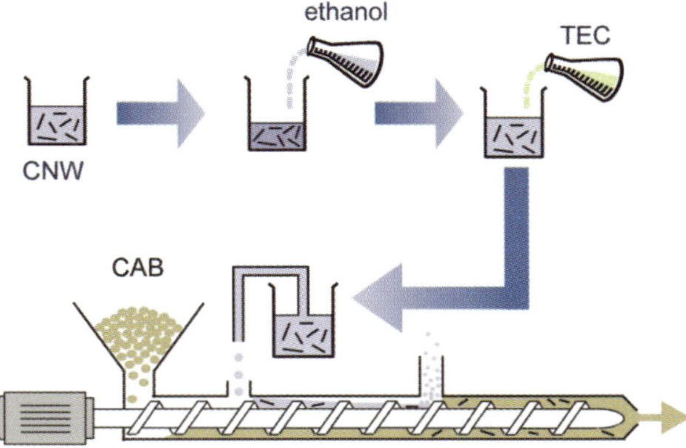

Fig. 5.7. Schematic view of the compounding processing route for cellulose nanocomposites, showing the liquid vehicle consisting the nanocrystals (CNC), ethanol and plasticizer (TEC), which is pumped into the melt CAB.

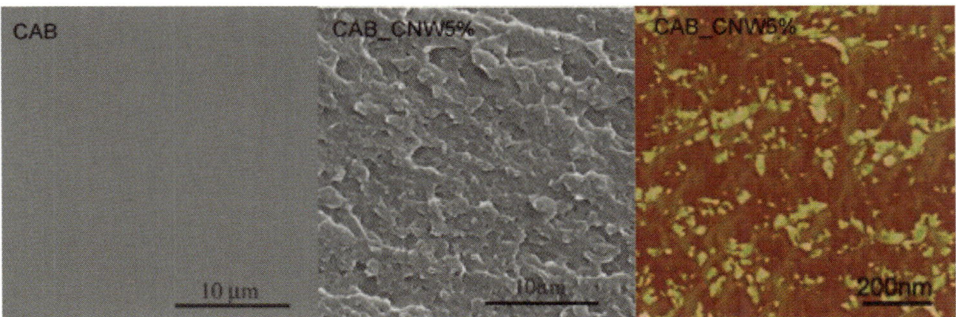

Fig. 5.8. Scanning electron micrographs of the fractured surfaces of the CAB matrix with and without 5 wt% nanocrystals, showing that an addition of CNC affected the fracture behavior of the CAB polymer. The AFM image shows the dispersion and distribution of the crystals.

based on renewable raw material, it is not as biodegradable as PLA and it needs to be plasticized, otherwise it is very brittle. The processing strategy in this case was that CNCs were dispersed in a mixture of biobased plasticizer (triethyl citrate (TEC)) and water/ethanol mixture and that liquid was fed to the extruder. The TEC is expected to act as plasticizer and also prevents agglomeration of the crystals. Figure 5.7 shows a schematic image of the processing route.

Fractured surfaces of matrix and nanocomposites showed that the addition of only 5% CNC had a large impact on the material's microstructure. The CAB matrix had smooth surface, while the composites have a rough surface with many small dots. The comparison of the microstructures can be seen in Fig. 5.8. Also, an atomic force microscopy (AFM) study on the materials showed that the nanocrystals were well dispersed and distributed in the CAB polymer.

The addition of CNC also showed a positive impact on the composite thermo-mechanical properties; the composite dynamic modulus increased with increased temperature, meaning that the operating temperature of this material increased

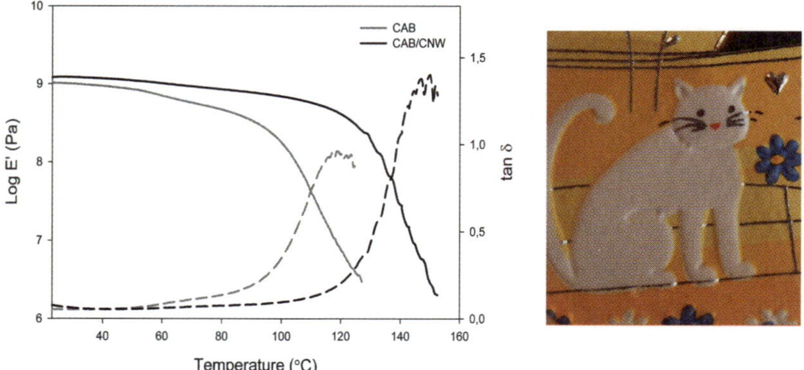

Fig. 5.9. Thermo-mechanical properties of CAB matrix and its nanocomposite with 5% nanocrystals, showing a positive effect of the addition on the modulus, especially at higher temperature and also the shift of the tan delta peak toward higher temperature. Also, the optical transparency of the nanocomposite is demonstrated.

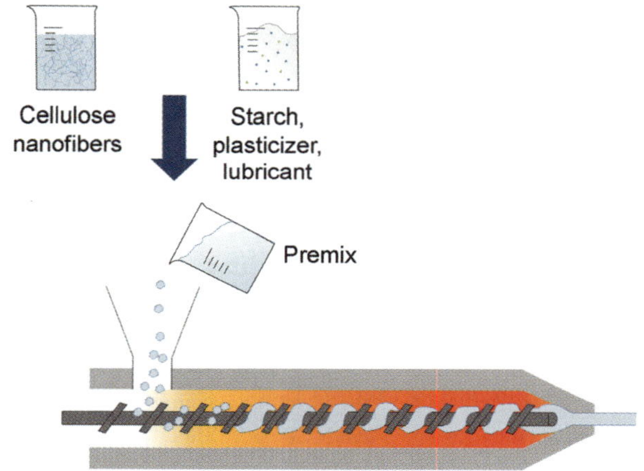

Fig. 5.10. Process setup for the starch–cellulose nanocomposite.

from approximately 100 to 140°. The nanocomposite was also very transparent, as seen in Fig. 5.9, where the cat is partly covered by the nanocomposite film.

The last example is starch nanocomposite, where the thermoplastic starch (TPS) and cellulose nanocomposite were produced in one step.[12] CNF gel with 2% CNF was premixed with starch powder and sorbitol plasticizer and fed to the extruder manually. Usually, the cooking process of TPS is done in the presence of water; therefore, the high water content of CNFs is not a problem with this polymer.[12]

The plasticizer content was 30 wt% based on the dry weight of starch. Stearic acid was also used as a processing aid, preventing the compound from sticking to the screws or clogging the die. The processing temperature was between 90 and 110°C and the water was removed during the process using the atmospheric ventings. The process setup is shown in Fig. 5.10.

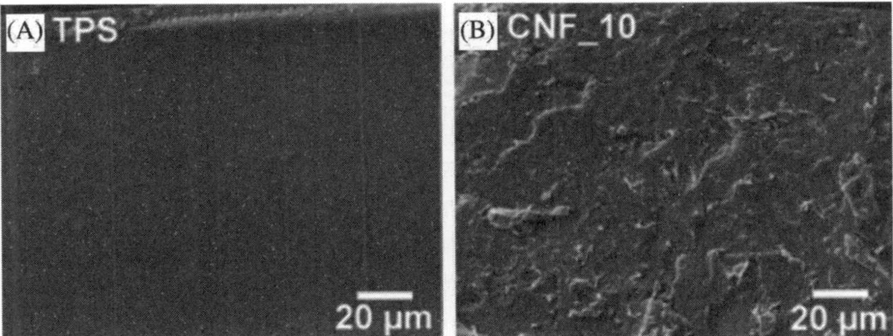

Fig. 5.11. Scanning electron microscopy (SEM) micrographs of fractured surface of TPS (A) and (B) the nanocomposite with 10 wt% CNF. The TPS has a smooth surface and the nanocomposite is rough with visible dots (CNF) relatively well distributed and dispersed in the matrix.

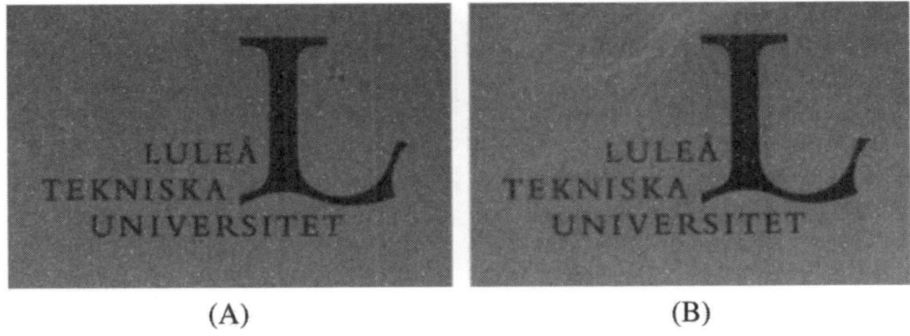

Fig. 5.12. Visual transparency of (A) TPS and (B) the nanocomposites with 10 wt% CNF showing that nanofibers decreased the transparency slightly.

Generally, it was no problem to compound the materials; but the high water content resulted in low melt viscosity and the extrudates were quite sticky. The moisture content of the extrudates was reduced after the process by drying the materials in room temperature and in a controlled humidity chamber for 2–3 days before compression molding to sheets. The fractured surfaces of the TPS and the composite are shown in Fig. 5.11. The pure TPS has a smooth surface, while the nanocomposite looks different. The fibers are well visible and seem to be relatively well dispersed and distributed in the TPS.[12]

Transparency can be used as an indirect measurement of the size and dispersion of CNFs in the matrix, since, if the reinforcement is not in nanoscale (non-fibrillated cellulose fibers, aggregated nanofibers), the light transmittance of the material decreases due to the increased light scattering. Figure 5.12 shows the optical transparency of the pure TPS film compared with the composite. It can be seen that the optical transparency of the neat TPS film (Fig. 5.12A) is very high and that it was reduced with the addition of cellulose fibers (Fig. 5.12B).

Table 5.2 shows the mechanical properties of the TPS and the nanocomposite. The mechanical properties of the TPS films improved with the addition of 10 wt% CNFs.[12]

Table 5.2. Mechanical properties of the TPS and its nanocomposite with 10 wt% NCF content.

Materials	Strength (MPa)	E-modulus (GPa)	Strain at break (%)
TPS	24.2 ± 1.3^{a}	1.36 ± 0.06^{a}	2.6 ± 0.3^{a}
TPS/10CNF	27.2 ± 2.0^{b}	1.83 ± 0.11^{b}	2.2 ± 0.4^{a}

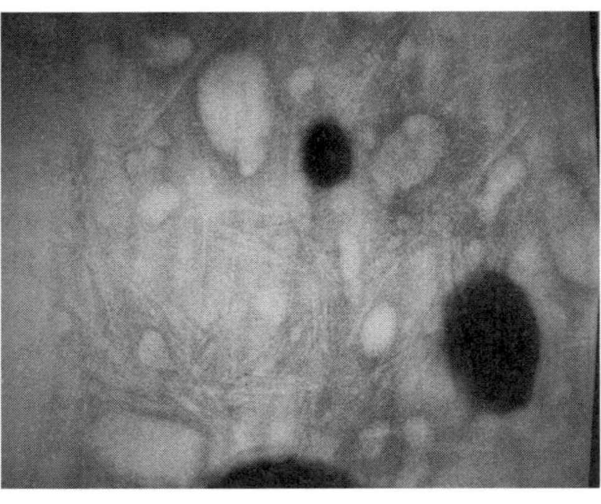

Fig. 5.13. Bionanocomposites of CNC in PVOH/PLA mix. Phase separation of the PLA and PVOH is seen as well as the CNC dispersed in a PVOH phase.

It can also be noticed that the pure TPS has high stiffness and strength compared to earlier results, where the stiffness values measured in MPa level have been 1000 times lower than in this study. The study shows that the extrusion process of TPS with sorbitol as a plasticizer is a promising way to prepare TPS and results in a polymer with mechanical properties very similar to those of polyethylene.

5.3.2 *Master-batch process*

The second approach to be able to make cellulose nanocomposites in extrusion is preparation of a master batch with higher nanocellulose content, which is then diluted during the compounding process to different nanocellulose concentrations.[14–22]

The first attempt to prepare a master batch was made with polyvinyl alcohol (PVOH), which is a water-soluble polymer reported by Bondeson and Oksman.[14] Dissolved PVOH and CNC in aqueous dispersion were mixed and dried. The dry mix of PVOH and CNC was crushed and fed to the melt compounding process, where PLA was used as matrix. This process worked well from a processing point of view, but it was observed that the PVOH phase separated in PLA and that the CNCs were well dispersed in a PVOH phase but not in PLA, which did not improve the composite mechanical properties as expected. Figure 5.13 shows the microstructure with phase-separated PVOH in PLA matrix and also a more detailed image in which well-dispersed crystals in PVAOH phase can be seen.[14]

Another example in which a master-batch process was tested employed polyvinyl acetate (PVAc), which is a water-dispersive polymer. In this case, the

Table 5.3. Tensile properties of PVAc and PVAc/CNF. (Adapted from Ref. 16.)

Materials	Tensile modulus (GPa)	Tensile strength (MPa)	Max strain (%)
PVAc	1.7 ± 0.1	39.3 ± 0.4	4.3 ± 0.3
PVAc/CNF1	2.5 ± 0.1	41.5 ± 1.3	2.9 ± 0.2
PVAc/CNF5	2.6 ± 0.1	42.6 ± 0.9	2.9 ± 0.1
PVAc/CNF10	2.7 ± 0.1	47.0 ± 0.2	2.4 ± 0.2

PVAc powder was dispersed in distilled water and mixed with an aqueous suspension of CNFs to reach a dry weight ratio between PVAc and CNF of 4:1.[16] After mixing, the dispersion was frozen at −20°C and then freeze-dried. This freeze-dried product was crushed into powder and used as the master batch (CNF content of 25 wt%). The master batch was compounded with the matrix polymer (PVAc) and a lubricant to nanocomposites using a Coperion ZSK 18 MEGAlab co-rotating twin-screw extruder in a temperature range of 120–150°C. The final nanocomposite compositions were 1, 5, and 10 wt% of CNF. The nanocomposites were named PVAc-CNF_1, PVAc-CNF_5, and PVAc-CNF_{10}. After compounding, the composites were compression molded to sheets of 0.2–0.6 mm thickness under pressure between 2.50 and 6.25 MPa at 130°C using a laboratory press. Dumbbell-shaped tensile bars (ASTM D638 Type V) were then punched from the sheets. All prepared specimens were conditioned in a desiccator at 50% relative humidity (RH) for 30 days to ensure equilibrium moisture content before testing. Results of the mechanical properties are shown in Table 5.3.[16]

The modulus and strength of PVAc were improved with the addition of CNFs; but it was noticed that the increment of modulus was almost independent of CNF content, whereas the strength was increased with increased CNF content. It was observed that the modulus and strength of PVAc/CNF10 were 59% and 20%, respectively, higher than those of neat PVAc. The average value of maximum strain was reduced with increased CNF content, revealing the limited deformation and decreased ductility of PVAc.

The DMA study showed a slight increase in the storage modulus as well as positive shift and broadening of the tan delta peak of PVAc with increased CNF content (see Fig. 5.14). This indicates the entanglement of CNFs and/or the CNF–PVAc interaction that restricts the segmental mobility of PVAc chains in the vicinity of CNFs. However, the reinforcing effect and interfacial adhesion can be reduced by bound moisture at the CNF–PVAc interface.[16]

The last example of master-batch development is a study in which CNFs were used to reinforce the PLA.[19,20] PLA was dissolved in a mixture of acetone and chloroform solvents. CNFs dispersed in water were solvent exchanged to acetone and mixed with the dissolved PLA. The mixture was allowed to evaporate at room temperature overnight followed by oven drying. The dry master batch with 10% CNF was powdered using a Warren blender. The powdered master batch was diluted with PLA to 1, 3, and 5 wt% nanofiber contents and pelletized using a co-rotating twin-screw extruder. The pellets were injection molded to test specimens and used for mechanical testing.[19,20]

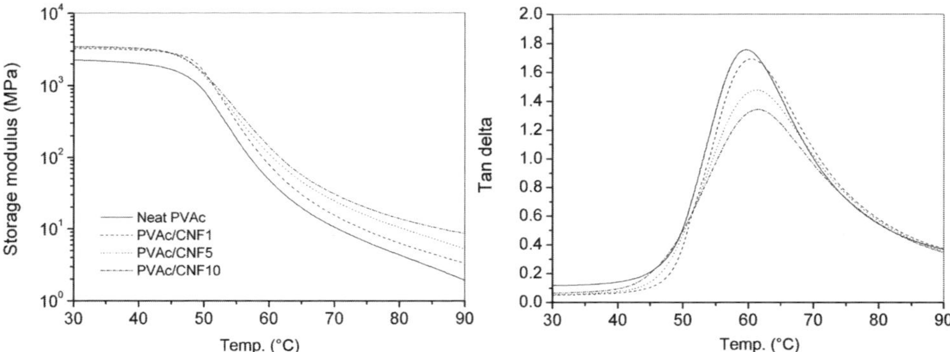

Fig. 5.14. Dynamic mechanical thermal properties of PVAc and the nanocomposites showing that the addition of nanofibers increased the storage modulus before and after the relaxation of PVAc resulted in a positive shift of the tan delta peak.[16]

Table 5.4. Mechanical properties of the PLA and its nanocomposites with different CNF contents. (Reproduced from Ref. 19.)

Materials	Young's modulus (GPa)	Tensile strength (MPa)	Strain at break (%)
PLA	2.9 ± 0.6	58 ± 0.5	3.4 ± 0.4
PLA-CNF1	3.3 ± 0.4	62 ± 0.9	2.8 ± 0.3
PLA-CNF3	3.4 ± 0.1	65 ± 0.6	2.7 ± 0.2
PLA-CNF5	3.6 ± 0.7	71 ± 0.6	2.7 ± 0.1

The microscopic studies of the nanocomposites indicated that the nanofibers were partly dispersed in the PLA matrix. The mechanical test results showed positive impact on the tensile strength and modulus, while the strain was slightly decreased compared with the neat PLA but remained similar with increased nanofiber content. The properties are presented in Table 5.4.

Figure 5.15 shows that the CNFs have large potential as reinforcement in polymers, especially if the fibers can be aligned as seen in the model by Halpin-Tsai. The composites' experimental values are closer to the prediction based on Krenchel's model, which was also expected because the fibers might be more randomly orientated in the PLA. The experimental values in Fig. 5.15 show that the composite modulus did not reach the experimental values; this may be explained by poor dispersion and distribution, especially in composites with higher nanofiber content.

Figure 5.16 shows the storage modulus and tan delta curves of the PLA and the nanocomposites. The storage modulus and tan delta peak position of the nanocomposites showed improvement when compared to the PLA matrix. The addition of nanofibers increased the thermal stability of the PLA, the storage modulus was up to 28 times higher than the pure PLA in 70°C and the tan peak position was moved by 7° to higher temperature, which indicates good interaction between the CNFs and PLA.

These examples of the master-batch processing route for manufacturing cellulose nanocomposites using extrusion demonstrate a promising way to feed

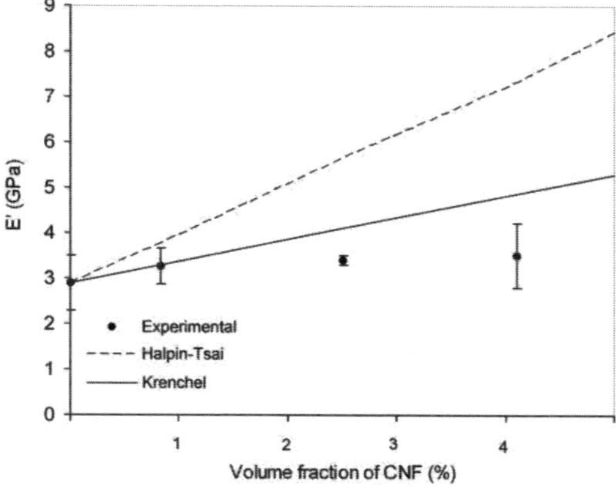

Fig. 5.15. Experimentally measured tensile modulus for the prepared nanocomposites compared to theoretical estimations by Halpin-Tsai and Krenchel. (Reproduced from Ref. 19 with permission from Elsevier.)

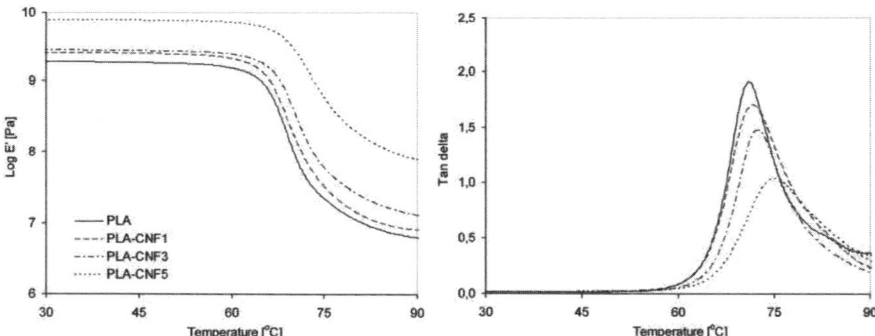

Fig. 5.16. Storage modulus and tan delta peak curves as a function of temperature, showing that 5% addition of CNF has a large reinforcing effect on the PLA also in a glassy state. The addition of CNFs also affects the tan delta peak position and restricts the PLA molecular motion. (Reproduced from Ref. 19 with permission from Elsevier.)

the cellulose nanomaterials into the extruder. There are some drawbacks with this method compared to liquid feeding. Making the master batch is more time consuming, and if the nanofibers are in contact with each other during drying, they will aggregate and are difficult to separate during extrusion. However, this is a more suitable way to make nanocomposites from CNFs than the liquid process.

5.3.3 *Challenges*

Although the melt compounding process for cellulose nanocomposites is a very promising route for taking these materials to a commercial manufacturing level, it presents several challenges. The major difficulties are (i) feeding the nanomaterials (both fibers and crystals) into the extruder, (ii) dispersing and distributing them in the polymer matrix, and (iii) avoiding thermal degradation of the materials.

The CNCs and CNFs have very high surface area and have a tendency to aggregate when dried. Aggregation can be avoided by feeding them in a suitable medium in which the nanomaterials are dispersible, which is compatible with the polymer matrix and does not cause any degradation at high temperatures.

The visual examination of the compounded materials showed that PLA_{DMAc} showed a slightly brown color compared to pure PLA. This color change may be due to some degradation that takes place due to the DMAc/LiCl. Therefore, the feeding medium is crucial when using liquid feeding.

5.4. Conclusions

Two different strategies have been used to prepare cellulose nanocomposites with thermoplastic polymers.

The first strategy, where CNCs and nanofibers were dispersed in a liquid vehicle and then compounded, indicated that nanocomposites can be prepared using a conventional compounding extruder and that liquid feeding of the nanocellulose materials into the extruder is a possible route to feed and disperse nanocelluloses in a polymer matrix.

The results of different studies showed that CNCs have a reinforcing capacity for polymers such as PLA and CAB. It is also evident that it is important to find a suitable liquid vehicle, a medium, that does not degrade the polymer or the crystals during the temperature process and that a co-rotating twin-screw extruder with high venting capacity has to be used. The use of this technique is more suitable for nanocrystal composites, since the nanofibers form very viscous gels already at low concentrations, thereby making pumping to the extruder more difficult.

The second strategy, where a master batch with higher nanocellulose content is first prepared, is also a possible route for manufacturing cellulose nanocomposites. This is especially suitable for manufacturing CNF composites because viscous gels cause problems with liquid feeding; however, with this method, it is more difficult to get well-dispersed and distributed nanofibers. The reason might be that the fibers should not be in contact with each other during drying; if they are, they will be aggregated and will not be separated during the extrusion.

It is shown that the compounding extrusion process is a possible way to produce cellulose nanocomposites, but more studies and efforts are needed to solve problems, especially problems associated with feeding of the nanomaterials into the extruder and the dispersion and distribution of the nanocellulose in the matrix polymer. If these issues are solved, then this method may be very useful, especially in large-scale manufacturing of cellulose biocomposites.

References

1. V. Favier, H. Chanzy and J.Y. Cavaille, Polymer nanocomposites reinforced with cellulose whiskers, *Macromolecules* **28** (1995) 6365–6367.

2. W. Helbert, J.Y. Cavaille and A. Dufresne, Thermoplastic nanocomposites filled with wheat straw cellulose whiskers. Part I: processing and mechanical behaviour, *Polym. Compos.* **17** (1996) 604–611.
3. A. Dufresne and J.Y. Cavaillé, Nanocomposite materials of thermoplastic polymers reinforced by polysaccharide, *ACS Symp. Ser.* **723** (1999) 39–54.
4. H. Matsumura, J. Sugiyama and W.G. Glasser, Cellulosic nanocomposites. I. thermally deformable cellulose hexanoates from heterogeneous reaction, *J. Appl. Polym. Sci.* **78** (2000) 2242–2253.
5. M. Grunert and W.T. Winter, Nanocomposites of cellulose acetate butyrate reinforced with cellulose nanocrystals, *J. Polym. Environ.* **10** (2002) 27–30.
6. T. Zimmermann, E. Pohler and P. Schwaller, Mechanical and morphological properties of cellulose fibril reinforced nanocomposites, *Adv. Eng. Mater.* **7** (2005) 1156–1161.
7. A. Bhatnagar and M. Sain, Processing of cellulose nanofibre-reinforced composites, *J. Reinforced. Plast. Compos.* **24** (2005) 1259–1268.
8. K. Oksman, A.P. Mathew, D. Bondeson and I. Kvien, Manufacturing process of polylactic acid (PLA) – cellulose whiskers nanocomposites, *Compos. Sci. Technol.* **66** (2006) 2776–2784.
9. K. Oksman, M. Skrifvars and J.-F. Selin, Natural fibres as reinforcement in polylactic acid (PLA) composites, *Compos. Sci. Technol.* **63** (2003) 1317–1324.
10. A.P. Mathew, A. Chacraborty, K. Oksman and M. Sain. The structure and mechanical properties of cellulose nanocomposites prepared by twin screw extrusion. In *Cellulose nanocomposites: processing, characterization and properties,* K. Oksman and M. Sain (Eds.), ACS Symposium Series, Vol. 938, Oxford University Press (2006).
11. D. Bondeson, P. Syre and K. Oksman Niska, All cellulose nanocomposites produced by extrusion, *J. Biomat. Bioenerg.* **1** (2007) 367–371.
12. M. Hietala, K. Kekäläinen, J. Niinimäki and K. Oksman, Bionanocomposites of thermoplastic starch and cellulose nanofibers manufactured using twin-screw extrusion, *J. Appl. Polym. Sci.* submitted 2013.
13. K. Oksman, A.P. Mathew and M. Sain, Novel bio-nanocomposites: processing properties and potential applications, *Plast. Rubber Compos: Macromol. Eng.* **38** (2009) 396–405.
14. D. Bondeson and K. Oksman, Polylactic acid/cellulose whisker nanocomposites modified by polyvinyl alcohol, *Comp. Part A* **38** (2007) 2486–2492.
15. G. Gong, A.P. Mathew and K. Oksman, Toughening effect of cellulose nanowhiskers (CNWs) on polyvinyl acetate (PVAc): fracture toughness and viscoelastic analysis, *Polym Compos.* **32** (2011) 1492–1498.
16. G. Gong, J. Pyo, A.P. Mathew and K. Oksman, Tensile behavior, morphology and viscoelastic analysis of cellulose nanofiber-reinforced (CNF) polyvinyl acetate (PVAc), *Comp. Part A* **42** (2011) 1275–1282.
17. A.P. Mathew, G. Gong, N. Björngrim, D. Wixe and K. Oksman, The effect of cellulose nanowhiskers (CNW) on the moisture absorption and mechanical properties of polyvinyl acetate (PVAc), *Polym. Eng. Sci.* **51** (2011) 2136–2142.
18. G. Gong, A.P. Mathew and K. Oksman, Efficient toughening and reinforcing of polyvinyl acetate, *SPE Plastic Research Online*, 15 February (2012).
19. M. Jonoobi, J. Harun, A.P. Mathew and K. Oksman, Mechanical properties of cellulose nanofiber (CNF) reinforced polylactic acid (PLA) prepared using twinscrew extrusion, *Compos. Sci. Technol.* **70** (2010) 1742–1745.
20. M. Jonoobi, A.P. Mathew, M.M. Abdi, M.D. Makinejad and K. Oksman, A comparison of modified and unmodified cellulose nanofiber reinforced polylactic acid (PLA) prepared by twin screw extrusion, *J. Polym. Environ.* (2012).

21. C. Eyholzer, P. Tingaut, T. Zimmermann and K. Oksman, Dispersion and reinforcing potential of carboxymethylated nanofibrillated cellulose powders modified with 1-hexanol in extruded poly(lactic acid) (PLA) composites, *J. Polym. Environ.* (2012).
22. M.L. Hassan, A.P. Mathew, E.A. Hassan, S.M. Fadel and K. Oksman, Chemical modification and maleated polypropylene to improve properties of polypropylene and bagasse nanofiber composites prepared by twin screw extrusion, Accepted for publication in *J. Reinforced Plast. Comp.* Accepted (2013).
23. K. Kohlgrüber and W. Wiedemann, *Co-rotating twin-screw extruders: fundamentals, technology and applications*, Munich, Germany: Carl Hanser Verlag (2008), pp. 59–89.

Chapter 6

In Situ Polymerization of Bionanocomposites

Youssef Habibi, Samira Benali and Philippe Dubois
Center of Innovation and Research in Materials and Polymers (CIRMAP), University of Mons Place du Parc Mons, Belgium

Bionanocomposites are a class of composites where both the polymer matrix and nanofillers are derived from renewable resources and/or based on biodegradable polymer matrices. Nevertheless, processing such systems faces many obstacles mainly related to interfacial incompatibility between the most industrially appealing biobased polymers that are hydrophobic and the most naturally abundant nanofillers that are hydrophilic. Among the methodologies developed to overcome this challenge, *in situ* polymerization constitutes one of the most viable and promising approaches because good filler dispersion and enhanced interfacial adhesion with the surrounding matrix can be achieved. This chapter collates key advances realized in the field of *in situ* polymerization of bionanocomposites, particularly based on polylactide and polycaprolactone as matrices while the fillers are nanoclays and nanocelluloses, and their related properties.

6.1. Introduction

Bionanocomposites refer to multiphase materials that are biodegradable and/or partially or fully derived from biobased resources where the dispersed constituent has at least one dimension in the nanoscale size range. In the burgeoning trend in developing sustainable economies, huge efforts are devoted to access high-performance bionanocomposite materials, including polymeric matrices and nanofillers, derived from renewable resources and/or biodegradable. Such technical innovations allowed, and still permitting further advances, to design and create new bionanocomposites and structures with unprecedented flexibility, improvements in their physical properties, and significant industrial relevancy. The resulting materials are indeed promised to new horizons in the field of engineering applications such as medicine, textiles, cosmetics, agriculture, food packaging, aerospace, construction, catalysis, etc., where they are intended to substitute their petroleum-based counterparts.[1]

Similarly to synthetic composites, the incorporation of biofillers into a biopolymer matrix can significantly affect the properties of the matrix resulting in bionanocomposites exhibiting improved thermal, mechanical, rheological, electrical, catalytic, fire retardancy, optical properties, and so on.[2] These properties are governed by the type of the incorporated nanofillers, their size and shape, their concentration, and most importantly their interactions with the polymer matrix.

These interactions, mainly linked to their specific surface area and volume effects, determine the quality of the dispersion of the nanofillers into the polymer matrix.[3] Preventing the nanofillers from aggregating and improving their dispersion in the polymeric matrix are not trivial tasks since there have been extensive works particularly in the field of synthetic polymers suggesting that surface modifications of the nanofillers constitute the most viable pathways. Among such modifications, the *in situ* synthesis of macromolecules, from their corresponding monomers, in the presence of (pre-functionalized) fillers presents many advantages including: (i) straightforward carrying out and processing, (ii) fine-tuning of the synthesized polymers through a proper selection of the species and synthesis conditions, and (iii) high dispersion of the fillers as a result of enhanced compatibility particularly when covalent bonding is created between the fillers and the matrix. This later case, commonly referred as "grafting", has attracted and still attracting, tremendous level of attention among the material scientific community owing to the outstanding performances that can be reached. It can be achieved through different strategies namely "grafting onto", "grafting from", or "grafting through". The "grafting onto" method involves attachment of pre-synthesized polymer chains, carrying reactive end groups to the surface of fillers. The polymers can be fully characterized before grafting to the surface, offering the possibility of controlling the properties of the resulting material. However, steric crowding for the attachment increases during the reaction, as the polymer chains have to diffuse through the layer of already attached brushes to reach the reactive sites on the surface. Thus, it is very difficult to achieve high grafting densities using the "grafting onto" method. In order to increase the grafting density of the polymer brush on the surface and to ensure its stability in different application conditions, the "grafting from" approach is often privileged. In this method, the polymer chains are formed by *in situ* surface-initiated polymerization from immobilized initiators on the filler. "Grafting through" offers intermediate benefits over the two above-mentioned approaches, but it is still marginally exploited in this field. The purpose of this chapter is to collate current knowledge and provide several key achievements realized in the *in situ* polymerization of bionanocomposites. As opposed to the flourishing literature abounding on the *in situ* synthesis of petroleum-based nanocomposites or their hybrids, published data focusing exclusively on biobased materials, where both polymer matrices and fillers are biosourced and/or biodegradable, are quite limited. Therefore, this chapter will focus on some biobased polymers, or being derived from bioresources and/or biodegradable, like polylactide (PLA) and poly(ε-caprolactone) (PCL). Consequently, there is no denying that ring-opening polymerization (ROP) is employed, in most cases in conjunction with the "grafting from" approach, to prepare such polymer-based bionanocomposites, will be mostly described. In regard to nanofillers, these building blocks are obtained from numerous sources leading to distinctive morphologies and characteristics thus implying the potential for diverse functionalities and uses as will become evident in this chapter. Because of their undeniable industrial potential, clays and nanocelluloses are amongst the most exploited biosourced nanofillers in this field; therefore, a systemic emphasis will be given to these substrates in this chapter providing therefore an "inorganic" and an "organic" model of nanofillers.

6.2. Biopolymer matrices

6.2.1 *Poly(ε-caprolactone)*

PCL is recognized not only as a biodegradable and non-toxic material but exhibits also a unique compatibilizing ability with a great variety of other polymers that can result in improved or modified properties.[4] Although PCL is well known as oil-derived polymers, it is actually being derived from renewable resources along with other derivatives polyesters.[56] PCL is in the rubbery state and exhibits high permeability to low molecular species at body temperature. PCL degradation proceeds through hydrolysis of backbone ester bonds as well as by enzymatic attack.[7] PCL can be obtained by anionic, cationic, or coordinated ROP of ε-caprolactone (CL), enzymatic, or radical polymerization.[7] PCL is a semicrystalline polymer with a degree of crystallinity around 50%. It has a rather low glass transition temperature T_g ($\sim -60°C$) and a relatively low melting temperature ($\sim 60°C$). PCL is a ductile polymer able to sustain large deformations. These properties allow considering PCL for use in many applications ranging from biomedical to commodity materials. Unfortunately, the elastic modulus is rather low making it useless for any application that requires higher rigidity. Thus, the addition of filler can contribute to improve its stiffness.

6.2.2 *Polylactide*

PLA belongs to the family of synthetic aliphatic polyesters and is considered as biodegradable and compostable. It is derived from renewable resources (mainly starch and sugar). The polymerization of PLA is largely described in the literature.[8–10] High molecular weight PLA is generally produced by the ROP of the lactide monomer. Catalytic ROP of the lactide intermediate results in the production of PLA with controlled molecular weights. Commercially available PLA grades are copolymers of poly(L-lactide) having different amounts of D-lactide. The amount of the D-enantiomers is known to affect the properties of PLA, such as melting temperature, degree of crystallinity, and so on. PLA is a thermoplastic, high strength, high modulus polymer that can be made from annually renewable resources. In the past four decades, focus has been on biomedical applications such as drug delivery, bioresorbable sutures, medical implants, and scaffolds for tissue engineering. More recently, PLA has garnered interest as a new environmentally friendly thermoplastic with wide applicability. Similar to many thermoplastic polymers, semicrystalline PLA exhibits T_g and T_m. Above T_g ($\sim 58°C$), PLA is rubbery, while below T_g, it becomes glassy which is still capable to creep until it is cooled to its β transition temperature at approximately $-45°C$, below which it behaves as a brittle polymer; PLA has relatively high T_g ($\sim 60°C$) and low T_m (~ 165–$180°C$) as compared to other thermoplastics.[9]

6.3. Nanofillers

6.3.1 *Type and properties of clays*

The "Clay Science" is a relatively young discipline which started with the visual macroscopic observations of clays in soils by geologists, followed by the chemical and structural analysis of clay minerals by mineralogists, and finally by other several physical-chemical determinations by different scientists.[11] In regard to *in situ* polymerization, the most used clays are montmorillonite, i.e., layered silicate clays and sepiolite, i.e., needle-like clays.

The montmorillonite is 2:1 phyllosilicate or smectite clay consisting of an octahedral sheet of alumina sandwiched between two opposing tetrahedral sheets of silica, negatively charged due to isomorphic substitution of a part of the trivalent aluminum cations for divalent cations such as Fe^{2+} or Mg^{2+}.[12] The isomorphic substitution within the layers generates negative charges that are counterbalanced by alkali or alkaline earth cations situated in the interlayer. In order to render these hydrophilic phyllosilicates more organophilic, the hydrated cations of the interlayer can be exchanged with cationic surfactant such as alkylammoniums or imidazoliums. The modified clay being organophilic, and therefore more compatible with organic polymers, is largely studied for enhancing biopolymer properties.

Concerning needle-like clays, Brigatti *et al.*[13] define sepiolite and attapulgite (or palygorskite) also as 2.1 phyllosilicate clays with two-dimensional tetrahedral sheet. However, they differ from layered in that they lack continuous octahedral sheets. A complete description is available in Bergaya *et al.*[12] The structural characteristics of these needle-like clays, with the presence of small channels, are responsible for their absorption and adsorption properties and the derived standard applications.

6.3.2 *Type and properties of nanocelluloses*

Cellulose constitutes the most abundant renewable polymer available in the biosphere. This fascinating and almost inexhaustible sustainable polymer possesses remarkable chemical and physical attributes allowing its use as raw material in different forms for the production of a wide range of materials and products. More recently, nanosized cellulosic substrates known as nanocelluloses are exceptionally emerging due to their unsurpassed quintessential physical and chemical properties in addition to the inherent attributes of cellulose such as low cost, availability, renewability, and lightweight. According to their morphology, cellulose fibers can be dissociated transversally at the amorphous parts present along their axis to yield defect-free rod-like nanoparticles referred hereafter as cellulose nanocrystals (CNCs), or laterally yielding to bundles of their elementary fibrils known as nanofibrillated cellulose (NFC) or cellulose nanofibers (CNFs).

CNCs are extracted from cellulose fibers by controlled acid hydrolysis with either sulfuric or hydrochloric acid. Acid molecules primarily degrade the less ordered, and thus more accessible regions along the microfibrils to finally leave intact nanometric and highly crystalline rod-like fragments (Fig. 6.1). CNCs can

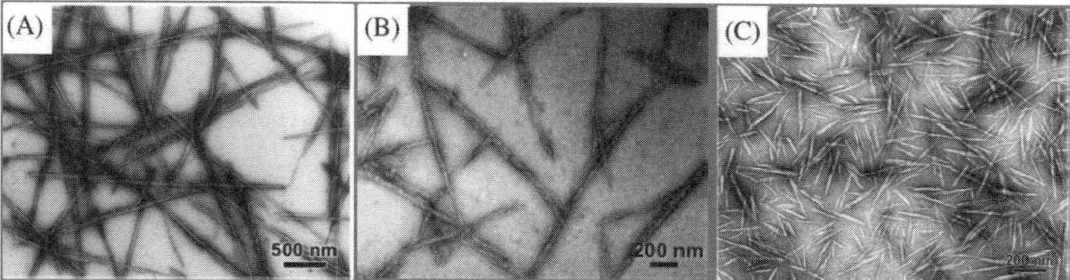

Fig. 6.1. Transmission electron microscopy (TEM) images of dried dispersion of CNCs derived from (A) tunicate,[15] (B) bacterial,[16] and (C) ramie.[17] ((A) Reprinted with permission from Ref. 15. Copyright (2008) American Chemical Society. (B) Reproduced from Ref. 16 with kind permission from Springer Science + Business Media. (C) Reproduced from Ref. 17 with permission from The Royal Society of Chemistry.)

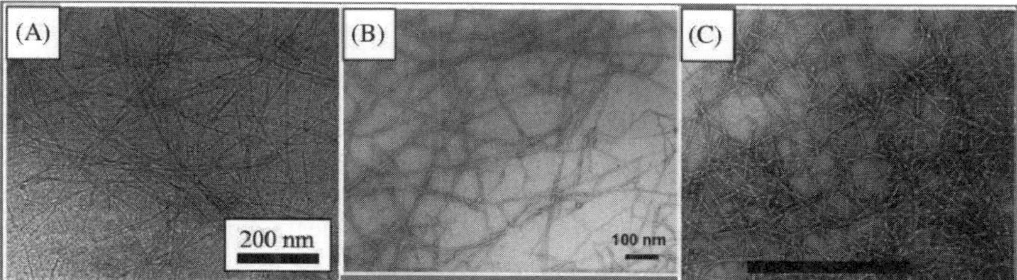

Fig. 6.2. Transmission electron micrographs from a dilute suspension of NFCs obtained from wood fibers by mechanical processing combined to (A) enzymatic,[23] (B) 2,2,6,6-tetramethylpiperidine-1-oxyl radical (TEMPO)-mediated oxidation,[24] (C) carboxylmethylation pretreatment (scale bar corresponds to 0.5 μm).[25] ((A) Reprinted with permission from Ref. 22. Copyright (2007) American Chemical Society. (B) Reprinted with permission from Ref. 23. Copyright (2007) American Chemical Society. (C) Reprinted with permission from Ref. 24. Copyright (2008) American Chemical Society.)

be prepared from a variety of sources and depending on the origin of the cellulosic fibers, CNCs with different morphologies and geometrical characteristics (length and width) can be obtained. The width is generally a few nanometers, but the length ranges from tenths of nanometers to several micrometers.[14]

Introduced by Turbak *et al.*[18] since 1980s, CNFs are long nanoparticles where the length can reach few microns while the width remains in the nanoscale range. They can be prepared by mechanical disintegration of the respective macrofibers. Manton Gaulin homogenizer, microfluidizer, and ultrafine friction grinder are the most commonly used technologies for such treatment.[19] The main issue with these processing methods is the very high-energy consumption involved when processing pure cellulosic fibers. Thus, combinations of enzymatic[20] or chemical pretreatments[21,22] with mechanical processing have been introduced in order to facilitate the fibrillation and mechanical shearing by decreasing the cellulose microfibril length and/or impairing their structural potential (breaking intra-fiber hydrogen bonds). Like CNCs, they have been prepared from numerous cellulosic sources (Fig. 6.2).

6.4. Clay-based bionanocomposites

6.4.1 *In situ polymerization of clay-based bionanocomposites*

In the case of biopolymers/clay nanocomposites, available published data reveals that the covalent bonding between organomodifiers from layered silicates and polymeric chains, so-called "grafting from" method, represents a much better way to finely disperse layered silicates within biopolymer matrix.[25–27] In this respect, the *in situ* intercalation of biopolymers chains into layered silicates has been investigated in order to reach fully exfoliated structures. In this method, the layered silicate is swollen by the liquid monomer or a monomer solution so the polymer formation occurs between the intercalated sheets. The polymerization can be then initiated either by heat or radiation, by the diffusion of a suitable initiator, or by an organic initiator or catalyst fixed through cation exchange inside the interlayer before the swelling step.[28] This strategy was successfully used in both types of clays, e.g., layered silicate and needle-like clays, where PCL and PLA were used as matrix biopolymers.

6.4.1.1 *PCL grafting from layered silicate clays*

The production of PCL/clay nanocomposites by *in situ* polymerization of CL is largely reported starting from organomodified layered silicate clays. The most common layered silicate used for the preparation of PCL grafting from clay nanocomposites is montmorillonite.[13,26] The process consists of the addition of layered silicate into CL prior to polymerization. The monomer intercalates into the clay galleries and, under appropriate conditions, polymerizes through ROP. The platelets are separated/expanded from each other by the action of the polymerization reaction, resulting eventually in exfoliation of the clay nanoplatelets within the so-formed PCL matrix.[26] The ability of CL to delaminate and disperse the layered silicate in the polymerization medium is related to a number of factors, including the exchange capacity of the silicate clays, the polarity of the medium, and the chemical nature of the interlayer cations.[4,29,30]

Messersmith *et al.* were the first to report on the intercalative polymerization of CL by thermal activation at 170°C in the presence of montmorillonite exchanged with Cr^{3+}cations.[29] Later, Haouas *et al.*[31] proposed a study demonstrating, by using solid-state nuclear magnetic resonance (NMR) characterization, that during the *in situ* formation of PCL/Maghnite (a partially protonated natural montmorillonite) nanocomposites, the grafting of the polyester chains onto the layered silicates took place from Brønsted/Lewis acidic sites located at the clay interface. Unlike the *in situ* polymerization of CL initiated by protonic acid without clay,[32] no linear dependence of the PCL molecular weight on the carboxylic acid was observed in the presence of this acidic montmorillonite.

Kubies *et al.*[33] developed a convenient route to control the *in situ* intercalative polymerization of CL initiated by metal alkoxides such as dibutyltin(IV) dimethoxide ($Bu_2Sn(OMe)_2$) or catalyzed by tin(II) octoate ($Sn(Oct_2)$) directly in presence of organomodified clays.

Dubois *et al.*[33–40] reported the effect of the nature of the catalyst/initiator on both the molecular parameters of the PCL (molar mass and polydispersity) and the degree of clay exfoliation, when *in situ* polymerization was combined with hydroxyl-containing ammonium cations or not. Various metal alkoxides, either preformed (e.g., $Bu_2Sn(OMe)_2$) or formed "*in situ*" by reaction of a metal catalyst (e.g., $Sn(Oct_2)$ or triethylaluminum ($AlEt_3$)) with alcohol-bearing ammonium modifiers were investigated. More recently, Chrissafis *et al.*[39] used a titanium(IV) butoxide as catalyst with non-functional alkylammonium modifier and Tarkin-Tas *et al.*[40] reported the effect of montmorillonite exchanged with an alkyl imidazolium ion bearing hydroxyl groups and activated by tin octoate for the preparation of exfoliated PCL/clay nanocomposites and its impact on crystallinity and thermal stability of the polyester matrix. Through all these studies, it was demonstrated that the presence of hydroxyl groups "ionically" anchored at the clay surface in combination with a controlled coordination–insertion ROP process allows for the formation of tethered polyester chains and largely favored the exfoliation of clay nanoplatelets.

Another method described by Liao *et al.*[41,42] can be pointed out as well, in which PCL/clay nanocomposites were prepared using hydroxyl-bearing organomodified clay via microwave-assisted *in situ* ROP of CL, using tin octoate as catalyst and montmorillonite modified with methyl, tallow, bis-2-hydroxyethyl as initiator.

6.4.1.2 *PCL grafting from needle-like clays*

As far as needle-like clays are concerned, Duquesne *et al.*[43] prepared PCL/sepiolite nanohybrids by ROP of CL and *in situ* PCL anchoring onto sepiolite surface. These nanohybrids were synthesized from the aminopropyl-functionalized sepiolite, the primary amino functions behaving as initiation sites for the ROP of CL catalyzed by $Sn(Oct)_2$. This sepiolite/PCL nanohybrids have been synthesized in order to further modify the sepiolite surface and to prepare PCL-grafted sepiolite to be used as "masterbatches". PCL not only interacts with the external surface of needle-like silicate, but it can also penetrate into the structural tunnels of the mineral. No properties were evaluated on sepiolite/PCL nanohybrids but the as-prepared nanocomposite materials (PCL/sepiolite-PCL) by twin-screw mini extrusion have been characterized in terms of thermal, morphological, and mechanical properties and compared to the pristine PCL matrix and a direct melt blending (PCL/sepiolite) at 3 wt% of inorganics. A significant improvement of the mechanical properties occurs if sepiolite–PCL nanohybrids are used without affecting morphological properties. The authors explain that the presence of covalent bonds between the polyester chains and the well-dispersed nanofiller allows for a strong enhancement of the interfacial adhesion as well as the overall stiffness of the nanocomposites materials.

6.4.1.3 *PLA grafting from layered silicate*

In the case of layered silicate/PLA nanocomposites, the *in situ* intercalation of PLA chains into layered silicates has been investigated in order to reach fully exfoliated structures. Some of us so-reported that exfoliated PLA/organomodified montmorillonite nanocomposites could be successfully prepared in bulk by triethylaluminum-catalyzed ROP of LA initiated from hydroxyl-functionalized ammonium cations as organomodifiers.[44] In this case, the active species in ROP of LA, i.e., aluminum alkoxide active species, were *in situ* generated, resulting from the reaction of the triethylaluminum with the hydroxyl groups available over the entire surface of clay.

6.4.1.4 *PLA grafting from needle-like silicate*

Only very recently, some authors[45–48] reported on the use of needle-like silicate clays in conjunction with PLA. All these studies have shown a strong potential for this type of clays to improve PLA properties. Unfortunately, these studies were not carried out using *in situ* polymerization route; therefore, they are out of the scope of this chapter.

6.4.2 *Properties of clay-based bionanocomposites*

6.4.2.1 *Microstructure*

The structure–property relationships of layered silicate bionanocomposites depend strongly on the nature of the (organo) clay used and their preparation method. Alike other polymer/layered silicate nanocomposites (see Ref. 26), various structures of PCL/layered silicate nanocomposites can be obtained: (i) microcomposites, (ii) intercalated, or (iii) exfoliated nanocomposites. X-ray diffraction (XRD) is mainly used to identify the microstructure of the nanocomposites since no change in the diffraction pattern corresponding to clays is observed for microcomposites while the diffraction peaks shift toward lower values in the presence of intercalated structures. However, a fully amorphous structure (no diffraction peaks in the low angle diffraction range) is distinctive in the case of exfoliated structure. In this case, TEM or atomic force microscopy (AFM) is then used to characterize their morphology. Yet, in most practical cases, PCL/clay nanocomposite preparation leads often to an intercalated–exfoliated intermediate organization impacting their structures and properties among which mechanical, barrier, flame-retardant properties, and crystallization behavior are often scrutinized.[4,49,50] PLA/layered silicate nanocomposites attract huge interest (see Ref. 27). Melt-intercalation of PLA into silicate layers has been widely described in the literature but surprisingly fully exfoliated structures were never achieved even through *in situ* polymerization.

6.4.2.2 *Thermal properties and crystallization*

The examination of the crystallization of organic polymer matrix in clay-based nanocomposites and more particularly at the interface of the large inorganic surface area is of particular interest. In fact, it has been found that the inorganic surface may favor nucleation and crystal growth or development of a different crystalline structure and/or morphology.[50] Such effects are important because the ultimate properties of the materials (e.g., mechanical and thermal) are greatly impacted.

So, in order to control the enhancement of the properties upon the incorporation of clays, the understanding of the thermal behavior and crystallization kinetics of PCL is essential. It is worth noting that it is well known that the silicate layers can either act as crystallization barriers distributing or completely stopping crystal growth, or hinder the polymer chain motion required for the crystallization.[51] With PCL/OMMT obtained by *in situ* polymerization, Pucciarello *et al.*[50] studied the isothermal kinetics of a series containing from 6 to 44 wt% in inorganics. They noted that the most enhanced thermal properties are obtained for exfoliated nanocomposites with smaller crystallites, observed by light microscopy, compared to those of neat PCL. In contrast, for intercalated nanocomposites, the crystallites have an irregular non-spherulitic shape, which is attributed to the interruption of the crystallite growth at the vicinity of the large tactoïds. Accordingly, authors concluded that the presence of OMMT does not prevent PCL from crystallizing in a spherulitic morphology, but the crystallization becomes slower when the OMMT content increases especially at higher temperatures and for intercalated nanocomposites. This delay can be explained by the fact that the nanoclay acts as nucleating agents for PCL but little or negatively affects the linear growth rate of PCL spherulites. Later, Tarkin-Tas *et al.*[40] studied the PCL/exfoliated silicate layer nanocomposite obtained by *in situ* polymerization of ε-CL from hydroxyl-containing ammonium cations used to organomodify the clay. The focus to the PCL crystallization behavior highlighted that the PCL degree of crystallinity increases with increasing PCL-grafted clay content, i.e., from 51 to 69% (for 5 wt% of inorganic content). The authors explained that highly exfoliated structure increases the surface area but with virtually all chains bound to the surfaces. This led to a pre-organization of the chains near the platelets enhancing the nucleation and consequently increasing the degree of crystallinity. In the case of PLA, the as-obtained PLA/OMMT nanocomposites presented enhanced thermal properties since the temperature corresponding to 50% of weight loss is shifted by $\sim$30°C toward higher temperature. The crystallinity was also affected since the mobility of the resulting grafted chains was restricted while T_g and T_m were not influenced by the addition of the clay.[52]

6.4.2.3 *Mechanical properties*

Mechanical properties of PCL are highly related to the crystalline state in turn linked to thermal history. Though, no study dedicated to the mechanical properties of PCL/clay (nano)composites fully produced by *in situ* polymerization has been reported except one illustration performed by Chrissafis *et al.*[39] for which no detailed investigation on the structure of the so-produced compositions was provided making difficult any conclusive statement. Otherwise, *in situ* polymerization has been combined to melt intercalation for producing higher performance nanocomposite materials (*vide infra*: masterbatch process). Indeed, melt intercalation of preformed polymers and *in situ* intercalative polymerization represent the most commonly used techniques to prepare polymer/clay nanocomposites. Melt blend intercalation requires the addition of the filler during the extrusion/kneading process, without making any important change on the industrial production line; while the *in situ* polymerization process allows to obtain exfoliated nanocomposites (when using the appropriate functionalized organomodified clay), displaying much enhanced mechanical properties. Accordingly, a two-step method, named masterbatch process, has been adopted for the preparation of PCL/layered silicate nanocomposites by combining the *in situ* intercalative polymerization and the melt blend intercalation process. In such process, a highly filled (organomodified) clay PCL is first prepared by *in situ* intercalation polymerization of ε-CL, followed by the blending of the resulting masterbatch with the polyester matrix. As it will be described, this method allows preparing PCL-based nanocomposites with a high degree of exfoliation, which cannot be achieved by direct mixing of PCL and clays. Lepoittevin *et al.*[53] and Pollet *et al.*[54] compared the mechanical properties for PCL/MMT nanocomposites (obtained by the two-step masterbatch process) and for microcomposites prepared by direct melt blending of PCL and MMT under the same processing conditions. The elastic modulus was improved from 217 MPa for unfilled PCL up to 365 MPa for the nanocomposites (filled with 10 wt% MMT) prepared by the masterbatch process. In contrast, the microcomposites prepared by direct melt blending with MMT showed only very limited increase in stiffness. All the PCL nanocomposites originated from the masterbatches remain rather ductile, with an elongation at break higher than 300%. The stress at break slightly drops when the clay content is increased but remains at an acceptable level. Similar observations have been reported for the two organomodified clays: dimethyl, dehydrogenated tallow 2-ethylhexyl ammonium (MMT-2MHTL8) and methyl, tallow, bis-2-hydroxyethyl ammonium (MMT-MT2EtOH), showing increased stiffness and preserved ultimate tensile properties.

6.5. Nanocellulose-based bionanocomposites

6.5.1 *In situ polymerization of nanocellulose-based bionanocomposites*

With the presence of large number of hydroxyl groups within the structure of nanocelluloses, these building blocks provide a unique platform for significant surface modification using various chemistries including *in situ* polymerization. Regarding their cellulosic nature, most of these (graft) polymerizations are realized similarly to cellulose fibers, that were collated in a critical review.[55] Polymers can be grown from nanocelluloses directly using the surface hydroxyl groups as initiating sites for ROP or the surface can be modified to introduce different initiator sites needed for controlled polymerization techniques such as reversible addition-fragmentation chain transfer (RAFT) polymerization, or atom transfer radical polymerization (ATRP). However, as far as the aspect biosourced is concerned, most of the synthesized or grafted biopolymers belong to polyesters mainly PLA and PCL, which limit the synthesis route to ROP. Carlmark *et al.*[56] recently reviewed key advances achieved using this polymerization route on cellulose and cellulose derivatives. Thus, polymerization methods, adopted to graft polymers from nanocelluloses, are a logic extending of those applied to cellulose. However, preserving the original morphology of nanocellulose, avoiding any polymorphic conversion and maintaining the integrity of their native crystalline structure remain a great challenge in conducting such synthesis. At the opposite of clays for which data are quite abundant, the field of nanocellulose is still at its infancy; therefore published studies are very limited.

6.5.1.1 *PCL grafting from CNCs*

Habibi *et al.*[17] were the first to report on the grafting of PCL from CNCs based on *in situ*-catalyzed ROP. $Sn(Oct)_2$ was used as the catalyst and the reaction was conducted at 95°C. Under these conditions, the CNCs kept their original crystalline structure and morphology as confirmed by the XRD and TEM (Fig. 6.3). The grafted PCL chains were long enough to be able to crystallize at the surface of CNCs allowing the processing of the corresponding PCL-*g*-CNC masterbatches by melt blending with virgin PCL matrix.[57] Later, Chen *et al.*[58] used the same strategy to modify CNCs derived from microcrystalline cellulose and the PCL-*g*-CNCs were incorporated into PCL matrix by thermoforming and, hence, produced molded sheets with good mechanical strength. In order to overcome the use of metal-based catalyst, Labet and Thielemans[59] recently used citric acid, a benign naturally available organic acid, as the catalyst. The reaction conditions used are within the range of normal conditions for metal-catalyzed ROP, so that additional energetic costs are not incurred which constitutes a major step toward truly green materials, as the incorporation of potentially harmful metal catalysts in polymers is a continuous concern.

6.5.1.2 *PCL grafting from CNFs*

Lönnberg *et al.* have successfully achieved the surface grafting of CNFs with PCL via ROP using $Sn(Oct)_2$ as the catalyst using two distinct procedures where: (i) NFCs were dispersed in the monomer and the reaction was conducted at 95°C without any solvent[60a] and (ii) the reaction was carried out in toluene where CNFs were previously dispersed.[61] Similarly to CNCs, the applied conditions in both procedures were not harmful to CNFs as they preserve their native morphology

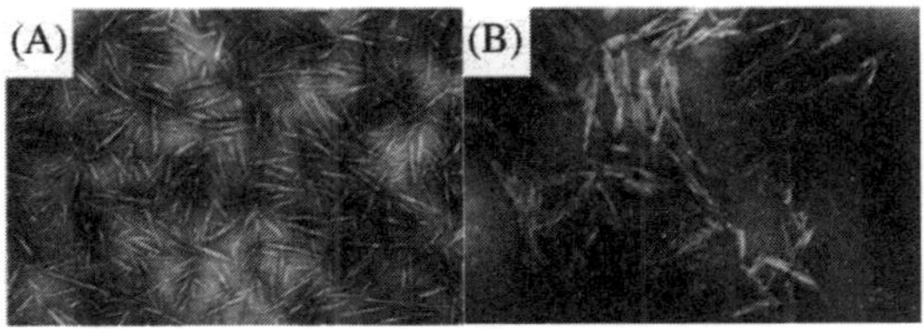

Fig. 6.3. Transmission electron micrograph of CNCs before (left) and after (right) ROP grafting. (Reproduced from Ref. 17 with permission from The Royal Society of Chemistry.)

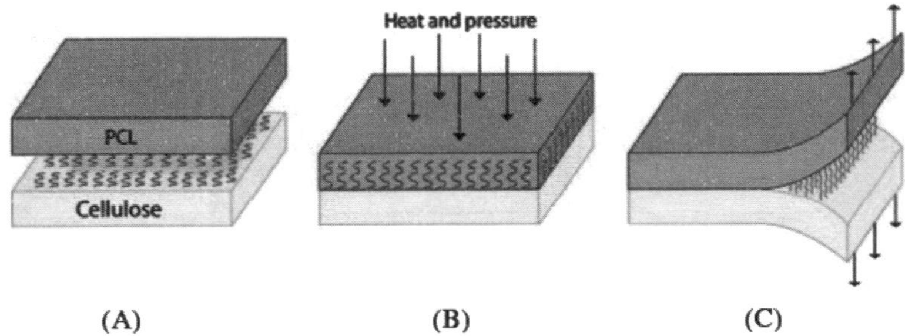

Fig. 6.4. Production of the PCL-grafted CNF and PCL bilayer laminates by hot pressing. (Reprinted from Ref. 56, Copyright (2012), with permission from Elsevier.)

and crystallinity throughout the grafting procedures. The grafting density as well as the length of the grafted PCL chains was tuned by varying the amount of the catalyst and the free sacrificial initiator added to the system.[62] Although their restricted mobility, the grafted PCL chains were able to crystallize prompting the authors to investigate, in a following work, the processing of the resulting PCL-grafted CNF films by thermoforming in the form of bilayer laminates (Fig. 6.4). A direct correlation between the length of the grafted PCL chains and the interfacial toughness of the resulting laminate was then evidenced.

6.5.1.3 *PLA grafting from CNCs*

Similarly to the strategy followed in the case of the grafting of PCL at the surface of CNCs, Goffin *et al.*[63] have successfully grafted PLLA chains from the surface of CNCs by ROP without impacting their morphological features nor their crystalline structure. However, the grafted chains were suspected to be very short according to their crystallization behavior. Therefore, and in order to achieve high molecular weight of grafted PLLA chains, Braun *et al.*[64] suggested the use of partially acetylated CNCs, obtained through Fischer esterification pathway (Fig. 6.5), to control the grafting density. A well-designed approach to produce PLLA-grafted CNCs is used to balance the competing effects; blocking a fraction of surface hydroxyl groups available for ROP initiation leads to high molecular weight chains covalently attached to the filler surface, while maintaining sufficient inter-particle contact for efficient stress transfer within the percolating network. In all these cases, the grafted PLLA chains were able to co-crystallize with free unbounded PLLA chains from the matrix. Furthermore, owing to the ability of the two stereoforms of PLA, e.g., PDLA and PLLA to form stereocomplexes having distinct thermal properties, few recent studies reported the grafting of one stereoform at the surface of CNCs using ROP and use the other one as the matrix.

6.5.1.4 *PCL-b-PLA grafting from CNCs*

In a singular work, Goffin *et al.*[65] employed ROP catalyzed by $Sn(Oct)_2$ to graft successively PCL and then PLLA from the surface of CNCs (Fig. 6.6). Despite the

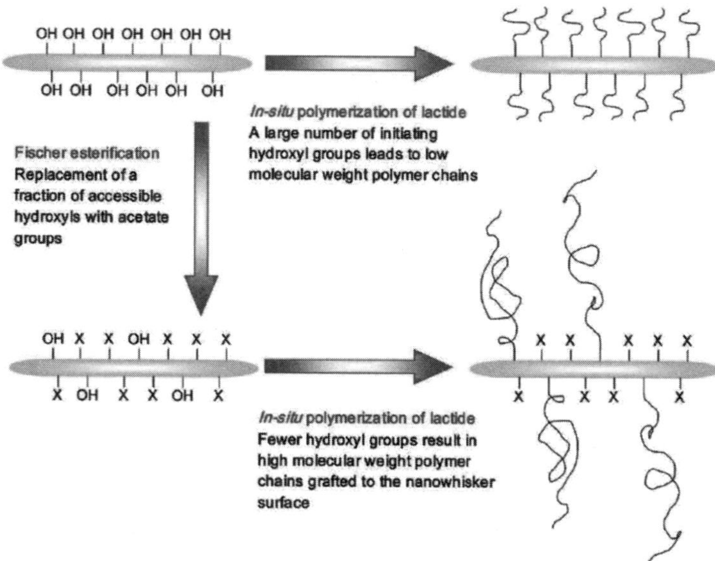

Fig. 6.5. Illustration of the synthesis approach to graft PLLA from the surface of acetylated CNCs. (Reprinted with permission from Ref. 64. Copyright (2012) American Chemical Society.)

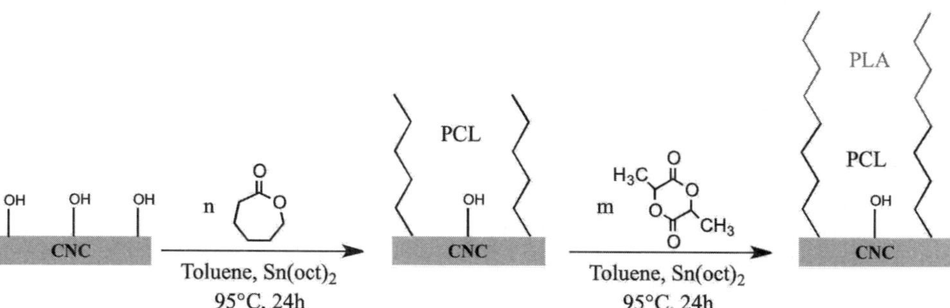

Fig. 6.6. Simplified schematic of the successive grafting of PCL and PLLA from the surface of CNCs.[65]

prolonged reaction time under metal-catalyzed ROP conditions, the morphology and crystalline structure of the CNCs were preserved. These CNCs grafted with PCL-*b*-PLA copolymer showed distinct crystallization behavior and dispersion ability when blended with a mixture of immiscible blend of PCL and PLA, as it will be detailed hereafter.

6.5.2 *Properties of nanocellulose-based bionanocomposites*

6.5.2.1 *Microstructure*

The final properties of nanocellulose-based bionanocomposites are strongly linked to the quality of the dispersion of nanocellulose substrates within the host matrices.

However, the hydrophilic nature of nanocellulose substrates makes them poorly dispersible in common non-polar solvents or in hydrophobic matrix such PLA or PCL and thus irreversible aggregations of nanocelluloses, predominantly stabilized by hydrogen bonds, occur. Surface chemical modifications, particularly *in situ* polymerization, are conducted to fine-tune their interfacial energy and enhance their dispersion in non-polar media. In the particular case of polyester (PCL or PLA)-grafted nanocellulose (both CNCs or CNFs), long-term dispersion tests carried out in toluene demonstrated their good ability to disperse in non-polar solvents. More importantly, when incorporated in the respective matrix, these polyester-grafted nanocelluloses finely and homogenously disperse within the host polymeric matrix as revealed by visual examination or microscopy observations.[17,57,63]

Interestingly enough is that PCL-*b*-PLA-grafted CNCs exhibited exceptional interfacial properties allowing their use to tune the compatibility of PCL/PLA immiscible blends and their related microstructures (Fig. 6.7) by preventing from coalescence the dispersed domains.[65]

6.5.2.2 *Thermal properties and crystallization*

The crystallization behavior of polyester-grafted nanocelluloses as prepared by *in situ* polymerization is linked directly to the length of the grafted polyester chains and their grafting density. Owing to the large number of hydroxyl groups available at the surface of nanocellulose and their stereoregularity, the grafting density is

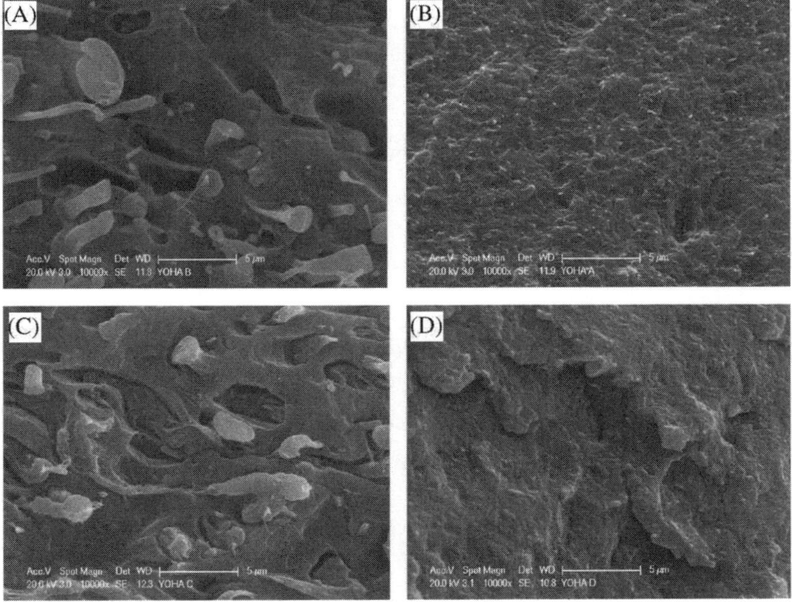

Fig. 6.7. Scanning electron microscopy (SEM) images of (A) binary PCL/PLA blend, and different ternary blends (B) PCL/PLA/unmodified CNCs, (C) PCL/PLA/PCL-*g*-CNC/PLA-*g*-CNC, and (D) PCL/PLA/PCL-*b*-PLA-*g*-CNC. (Reprinted with permission from Ref. 65. Copyright (2012) American Chemical Society.)

often sufficient to the grafted polyester to crystallize at the surface of nanocellulose. However, the melting temperatures recorded are usually lower than those for the corresponding polymers attesting for the formation, under restricted mobility, of small crystallites. Furthermore, when loaded into the respective polymer, this crystallization ability plays a key role in the alteration of the thermal and crystalline properties of the corresponding nanocomposites. Indeed, the grafted polyester chains can co-crystallize with free unbounded ones from the matrix, which affect the crystallization behavior. In the case of PCL filled with PCL–CNCs processed via solvent casting, glass transition temperatures T_g did not change regardless the amount of PCL–CNCs suggesting that these nanofillers do not affect the onset of translational and rotational backbone motions in the composite matrix. The melting temperatures changed slightly toward lower values while the overall crystallinity increased with increasing the PCL-grafted CNCs content, which suggests better filler–matrix compatibility.[17] Similar observations were also observed when PCL-*g*-CNCs/PCL were processed by extrusion.[57] Significant difference was observed in melting and crystallization behavior in the case of PCL-*g*-CNFs-based PCL laminate composites in comparison to neat PCL probably because of the restricted mobility of the PCL chains when anchored to the CNF surface. In addition, the non-isothermal crystallization shows that a much longer time is required for complete crystallization of PCL grafted to the surface of CNFs than for free PCL.[62]

More pronounced alterations were however obtained in the case of PLA-based nanocomposites for which PLA-*g*-CNCs modify substantially their crystalline and thermal behaviors.[63] In fact PLA-*g*-CNCs act as heterophase crystal nucleation agent more efficient than unmodified CNCs affecting the crystallization processes of the PLA matrix and therefore increasing the crystallization degree of the nanocomposites and triggers a decrease in the crystallization half-time recorded under isothermal conditions in DSC. Moreover, the incorporation of PLA-*g*-CNCs induced a splitting of the peak corresponding to the melting attesting for the modification of the PLA crystal growth. This phenomenon is often ascribed to the coexistence of two crystalline structures: less perfect crystals (α'-form crystals), which have enough time to melt and to reorganize into crystals with higher structural perfection (α-form crystals), before they re-melt at higher temperature. These results were much more improved when long PLA chains were grafted using partially acetylated CNCs. Indeed, the resulting nanocomposites displayed great improvements on heat distortion temperature and the nucleation by CNCs reduces the crystallization half-time to 15 s compared with 90 s for PLA.[64] In addition, a more striking result was however observed when CNCs were grafted with PDLA and incorporated into PLLA matrix. In fact in this case, the melting temperatures of the resulting nanocomposites shifted around 10°C toward higher values.

6.5.2.3 *Mechanical properties*

Nanoscale dimensions and impressive intrinsic mechanical properties make nanocellulose ideal candidates to improve the mechanical properties of host material. This lies to the fact that their axial Young's modulus is potentially stronger than steel and similar to Kevlar. For CNCs, the theoretical value of Young's modulus was estimated to 167.5 GPa.[66] Favier *et al.*[67,68] reported the first demonstration of the reinforcing effect of nanocelluloses, e.g., CNCs extracted from tunicate, in a polymeric nanocomposite. Since then great interest has risen on investigating the use of nanocellulose as a reinforcing phase in different polymeric matrices, including biopolymers offering the opportunity to access fully biobased materials. However, the interfacial strength and adhesion between hydrophilic nanocelluloses and hydrophobic biopolymers are usually weak and therefore the performance of the final bionanocomposite is limited by filler pullout rather than the nanofiller break (as a consequence of stress transfer facilitated through the filler–matrix interface). By using *in situ* polymerization approach to graft polymer chains, the obtained nanocomposites exhibit improved mechanical performances resulting for the maximized adhesion between the matrix and fillers. Indeed, the grafted polymer chains acted as long tails protruding in the matrix thereby creating a co-continuous phase between the nanocellulose and the polymeric matrix. Moreover, entanglements and/or co-crystallization between grafted and ungrafted biopolymer chains occurred when the molar weight of the grafted chains is high enough as observed in the case of PLA and PCL. For example, the addition of PCL-*g*-CNCs into the PCL matrix improved considerably the Young's modulus, which increased from 231 to 582 MPa and more importantly the ultimate elongation reduction was limited (561% for PCL-*g*-CNCs versus 391% CNCs for 10% CNCs content while for unfilled PCL 640%).[17] This unusual behavior clearly revealed the distinctive reinforcing effect of polymer-grafted CNCs resulting from the good adhesion ensuing from the entanglement and co-crystallization of polymer chains. Moreover, a solid-like behavior was observed by rheological investigations most likely due to the formation of a polymer physical network resulting from chains entanglement.[57] Lönnberg *et al.*[62] were inspired from these advances and used the same procedure to *in situ* graft PCL chains on CNFs and they were able to process the resulting materials using hot pressing in the form of laminates. These laminates exhibit great interfacial toughness resulting from the physical entanglements between PCL grafts and PCL matrix, which promote significant plastic deformation in the PCL bulk, thus increasing interfacial peeling energy.

6.6. Conclusions and outlooks

Bionanocomposites processed through *in situ* polymerization display very interesting thermomechanical properties, e.g., strength, modulus and dimensional stability, thermal stability, and heat distortion temperature, in addition to their permeability to gases and water, surface appearance, and optical clarity, etc., allowing their application in different fields where they are devoted to replace conventional nanocomposites derived from fossil resources. Nevertheless, there are still significant scientific and technological challenges to take up. To exploit their potential fully, research and development investments must be made to develop new processing and production technologies. Likely, future trends will include the investigation of novel types of biosourced nanofillers like chitin nanocrystals, crystalline starch nanoplatelets, and the like. Clearly, other biobased matrices will be developed and industrialized such as polycondensates based on polyester and/or polyamides. Combination of such biobased matrices and nanofillers will represent important challenges particularly in their interfacial compatibilization. Undoubtedly, *in situ* polymerization as promoted at the nanofiller surface will continue to represent one of the most effective synthesis strategy to reach this goal.

References

1. L. Yu, K. Dean and L. Li, Polymer blends and composites from renewable resources, *Prog. Polym. Sci.* **31** (2006) 576–602.
2. S.S. Ray and M. Bousmina, *Handbook of biodegradable polymeric materials and their applications*, American Scientific Publishers (2006) pp. 1–53.
3. E.T. Thostenson, C. Li and T.-W. Chou, Nanocomposites in context, *Compos. Sci. Technol.* **65** (2005) 491–516.
4. P.B. Messersmith and E.P. Giannelis, Synthesis and barrier properties of poly (ε-caprolactone)-layered silicate nanocomposites, *J. Polym. Sci., Part A: Polym. Chem.* **33** (1995) 1047–57.
5. T. Buntara, S. Noel, P.H. Phua, I. Melián-Cabrera, J.G. de Vries and H.J. Heeres, Caprolactam from renewable resources: catalytic conversion of 5-hydroxymethylfurfural into caprolactone, *Angew. Chem. Int. Ed.* **50** (2011) 7083–7087.
6. J.R. Lowe, M.T. Martello, W.B. Tolman and M.A. Hillmyer, Functional biorenewable polyesters from carvone-derived lactones, *Polym. Chem.* **2** (2011) 702–708.
7. A.C. Albertsson, Degradable aliphatic polyesters, *Adv. Polym. Sci.* **157** (2002) 1–179.
8. D. Garlotta, A literature review of poly (lactic acid), *J. Polym. Environ.* **9** (2001) 63–84.
9. L.T. Lim, R. Auras and M. Rubino, Processing technologies for poly (lactic acid), *Prog. Polym. Sci.* **33** (2008) 820–852.
10. L. Avérous, *Monomers, Polymers and Composites from Renewable Resources*, Oxford: Elsevier Ltd (2008) pp. 433–450.
11. F. Bergaya, G. Lagaly and K. Beneke, *Handbook of Clay Science*, Oxford: Elsevier Ltd (2006), pp. 1163–1181.
12. F. Bergaya, M. Jaber and J.F. Lambert, *Environmental Silicate Nano-Biocomposites*, London: Springer (2012) pp. 41–75.
13. M.F. Brigatti, E. Galan and B.K.G. Theng, *Handbook of Clay Science*, Oxford: Elsevier Ltd (2006) pp. 19–86.
14. Y. Habibi, L. A. Lucia and O. J. Rojas, Cellulose nanocrystals: chemisty, self-assembly and applications, *Chem. Rev.* **110** (2010) 3479–3500.
15. S. Elazzouzi-Hafraoui, Y. Nishiyama, J.-L. Putaux, L. Heux, F. Dubreuil and C. Rochas, The shape and size distribution of crystalline nanoparticles prepared by acid hydrolysis of native cellulose, *Biomacromolecules* **9** (2008) 57–65.
16. M. Grunert and W.T. Winter, Nanocomposites of cellulose acetate butyrate reinforced with cellulose nanocrystals, *J. Polym. Environ.* **10** (2002) 27–30.
17. Y. Habibi, A.L. Goffin, N. Schiltz, E. Duquesne, P. Dubois and A. Dufresne, Bionanocomposites based on poly (3-caprolactone)-grafted cellulose nanocrystals by ring-opening polymerization, *J. Mater. Chem.* **18** (2008) 5002.
18. A-F. Turbak, F.W. Snyder and K.R. Sandberg, Microfibrillated cellulose, a new cellulose product: properties, uses and commercial potential, *J. Appl. Polym. Sci. Appl. Polym. Symp.* **37** (1983) 815–827.

19. I. Siró and D. Plackett, Microfibrillated cellulose and new nanocomposite materials: a review, *Cellulose* **17** (2010) 459–494.
20. M. Henriksson, G. Henriksson, L.A. Berglund and T. Lindström, An environmentally friendly method for enzyme-assisted preparation of microfibrillated cellulose (MFC) nanofibers, *Eur. Polym. J.* **43** (2007) 3434–3441.
21. T. Lindström, M. Ankerfors and G. Henriksson, *Method for treating chemical pulp for manufacturing microfibrillated cellulose*, STFI-Packforsk AB: WO Patent 2007091942 (2007) p. 14.
22. S. Miyawaki, S. Katsukawa, H. Abe, Y. Iijima and A. Isogai, *Process for Producing Cellulose Nanofibers*, Jujo Paper Co LTD: WO 2010 116826 (2009).
23. M. Pääkkö, M. Ankerfors, H. Kosonen, A. Nykänen, S. Ahola, M. Österberg, J. Ruokolainen, J. Laine, P.T. Larsson, O. Ikkala and T. Lindström, Enzymatic hydrolysis combined with mechanical shearing and high-pressure homogenization for nanoscale cellulose fibrils and strong gels, *Biomacromolecules* **8** (2007) 1934–1941.
24. T. Saito, S. Kimura, Y. Nishiyama and A. Isogai, Cellulose nanofibers prepared by TEMPO-mediated oxidation of native cellulose, *Biomacromolecules* **8** (2007) 2485–2491.
25. L. Wagberg, G. Decher, M. Norgren, T. Lindstrom, M. Ankerfors and K. Axnas, The build-up of polyelectrolyte multilayers of microfibrillated cellulose and cationic polyelectrolytes, *Langmuir* **24** (2008) 784–795.
26. M. Alexandre and P. Dubois, Polymer-layered silicate nanocomposites: preparation, properties and uses of a new class of materials, *Mater. Sci. Eng. R* **28** (2000) 1–63.
27. P. Bordes, E. Pollet and L. Avérous, Nano-biocomposites: biodegradable polyester/nanoclay systems, *Prog. Polym. Sci.* **34** (2009) 125–155.
28. S. Sinha Ray and M. Bousmina, Biodegradable polymers and their layered silicate nanocomposites: in greening the 21st century materials world, *Prog. Mater. Sci.* **50** (2005) 962–1079.
29. P.B. Messersmith and E.P. Giannelis, Polymer-layered silicate nanocomposites: in situ intercalative polymerization of ε-caprolactone in layered silicates, *Chem. Mater.* **5** (1993) 1064–1066.
30. F. Gardebien, A. Gaudel-Siri, J.-L. Bredas and R. Lazzaroni, Molecular dynamics simulations of intercalated poly (ε-caprolactone)-montmorillonite clay nanocomposites, *J. Phys. Chem. B* **108** (2004) 10678–10686.
31. M. Haouas, A. Harrane, M. Belbachir and F. Taulelle, Solid state NMR characterization of formation of poly(ε-caprolactone)/maghnite nanocomposites by in situ polymerization, *J. Polym. Sci., Part B: Polym. Phys.* **45** (2007) 3060–3068.
32. D. Mecerreyes, R. Jérôme and P. Dubois, Novel macromolecular architectures based on aliphatic polyesters relevance of the 'coordination-insertion' ring-opening polymerization, *Adv. Polym. Sci.* **147** (1999) 1–59.
33. D. Kubies, N. Pantoustier, P. Dubois, A. Rulmont and R. Jerome, Controlled ring-opening polymerization of ε-caprolactone in the presence of layered silicates and formation of nanocomposites, *Macromolecules* **35** (2002) 3318–3320.
34. B. Lepoittevin, N. Pantoustier, M. Alexandre, C. Calberg, R. Jerome and P. Dubois, Layered silicate/polyester nanohybrids by controlled ring-opening polymerization, *Macromol. Symp.* **183** (2002) 95–102.
35. B. Lepoittevin, N. Pantoustier, M. Alexandre, C. Calberg, R. Jerome and P. Dubois, Polyester layered silicate nanohybrids by controlled grafting polymerization, *J. Mater. Chem.* **12** (2002) 3528–3532.
36. G. Gorrasi, M. Tortora, V. Vittoria, E. Pollet, B. Lepoittevin, M. Alexandre and P. Dubois, Vapor barrier properties of polycaprolactone montmorillonite nanocomposites: effect of clay dispersion, *Polymer* **44** (2003) 2271–2279.

37. P. Viville, R. Lazzaroni, E. Pollet, M. Alexandre, P. Dubois, G. Borcia and J.-J. Pireaux, Surface characterization of poly (ε-caprolactone)-based nanocomposites, *Langmuir* **19** (2003) 9425–9433.
38. O. Gain, E. Espuche, E. Pollet, M. Alexandre and P. Dubois, Gas barrier properties of poly(ε-caprolactone)/clay nanocomposites: influence of the morphology and polymer/clay interactions, *J. Polym. Sci., Part B: Polym. Phys.* **43** (2004) 205–214.
39. K. Chrissafis, G. Antoniadis, K.M. Paraskevopoulos, A. Vassiliou and D.N. Bikiaris, Comparative study of the effect of different nanoparticles on the mechanical properties and thermal degradation mechanism of in situ prepared poly (ε-caprolactone) nanocomposites, *Compos. Sci Technol.* **67** (2007) 2165–2174.
40. E. Tarkin-Tas, S.K. Goswami, B.R. Nayak and L.J. Mathias, Highly exfoliated poly(ε-caprolactone)/organomontmorillonite nanocomposites prepared by in situ polymerization, *J. Appl. Polym. Sci.* **107** (2008) 976–984.
41. L. Liao, C. Zhang and S. Gong, Preparation of poly (ε-caprolactone)/clay nanocomposites by microwave-assisted in situ ring-opening polymerization, *Macromol. Rapid Commun.* **28** (2007) 1148–1154.
42. L. Liao, C. Zhang and S. Gong, Microwave-assisted synthesis and characterization of poly (ε-caprolactone)/montmorillonite nanocomposites, *Macromol. Chem. Phys.* **208** (2007) 1301–1309.
43. E. Duquesne, S. Moins, M. Alexandre and P. Dubois, How can nanohybrids enhance polyester/sepiolite nanocomposite properties? *Macromol. Chem. Phys.* **208** (2007) 2542–2550.
44. M.-A. Paul, M. Alexandre, P. Degée, C. Calberg, R. Jérôme and P. Dubois, Exfoliated polylactide/clay nanocomposites by in-situ coordination–insertion polymerization, *Macromol. Rapid Commun.* **24** (2003) 561–566.
45. K. Fukushima, A. Fina, F. Geobaldo, A. Venturello and G. Camino, Properties of poly (lactic acid) nanocomposites based on montmorillonite, sepiolite and zirconium phosphonate, *eXPRESS Polym. Lett.* **6** (2012) 914–926.
46. K. Fukushima, D. Tabuani and G. Camino, Nanocomposites of PLA and PCL based on montmorillonite and sepiolite, *Mater. Sci. Eng. C* **29** (2009) 1433–1441.
47. A. Dasari, J. Quirós, B. Herrero, K. Boltes, E. García-Calvo and R. Rosal, Antifouling membranes prepared by electrospinning polylactic acid containing biocidal nanoparticles, *J. Membr. Sci.* **405–406** (2012) 134–140.
48. M. Liu, M. Pu and H. Ma, Preparation, structure and thermal properties of polylactide/sepiolite nanocomposites with and without organic modifiers, *Compos. Sci. Technol.* **72** (2012) 1508–1514.
49. M. Tortora, V. Vittoria, G. Galli, S. Ritrovati and E. Chiellini, Transport properties of modified montmorillonite-poly (ε-caprolactone) nanocomposites, *Macromol. Mater. Eng.* **287** (2002) 243–249.
50. R. Pucciariello, V. Villani, S. Belviso, G. Gorrasi, M. Tortora and V. Vittoria, Phase behavior of modified montmorillonite-poly (ε-caprolactone) nanocomposites, *J. Polym. Sci., Part B: Polym. Phys.* **42** (2004) 1321–1332.
51. D. Homminga, B. Goderis, I. Dolbnya and G. Groeninckx, Crystallization behavior of polymer/montmorillonite nanocomposites. Part II. Intercalated poly(ε-caprolactone)/montmorillonite nanocomposites, *Polymer* **47** (2006) 1620–1629.
52. M.-A. Paul, C. Delcourt, M. Alexandre, P. Degée, F. Monteverde, A. Rulmont and P. Dubois, (Plasticized) Polylactide/(organo-)clay nanocomposites by in situ intercalative polymerization, *Macromol. Chem. Phys.* **206** (2005) 484–498.
53. B. Lepoittevin, N. Pantoustier, M. Devalckenaere, M. Alexandre, C. Calberg, R. Jerome, C. Henrist, A. Rulmont and P. Dubois, Polymer/layered silicate nanocomposites by combined intercalative polymerization and melt intercalation: a masterbatch process, *Polymer* **44** (2003) 2033–2040.

54. E. Pollet, M.-A. Paul and P. Dubois, *Biodegradable polymers and plastics*, New-York, USA: Klumer Academic Plenum Publishers (2003) pp. 327–354.
55. D. Roy, M. Semsarilar, J.T. Guthrie and S. Perrier, Cellulose modification by polymer grafting: a review, *Chem. Soc. Rev.* **38** (2009) 2046–2064.
56. A. Carlmark, E. Larsson and E. Malmström, Grafting of cellulose by ring-opening polymerisation–a review, *Eur. Polym. J.* **48** (2012) 1646–1659.
57. A.L. Goffin, J.M. Raquez, E. Duquesne, G. Siqueira, Y. Habibi, A. Dufresne and P. Dubois, Poly(ε-caprolactone) based nanocomposites reinforced by surface-grafted cellulose nanowhiskers via extrusion processing: morphology, rheology, and thermomechanical properties, *Polymer* **52** (2011) 1532–1538.
58. G. Chen, A. Dufresne, J. Huang and P.R. Chang, *Macromol. Mater. Eng.* **294** (2009) 59.
59. M. Labet and W. Thielemans, Citric acid as a benign alternative to metal catalysts for the production of cellulose-grafted-polycaprolactone copolymers, *Polym. Chem.* **3** (2012) 679–684.
60. H. Lönnberg, L. Fogelström, L. Berglund, E. Malmström and A. Hult, Surface grafting of microfibrillated cellulose with poly(ε-caprolactone) — synthesis and characterization, *Eur. Polym. J.* **44** (2008) 2991–2997.
61. H. Lönnberg, K. Larsson, T. Lindström, A. Hult and E. Malmström, Synthesis of polycaprolactone-grafted microfibrillated cellulose for use in novel bionanocomposites–influence of the graft length on the mechanical properties, *ACS Appl. Mater. Interfaces* **3** (2011) 1426–1433.
62. H. Lönnberg, L. Fogelström, Q. Zhou, A. Hult, L. Berglund and E. Malmström, Investigation of the graft length impact on the interfacial toughness in a cellulose/poly (ε-caprolactone) bilayer laminate, *Compos. Sci. Technol.* **71** (2011) 9–12.
63. A.-L. Goffin, J.-M. Raquez, E. Duquesne, G. Siqueira, Y. Habibi, A. Dufresne and P. Dubois, From interfacial ring-opening polymerization to melt processing of cellulose nanowhisker-filled polylactide-based nanocomposites, *Biomacromol.* **12** (2011) 2456–2465.
64. B. Braun, J.R. Dorgan and L.O. Hollingsworth, Supra-molecular ecoBioNanocomposites based on polylactide and cellulosic nanowhiskers: synthesis and properties, *Biomacromol.* **13** (2012) 2013–2019.
65. A.-L. Goffin, Y. Habibi, J.-M. Raquez and P. Dubois, Polyester-grafted cellulose nanowhiskers: a new approach for tuning the microstructure of immiscible polyester blends, *ACS Appl. Mater. Interfaces* **4** (2012) 3364–3371.
66. K.Tashiro and M. Kobayashi, Theoretical evaluation of three-dimensional elastic constants of native and regenerated celluloses: role of hydrogen bonds *Polymer* **32** (1991) 1516–1526.
67. V. Favier, G.R. Canova, J.Y. Cavaillé, H. Chanzy, A. Dufresne and C. Gauthier, Nanocomposite materials from latex and cellulose whiskers, *Polym. Adv. Technol.* **6** (1995) 351–355.
68. V. Favier, H. Chanzy and J.Y. Cavaille, Polymer nanocomposites reinforced by cellulose whiskers, *Macromolecules* **28** (1995) 6365–6367.

Chapter 7

Characterization of Nanocomposites Structure

Kristiina Oksman[1] and Robert J Moon[2,3]
[1] *Composite Center Sweden, Division of Materials Science, Luleå University of Technology, Luleå, Sweden*
[2] *USDA-Forest Service, Forest Products Laboratory, Madison, WI, USA*
[3] *School of Materials Engineering and Birck Nanotechnology Center, Purdue University, West Lafayette, IN, USA*

This chapter summarizes several techniques that have been used in the characterization of cellulose nanocomposites, in particular cellulose nanomaterials (CNs) dispersion, distribution, and orientation within polymer matrixes. The microscopy techniques described are optical microscopy, scanning electron microscopy, transmission electron microscopy, and atomic force microscopy. Also, the use of X-ray diffraction for quantification of CN orientation is discussed. The characterization of bionanocomposites is challenging because these materials are soft, moisture sensitive, non-conductive, and usually both the matrix phase and the reinforcement phase primarily consist of low atomic number elements (making differentiation difficult). Different sample preparation techniques for CN composite materials are also discussed.

7.1. Introduction

To take advantage of nanoscale features and create nanocomposites based on plant biomass, reliable characterization measurements are needed to resolve nanosized-scale features so that the "nanoeffect" can be elucidated. Although nanoscale measurement methods have expanded in recent years, not all these techniques are useful for soft, hydrophilic, non-conducting biomass specimens. This chapter summarizes several methods that have been shown to be particularly useful in nanoscale characterization of cellulose nanocomposite surfaces, cellulose nanomaterial (CN) distribution, dispersion, and their orientation within the polymer matrix.

The structure of polymer nanocomposites is traditionally characterized by a combination of transmission electron microscope (TEM) and wide-angle X-ray diffraction (WAXD).[1] This combination is, however, convenient only for layered silicate-based nanocomposites because of the ordered stacking of the silicate layers. For cellulose, only the 3D arrangement of the cellulose chains in the crystallites is detectable in WAXD and no peaks corresponding to the stacking of the crystallites can be observed in the nanocomposites.

Table 7.1. Resolution of the different microscopy techniques.

Technique	OM	SEM	FESEM	TEM	AFM
Resolution	0.2–1.3 μm	5 nm	1 nm	0.2 nm	1 nm (x,y) 0.1 nm (z)

For observation of the nanostructure, different microscopy techniques can be utilized. In general, bionanocomposites are non-conductive, soft, and moisture-sensitive materials and therefore the sample preparation and microscopic characterization are challenging. For example, the use of electron microscopes will, in particular, require special attention to electron dose, contrast, and methods to assess the bulk structure without affecting the nanocomposites morphology.

Optical microscopy (OM), scanning electron microscopy (SEM), TEM, and atomic force microscopy (AFM) have been applied to study the structure, size, and morphology of cellulose nanocrystals (CNCs) and cellulose nanofibrils (CNFs).[2–6] Typically, SEM has been used most for the characterization of nanocomposites structure[7–14] even though the resolution is limited, as compared to AFM or TEM, and detailed information of the CN distribution within the polymer matrix is difficult to obtain. There are, however, SEM microscopes and field emission gun scanning electron microscopes (FESEM) with higher resolution comparable with TEM, but operate at very low voltages, making it possible to observe organic materials without conductive coating. AFM[14–17] and TEM[12,13,18–20] have also been utilized in studies on nanocomposites structure, especially the distribution and dispersion of cellulose or chitin nanomaterials. Table 7.1 shows the maximum resolution of the different microscopy techniques.[21,22]

The following sections describe the common sample preparation for microscopy studies of nanocomposites and some examples of micro- and nanostructures of the composite materials are shown which are viewed using different microscopy techniques.

7.2. Sample preparation

The most commonly used way to study the microstructure of nanocomposites is to examine fractured surfaces. The problem with this is that polymer nanocomposites are non-conducting and when imaged using SEM, the surfaces are usually coated with a thin layer (<10 nm) of conducting material (typically gold or platinum), otherwise the high intensive electron beam, that is focussed on the sample surface, can damage the surface. However, this coating will cover the finer details on the surface and may also create other artifacts (cracks, altered surface roughness, etc.). Figure 7.1 shows a platinum-coated cellulose nanocomposite where the coating has cracked and the underlying structure is visible.

Another sample preparation method for studies of cellulose nanocomposites structure is ultramicrotoming. This is a method where ultrathin (<100 nm) samples are produced and these are usually used for TEM imaging or when very smooth

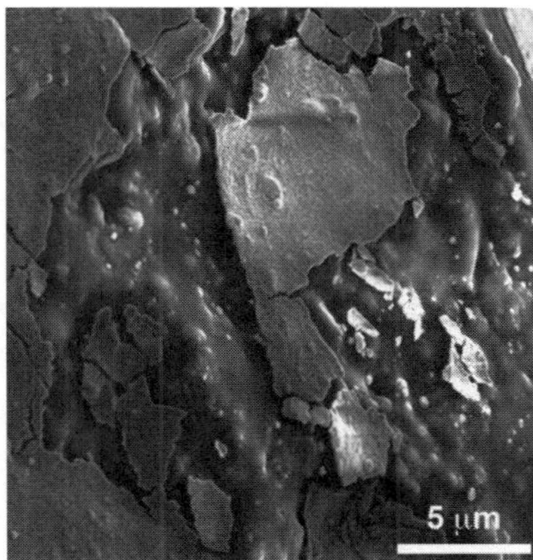

Fig. 7.1. FESEM image of platinum-coated nanocomposites of CNCs and polylactic acid (PLA), the coating has cracked and the underlying structure is visible.

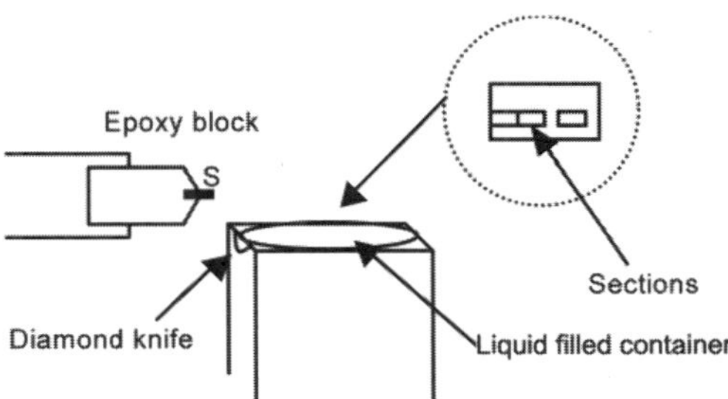

Fig. 7.2. Ultramicrotoming of the thin slices. S is the sample embedded in epoxy and sliced using a diamond knife and collected from a liquid-filled container. (Figure is based on Ref. 21.)

surfaces are needed for AFM studies. This is a common sample preparation method for polymeric materials. Ultramicrotomed samples will give information of actual "bulk" structure of the composite material. The sample is prepared using the following three steps: (i) the specimen is first embedded in a thermosetting resin and cured, (ii) then it is trimmed and sliced using microtome, and finally (iii) the thin slices can be stained if needed.

To obtain good sectioning, it is important that the hardness of the embedding polymer matches the hardness of the sample. Epoxies are typically used as the embedding polymer because they are stable in the electron beam. The schematic of the ultramicrotoming is shown in Fig. 7.2.

After the microtoming procedure, the slices are mounted on grids. The thickness of the slices is usually less than 100 nm, and in TEM imaging allows sufficient electron transmission through the sample to obtain a signal for imaging. It is important to slice the samples to an even thickness to avoid artifacts such as mass-thickness contrast.

Polymers with glass transition temperature below room temperature can be too soft for sectioning at room temperature. In this case, ultramicrotoming is performed at low temperature using liquid nitrogen. The method is called cryo-ultramicrotoming. The advantages of cryo-technique are that embedding is not needed and soft polymers can be sectioned. Disadvantages are that this technique is more time consuming, it is also difficult to collect the sectioned samples, and frost could build-up.[23]

Freeze-etching method is an alternative to cryo-ultramicrotomy and can be used for water-soluble or soft polymers and this method is especially useful for beam-sensitive materials. The advantage of the freeze-etching technique is that there is no need for chemical fixation or staining of the sample. The freeze-etching method involves the following steps:

- rapid freezing,
- freeze fracturing or etching,
- shadowing, and
- replication and replica cleaning by dissolving the specimen.[24]

For the rapid freezing step, the polymer sample is mounted on a support and rapidly immersed in liquid nitrogen and placed in a cooled chamber and fractured. The freshly cleaved surface is either etched or fractured and shadowed with carbon or metallic deposition and a replication is made with a thin layer of carbon. The replica is taken off the sample, washed, and gathered onto TEM grids. The replica is an inverted copy of the original sample surface topography and is very stable in the electron beam[24] allowing imaging of the surface topography. Note that this "inversion" of the surface topography needs to be accounted for when interpreting the imaging.

Figure 7.3 shows an example of nanocomposite structure when the sample is prepared using the freeze-etching technique. This is a replica of a nanocomposite with CNCs embedded in a thermoplastic starch (TS). The freeze-etching method was used because of the difficulty to section the starch matrix.[20] The CNCs can be detected in which the contrast is created by platinum metal shadowing. The CNCs appear to be very wide (darker dots in the image), and it is difficult to determine

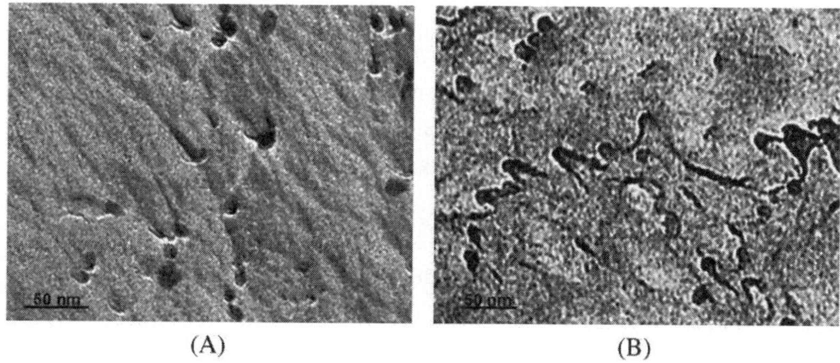

Fig. 7.3. TEM images of freeze-etched replicas of TS-CNC nanocomposite (a) parallel view and (b) perpendicular view to the film surface. (Adapted from Ref. 20 with permission from Springer.)

if the crystals are individual, but they broaden due to metal shadowing or if they are aggregated.

7.3. Characterization of the nanocomposites structure

7.3.1 *Optical microscopy*

OM is not a suitable tool for characterization of nanosize materials such as CN within nanocomposites, but is useful in characterizing the macroscopic composite structure in cases when the matrix is transparent or the sample is very thin. For example, polarized OM can be used to study crystallization behavior of CN composites. Pei *et al.*[25] studied the effect of CNC additions on the crystallization nucleation rate of a PLA polymer. Unmodified CNCs and silylated functionalized CNCs were tested and it was seen that CNC without surface modification aggregated and was not as effective nucleation agent as the silylated CNCs. Figure 7.4 shows the polarized OM images of pure PLA, PLA with 1% CNC, and PLA with 1% silylated crystals in melt at 210°C and how the crystallites nucleate during the cooling, after 0, 5, and 10 min. It is possible to see that the PLA with unmodified CNC has some aggregates compared to the PLA with silylated crystals (0 min).

Kvien and Oksman[26] used polarized OM to study alignment of the CNCs in a PVA matrix using magnetic field. The films were thin and transparent and when viewed in OM, the film was bright at 45° and dark at 0° and 90° between the magnetic field direction and the polarization plane (Fig. 7.5). This OM indicated

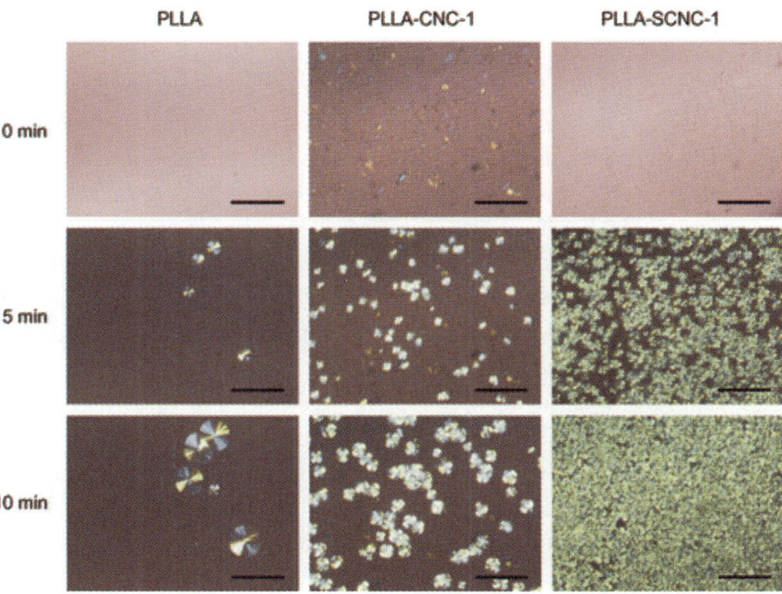

Fig. 7.4. Polarized optical microscope images of PLLA, PLLA–CNC1, and PLLA–SCNC1 after cooling from melt at 210°C showing the ability of CNCs as nucleation agent for PLLA. Scalebar, 200 μm. (The figure is adapted from Ref. 25, with permission from Elsevier.)

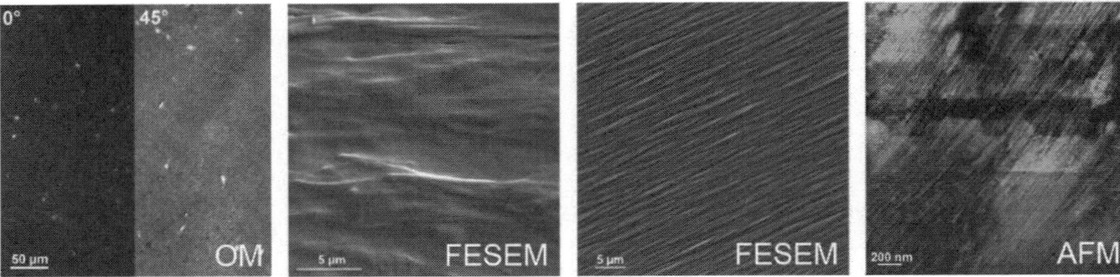

Fig. 7.5. Polarized optical microscope images of the nanocomposite showing reflected light at 45° in polarized light indicating alignment of CNCs, the alignment of the CNC is further strengthened by studies with FESEM (fractured and etched surface) and with AFM where highly oriented structures can be seen.

that the CNCs were aligned in the PVA matrix, either parallel or transverse to the field direction. Also, the alignment of the nanocrystals was further studied by FESEM and AFM as seen in Fig. 7.5. The FESEM images of the fractured surface and etched nanocomposites showed a highly oriented structure, similar orientation was seen also in AFM.

7.3.2 *Electron microscopy*

7.3.2.1 *Scanning electron microscopy*

SEM is one of the most used microscopy for materials structure and surface characterization. These microscopes have a resolution between 1 and 5 nm. In addition to the high resolution, they also have a large depth of field, which is the reason that the images appear three dimensional. The principle of SEM is that an electron gun generates electrons and accelerates them through lenses which focus the beam with very small spot size. These electrons interact with the specimen to a depth of about 1 μ and generate signals that are used to form the image. The three most important signals are backscattered electrons, secondary electrons, and X-rays. The backscattered electrons are elastically scattered electrons and give compositional contrast depending on the atomic number of the specimen. These electrons have high energy and they come from the depth of the specimen (1 μm or more). Secondary electrons are low energy electrons and come from the top surface of the specimen (a few nm), and are mainly used for topography imaging of the sample.[27]

Cellulose nanocomposites or bionanocomposites contain polymers, where both the matrix and the reinforcing phases are polymeric, which means they are nonconductive and consist of low atomic number elements and, in topography imaging, mainly used for these materials.

SEM images of two different cellulose nanocomposites are shown in Figs. 7.6 and 7.7. The nanocomposite in Fig. 7.6 shows the fractured surface of PLA nanocomposites with CNC and CNF.[9]

This material was fractured using liquid nitrogen and coated with platinum to avoid charging of the electron beam. It is possible to see that nanocrystals and

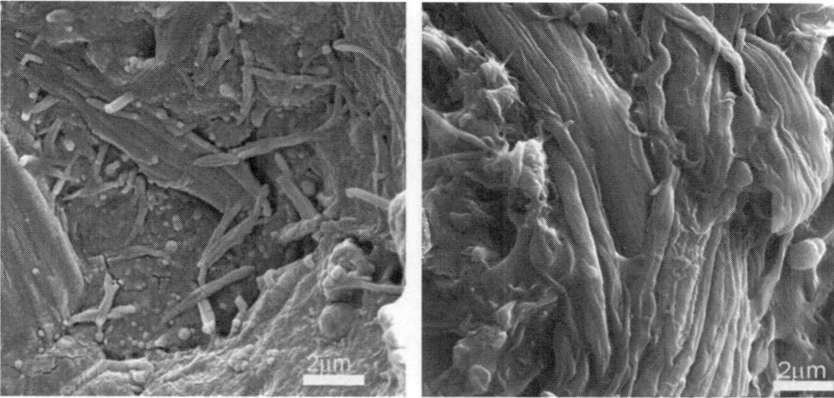

Fig. 7.6. CNCs (left) and cellulose nanofibers (right) in a PLA matrix. It is seen that none of the nanomaterials are well dispersed in the PLA but the nanocrystals are of smaller size and better dispersed and distributed than the nanofibers.

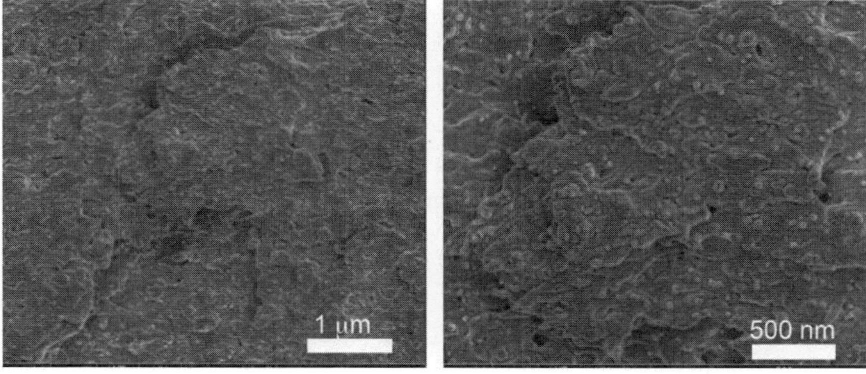

Fig. 7.7. Fractured surface of nanocomposites with CNC (12 wt%) in a CAB matrix. It is seen that CNCs are well distributed and are also dispersed, small dots are seen which are believed to be CNC and no large agglomerate can be seen.

nanofibers are not well dispersed in the PLA matrix, agglomerates in micrometer size are seen.

The second image, Fig. 7.7, shows the fractured structure of nanocomposite with cellulose acetate butyrate (CAB) as matrix and CNC as the reinforcing phase. This sample was not coated, but it was still possible to obtain reasonable resolution using a low voltage of 1 kV in the FESEM. In this case, the CNC was better dispersed in the polymer matrix. There were no visible agglomerates of the crystals in this composite; however, the small spherical particles in the matrix might be cross-sections of small clusters of cellulose crystals in the matrix.[10] In this case, a further investigation of the material in AFM or TEM would give more information of the distribution of CNC in the matrix.

7.3.2.2 *Transmission electron microscopy*

The principle of TEM is that the high energy electrons are transmitted through an ultrathin section of the specimen. The image is formed due to electron scattering when the beam hits the specimen.[28]

The electrons are emitted from an electron gun (filament). Below the electron gun are two or more condenser lenses which demagnify the beam emitted by the gun and control its diameter as it hits the specimen, which is held inside an objective lens just below the condenser lenses. There are two lenses after the objective lens, the intermediate lens and the projector lens. Each produces a real and magnified image, which then produces the image on the fluorescent imaging screen or film. TEM image contrast is due to electron scattering. Bright field (BF) is an imaging mode where an objective aperture is inserted so that the direct unscattered electrons form the image. Regions in the specimen which are thicker or of higher density will scatter more strongly and will appear darker in the image because highly scattered electrons are stopped by the objective aperture. An image field with no specimen in it is bright in BF. In TEM, there are three basic contrast mechanisms, which may contribute to the formation of image:[23,28]

- diffraction contrast,
- mass-thickness contrast, and
- phase contrast.

The cellulose nanocomposites are composed of low atomic number elements and therefore scatter electrons weakly, giving poor contrast in the TEM. For these materials, the mass-thickness contrast mechanism can be exploited by deliberately staining the thin specimen with a heavy metal which highlights specific features of interest. For example, uranyl acetate is one suitable staining agent for CNs to enhance better contrast.[21]

The properties of nanoparticle-reinforced polymer composites are directly affected by the distribution of the nanoparticle phase within the matrix phase. The ability to characterize the nanoparticle distribution along a surface or when embedded within a polymer matrix is critical when relating the resulting composite properties to mechanism of property enhancement.

Transmission electron microscopy is used to analyze the thought thickness of the nanocomposites material and is therefore the most suitable microscopy method, provided the CN distribution and dispersion are of interest.

Bondeson and Oksman[13] studied how the addition of a surfactant affected CNC dispersion and distribution in the PLA matrix. Figure 7.8 shows the FESEM images as well as a TEM image of the CNC composites. The first image is a fractured surface of the PLA nanocomposite where no surfactant was used showing aggregates of CNCs. The second image is also a fractured surface of the nanocomposites with surfactant showing well dispersed and distributed nanocrystals. The last image is TEM of the surfactant sample showing individual CNCs dispersed in the PLA matrix.

Fig. 7.8. The structure of PLA–CNC nanocomposites viewed with FESEM and TEM. The first image is a composite with poor dispersion and distribution and the second and third images show that the dispersion is affected by a surfactant. The FESEM confirms a better dispersion and distribution of CNC in a PLA and this is further shown by the TEM with individual crystals in PLA.

7.4. Atomic force microscopy

AFM is a scanning probe microscopy technique that can characterize nanometer-scale features of surfaces,[29,30] and has been extensively used in characterizing nanocomposites[31]. A detailed description of the AFM system and technique can be found in Refs. 29, 30, and 31, a brief description is given in Vol. 1.

Topography imaging is usually completed by scanning the AFM probe (typical radius of curvature of 10 nm) over the composite surface while the interaction response between the probe and the sample surface is monitored. Two operating modes can be used (contact and intermittent-contact), and measurements can be completed under different environments (vacuum, vapor, fluid). In the contact mode, the AFM probe tip is scanned across the surface and feedback is used to maintain a constant force between the tip and the sample (see Fig. 7.9). For the intermittent-contact mode (or "tapping" mode), the AFM probe is vibrated near its resonance frequency and feedback is used to maintain a constant force in some aspects of the probe's vibration (such as amplitude). Advantages of intermittent-contact mode are

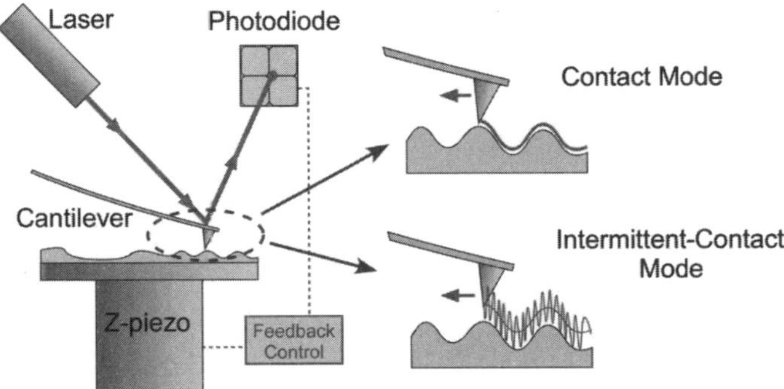

Fig. 7.9. Schematic of AFM setup and the representative configurations of the AFM cantilever/tip and surface for the different imaging modes: contact and intermittent-contact.

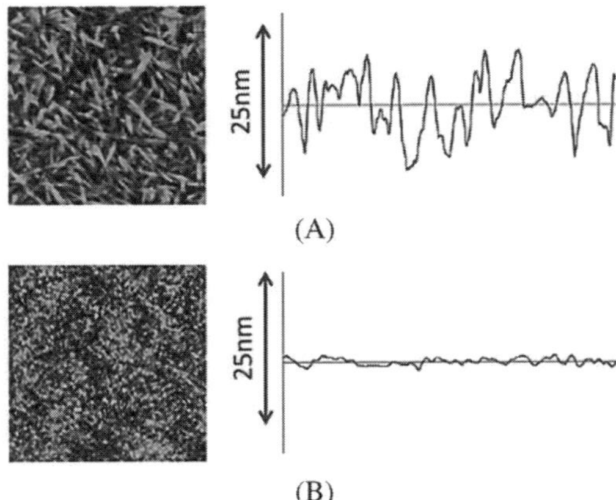

Fig. 7.10. Topography images (3 by 3 μ scans) and roughness profiles (1 μ trace) measured by AFM intermittent-contact mode imaging of cellulose films on silica: (A) CNC film and (B) Langmuir–Schaefer cellulose films. (Adapted with permission from Ref. 34. Copyright (2009) American Chemical Society.)

that it has lower lateral forces compared to contact mode and offers the possibility to acquire additional contrast channels such as a phase image.

AFM topography imaging of cellulose nanocomposites has been used to characterize the surface roughness.[32–34] The sub-nanometer quantitative height resolution and qualitative nanometer lateral resolution (depending on tip radius) have allowed for comparative studies on cellulose nanocomposite processing on the resulting surface roughness/finish[32,34] (see Fig. 7.10).

In the intermitted-contact mode, and when feedback is maintained on the amplitude of the probe vibration, information about nanoparticle distribution within polymer matrix composites can be obtained using phase imagining. Phase imaging refers to recording the phase lag (i.e., the "delay") of the cantilever oscillation, relative to the signal sent to the cantilever's piezo driver. The phase lag is sensitive to variations in material properties (e.g., adhesion, viscoelasticity, etc.), and thus can create contrast between the different material components within nanocomposites. Additionally, phase imaging highlights edges and is not affected by large-scale height differences, providing clearer observation of fine features, such as grain edges. These characteristics of phase imaging make this technique useful for investigating nanoparticle distribution within polymer composites, which has been completed for various CN–polymer composites[15,16,35] (see Fig. 7.11). Since AFM is a surface measurement technique, to characterize the nanoparticle distribution through the composite thickness, it is necessary to image multiple sections through the thickness of the composite.

7.4.1 *AFM — surface chemistry*

AFM techniques have been used to investigate surface chemistry of nanocomposites. The most common technique for these studies is the use of AFM force

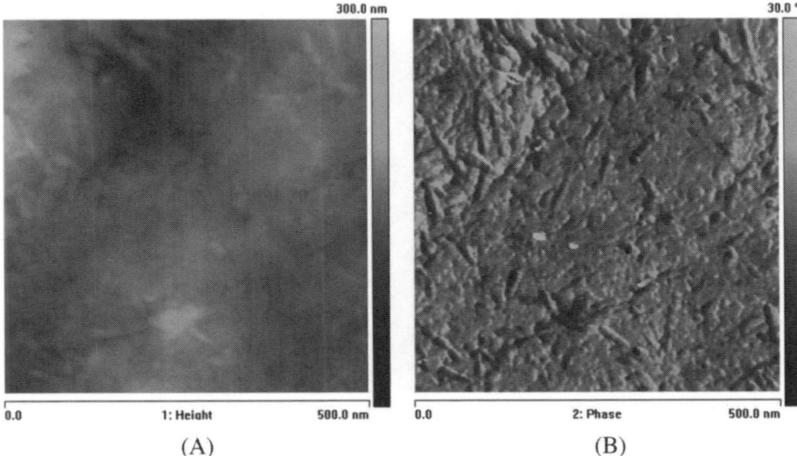

Fig. 7.11. AFM images of 5 wt% CNC–PHBV (poly(3-hydroxybutyrate-co-3-hydroxyvalerate)) nanocomposite: (A) topography and (B) phase. In the phase image, the darker colored rod-like CNC particles have increased contrast from the matrix polymer, allowing greater discrimination of the CNC particles from the polymer matrix as compared to that in the topography image. (Reprinted from Ref. 35, Copyright (2010) with permission from Elsevier.)

spectroscopy.[36] A detailed description of this technique can be found in Refs. 4, 37, 38, and 39. A brief description is given here. One method of monitoring the interaction between AFM probe tip and that of the sample surface is measuring force–distance (F–D) curves, which summarize the vertical force acting on the AFM tip as it approaches and withdraws from the sample surface (i.e., the distance between the tip and the surface, D).

An idealized F–D plot is given in Fig. 7.12 which shows five general zones: (1) approach, (2) jump-to-contact, (3) indentation, (4) jump-off-contact, and (5) withdrawal. During a typical test, the AFM tip is initially far away from the sample surface that there is minimal interaction (i.e., no force acting on the tip). Jump-to-contact occurs once the tip moves sufficiently close to the surface where the tip–surface attractive force gradient is greater than the cantilever stiffness. Additional applied forces result in the AFM tip indenting into the sample surface, which can be used to extract mechanical properties of the surface. Jump-off-contact happens during the withdrawal, giving information about the adhesive forces (hydrogen, van der Waals, Coulomb, etc.). After this, the tip–sample distance is sufficiently far that no force acts on the tip. The discontinuity of the jump-off-contact can be used to define the force of adhesion (F_{ad}), which can be used to assess the tip–surface interaction.

A single F–D plot provides adhesion/chemical information at the given contact point, whereas to map the adhesion/chemical information for a given area on the surface, multiple F–D plots are needed. Measurements can be completed in vacuum, vapor, or fluids. When testing within vapor, in particular, in ambient conditions with moisture, meniscus formation at the tip–sample contact may occur and the resulting capillary force will contribute to the force of adhesion. By adjusting the chemistry

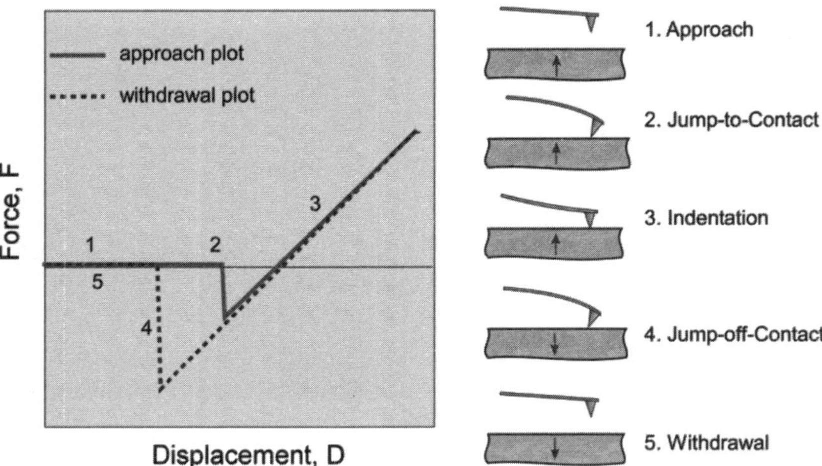

Fig. 7.12. Idealized AFM force–displacement (F–D) plot of AFM tip showing the five stages: (1) approaching, (2) jump-to-contact, (3) indentation, (4) jump-off-contact, and (5) withdrawing from a surface.

on the AFM tip and the measurement, fluid-specific chemical interactions with a given surface can be investigated.[37,38]

7.5. Cellulose nanoparticle orientation

The mechanical properties of crystalline cellulose (and thus CNCs) are expected to be anisotropic because of the parallel alignment of the cellulose chains along the length of the cellulose crystal. Properties along the crystal length (elastic modulus, $E = \sim 145\,\text{GPa}$ and ultimate tensile strength, $\sigma f = 7.5\,\text{GPa}$) are higher than in perpendicular directions ($E = 3 - 50\,\text{GPa}$).[40] Owing to the high axial properties and particle aspect ratio of CNCs, preferentially aligning the particles within a composite to increase properties in the aligned direction has been investigated. Several methods have been used to induce CNC alignment within composites: magnetic fields,[26,41] electric fields,[42,43] mechanical shearing of CNC suspensions,[44,45] combined field and mechanical shearing,[46] drawing of as-cast BC films[47] and cellulose–cellulose composites,[48] and the wet-spinning of CNC composite fibers.[49] Because of the strong influence of CNC orientation on composite properties, this is an important parameter to characterize.

7.5.1 *Wide-angle X-ray diffraction*

The alignment of the CNC within the composites can be characterized via WAXD,[42,44,47–50] in which differences in diffracted X-ray intensities for a given diffraction plane within the cellulose crystal structure are measured as a function of orientation within the composite sample geometry. WAXD measures the diffracting X-rays of a given sample as a function of the diffraction angle (2θ) with

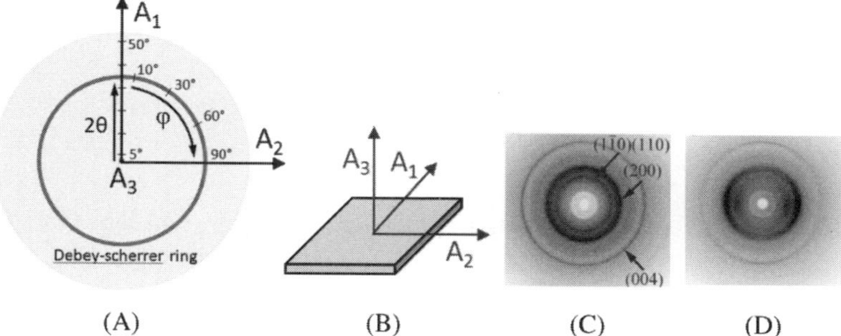

Fig. 7.13. 2D WAXD diffractograms: (A) schematic diffractogram showing *Debye*–Scherrer ring, θ, and φ. (B) WAXD can be measured from different composite axis giving additional information on CNC alignment within the composite. (C and D) are diffractograms of a CNC composite,[49] for (C) random CNC alignment showing uniform ring intensity, $I(\varphi)$, and (D) high CNC alignment showing variation in $I(\varphi)$ within a given ring. (Adapted with permission from Ref. 49. Copyright (2011) American Chemical Society.)

respect to the direction of the primary X-ray beam (e.g., monochromatic X-ray, $\gamma = 0.1541$ nm, from CuKα radiation generated at a given accelerating voltage and current). Diffracting X-rays occur only at specific conditions based upon the internal crystalline structure of the solid. For cellulose I, diffraction peaks are at 2θ of ~14.5°, ~16.6°, ~20.4°, ~22.7°, and ~34.4°, corresponding to the miller indices of the crystallographic planes for Iβ ($1\bar{1}0$), (110), (102), (200), (004), or Iα (100), (010), ($\bar{1}\bar{1}0$), (110), and ($\bar{1}\bar{2}3$ & $\bar{1}\,\bar{1}4$), respectively.[51,52] Peaks assignments use the Iα and Iβ unit cells of Sugiyama *et al.* in Ref. 53.

The degree of CNC alignment within a composite can be measured from two-dimensional (2D) WAXD diffraction patterns that give diffraction intensities for both 2θ (typically 5–50°), and the azimuthal angle, φ, which is the 360° rotation about one of the axis of the composite sample. The resulting 2D diffraction patterns consist of *Debye*–Scherrer rings which give the azimuthal intensity distribution for each 2θ diffraction peaks (Fig. 7.13A). Since the resulting diffraction patterns are linked with the composite sample, choosing which axis of the composite sample to scan can give additional information regarding the CNC alignment within the composite.[49] Typically for CNC composite films or plates, diffraction patterns are measured in the A_3 direction (see Fig. 7.13B), and for composites with CNC alignment, attempts are made to arrange the composite test sample such that the CNC alignment runs parallel to the A_1 or A_2 axis. For assessing CNC alignment, the (200), (110), or ($1\bar{1}0$) crystallographic planes (Iβ crystal structure) are measured. The (200) is used to assess the orientation of the CNC fiber axis, whereas the (110) and ($1\bar{1}0$) can be used to assess the rotation about the CNC fiber axis. The integrated intensity for a given Debye–Scherrer ring is used to calculate the orientation factor:

$$\{\cos^2\varphi\} = \frac{\int_0^{\pi/2} I(\varphi)\cos^2\varphi\sin\varphi d\varphi}{\int_0^{\pi/2} I(\varphi)\sin\varphi d\varphi}, \quad (7.1)$$

where $I(\varphi)$ is the diffracted intensity as a function of azimuthal angle. The more commonly used Herman's order parameter, S, is calculated using the equation

$$S = \frac{1}{2}(3\{\cos^2\varphi\} - 1), \tag{7.2}$$

in which, $S = 0$, for random CN alignment and $S = 1$ for complete CN alignment.

7.5.2 *Atomic force microscopy*

AFM imaging (topography, phase imaging, etc.) has been used to characterize the CNC alignment within nanocomposites[26,33,41,45] in which the degree of CNC alignment can be qualitatively described using various image analysis methods.[45] Typically, the raw AFM image undergoes some post-processing to highlight individual particles and some feature is used to identify the long axis of the CNC. The angle of the long axis of the CNC with respect to a given axis (e.g., sample geometry, or the direction of mechanism that caused CNC alignment) is measured. A histogram is produced that summarizes the percent of CNCs for a given angle from 0 to 90°. Note that such techniques describe a 2D CNC alignment, assuming that all the CNCs lie parallel to the imaging plane of the AFM. To account for this additional degree of freedom in the CNC orientation, a method for characterizing the three-dimensional alignment of short fiber composites using the geometry of the elliptical cross-section shape on a polished surface may be applicable to CNC composites, and for a description of this methodology, the reader can refer to Ref. 54.

7.6. Conclusions

This chapter summarizes several of the more prevalent methods used to date to characterize cellulose nanocomposites (OM, SEM, FESEM, TEM, AFM), and the CNC orientation within the polymer matrix (WAXD, TEM, AFM). For several of these characterization methods, the small size and low atomic number result in some unique challenges for quantitative measurements and require special attention as briefly described. Progress is being made for improved characterization of cellulose nanocomposites, and with this, our understanding of these materials is also growing.

References

1. S.S. Ray and M. Okamoto, Polymer/layered silicate nanocomposites: a review from preparation to processing. *Prog. Polym. Sci.* **28** (2003) 1539–1641.
2. D. Bondeson, A.P. Mathew and K. Oksman, Optimization of the isolation of nanocrystals from microcrystalline cellulose by acid hydrolysis. *Cellulose* **13** (2006) 171–180.
3. G. Gong, A.P. Mathew and K. Oksman, Strong aqueous gels of cellulose nanofibers and nanowhiskers isolated from softwood flour, *TAPPI J.* **10** (2011).
4. R.R. Lahiji, X. Xu, R. Reifenberger, A. Raman, A. Rudie and R.J. Moon, Atomic force microscopy characterization of cellulose nanocrystals, *Langmuir* **26** (2010) 4480–4488.
5. T. Saito, S. Kimura, Y. Nishiyama and A. Isogai, Cellulose nanofibers prepared by TEMPO-mediated oxidation of native cellulose, *Biomacromol.* **8** (2007) 2485–2491.

6. M.A.S. Azizi Samir, F. Alloin and A. Dufresne, Review of recent research into cellulosic whiskers, their properties and their application in nanocomposite fields. *Biomacromol.* **6** (2005) 612–626.
7. B.S. Tanem, I. Kvien and K. Oksman, *Morphology of cellulose and its nanocomposites.* In *Cellulose nanocomposites: processing, characterization and properties,* K. Oksman and M. Sain (Eds.) Oxford Press (2006), ACS Symposium Series, Vol. 938.
8. E. Trovatti, L. Oliveira, C.S.R. Freire, A.J.D. Silvestre, C. Pascoal Neto, J.J.C. Cruz Pinto and A. Gandini, Novel bacterial cellulose–acrylic resin nanocomposites, *Compos. Sci. Technol.* **70** (2010) 1148–1153.
9. A.P. Mathew, A. Chacraborty, K. Oksman and M. Sain, The Structure and mechanical properties of cellulose nanocomposites prepared by twin screw extrusion. In *Cellulose nanocomposites: processing, characterization and properties.* K. Oksman and M. Sain, (Eds.) Oxford Press (2006), ACS Symposium Series, Vol 938.
10. G. Siqueira, A.P. Mathew and K. Oksman, Processing of cellulose nanowhiskers/ cellulose acetate butyrate nanocomposites using sol-gel process to facilitate dispersion, *Comp. Sci. Technol.* **71** (2011), 1886–1892.
11. A.J. Svagan, P. Jensen, S.V. Dvinskikh, I. Furo and L.A. Berglund, Towards tailored hierarchical structures in cellulose nanocomposites biofoams prepared by freezing/freeze-drying, *J. Mater. Chem.* **20** (2010) 6646–6654.
12. L. Petersson, I. Kvien and K. Oksman, Structure and thermal properties of polylactic acid/cellulose whiskers nanocomposites, *Compos. Sci. Technol.* **67** (2007) 2535–2544.
13. D. Bondeson and K. Oksman, Dispersion and characteristics of surfactant modified cellulose whiskers nanocomposites, *Comp. Interf.* **14** (2007) 617–630.
14. I. Kvien, B.S. Tanem and K. Oksman, Characterization of cellulose whiskers and its nanocomposites by atomic force and electron microscopy, *Biomacromol.* **6** (2005) 3160–3165.
15. L. Goetz, A. Mathew, K. Oksman, P. Gatehnolm and A.J. Ragauskas, A novel nanocomposite film prepared from crosslinked cellulosic whiskers, *Carbohydr. Polym.* **75** (2009) 85–89.
16. J.E. Ayuk, A.P. Mathew and K. Oksman, The effect of plasticizer and cellulose nanowhisker content on the dispersion and properties of cellulose acetate butyrate nanocomposites, *J. Appl. Polym. Sci.* **114** (2009) 2723–2730.
17. A.P. Mathew, M-P.G. Laborie and K. Oksman, Cross-linked chitosan/chitin crystal nanocomposites with improved permeation selectivity and Ph stability, *Biomacromol.* **10** (2009) 1627–1632.
18. K. Oksman, A.P. Mathew, D. Bondeson and I. Kvien, Manufacturing process of polylactic acid (PLA) — cellulose whiskers nanocomposites, *Compos. Sci. Technol.* **66** (2006) 2776–2784.
19. D. Wang, J. Yu, J. Zhang, J. He and J. Zhang, Transparent bionanocomposites with improved properties from poly (propylene carbonate) (PPC) and cellulose nanowhiskers (CNWs), *Compos. Sci. Technol.* **85** (2013) 83–89.
20. I. Kvien, J. Sugiyama, M. Vortrubec and K. Oksman, Characterization of starch based nanocomposites, *J. Mater. Sci.* **42** (2007) 8163–8171.
21. I. Kvien and K. Oksman, Microscopic examination of cellulose whiskers and their nanocomposites. In *Characterization of lignocellulosic materials,* T.Q. Hu (Ed.) Oxford, UK: Blackwell Publishing (2008), Ch. 20, p.342–356.
22. NIKON Microscopy U, the source for microscopy education. Available at: http://www.microscopyu.com/articles/formulas/formulasresolution.html
23. L.C. Sawyer and D.T. Grubb, *Polymer microscopy,* 1st edn. London: Chapman and Hall (1987).
24. J.H.M. Willison and A.J. Rowe. Replica, shadowing and freeze-etching techniques. In *Practical methods in electron microscopy,* M. Audrey Glauert (Ed.) Elsevier/North-Holland: Biomedical Press (1980).

25. A. Pei, Q. Zhou and L.A. Berglund, Functionalized cellulose nanocrystals as biobased nucleation agents in poly (L-lactide) (PLLA) — crystallization and mechanical property effects, *Compos. Sci. Tech.* **70** (2010) 815–821.
26. I. Kvien and K. Oksman, Orientation of cellulose nanowhiskers in polyvinyl alcohol, *Appl. Phys. A.* **87** (2007) 641–643.
27. J. Goldstein, D. Newbury, D. Joy, D. Lyman, P. Echlin, E. Lifshin, L. Sawyer and J. Michael, *Scanning electron microscopy and X-ray microanalysis*, 3rd edn. New York: Kluwer Academic/Plenum Publishers (2003).
28. P.J. Goodhew, J. Humphreys and R. Beanland, *Electron microscopy and analysis,* London: Taylor & Francis (2001).
29. P.C. Braga and D. Ricci, *Atomic force microscopy: biomedical methods and applications*, Springer **242** (2004).
30. P. Eaton and P. West, *Atomic force microscopy,* Cary, NC, USA: Oxford University Press, (2010).
31. S. Bandyopadhyay, S.K. Samudrala, A.K. Bhowmick and S.K. Gupta. Applications of atomic force microscope (AFM) in the field of nanomaterials and nanocomposites. In *Functional nanostructures: processing, characterization, and applications*, S. Seal, (Ed.) Springer, 2007. pp. 504–568.
32. E.E. Brown and M-P.G. Laborie, Bioengineering bacterial cellulose/poly(ethylene oxide) nanocomposites, *Biomacromol.* **8** (2007) 3074-3081.
33. B. Jean, F. Dubreuil, L. Heux and F. Cousin, Structural details of cellulose nanocrystals/polyelectrolytes multilayers probed by neutron reflectivity and AFM, *Langmuir* **24** (2008) 3452–3458.
34. C. Aulin, S. Ahola, P. Josefsson, T. Nishino, Y. Hirose, M. Osterberg and L. Wagberg, Nanoscale cellulose films with different crystallinities and mesostructures-their surface properties and interaction with water, *Langmuir* **25** (2009) 7675–7685.
35. E. Ten, J. Turtel, D. Bahr, L. Jiang and M. Wolcott, Thermal and mechanical properties of poly(3-hydroxybutyrate-co-3-hydroxyvalerate)/cellulose nanowhiskers composites, *Polymer* **51** (2010) 2652–2660.
36. H. Butt, B. Cappella and M. Kappl, Force measurements with the atomic force microscope: technique, interpretation and applications, *Surf. Sci. Rep.* **59** (2005) 1–152.
37. X. Zhang and R. Young, Adhesion properties of cellulose films, *Mater. Res. Soc. Symp. Proc.* **586** (2000) 157–162.
38. J.C. Bastidas, R. Venditti, J. Pawlak, R. Gilbert, S. Zauscher and J.F. Kadla, Chemical force microscopy of cellulosic fibers, *Carbohydr. Polym.* **62** (2005) 369–378.
39. R.R. Lahiji, Y. Boluk and M. McDermott, Adhesive surface interactions of cellulose nanocrystals from different sources, *J. Mater Sci.* **47** (2012) 3961–3970.
40. R.J. Moon, A. Martini, J. Nairn, J. Simonsen and J. Youngblood, Cellulose nanomaterials review: structure, properties and nanocomposites, *Chem. Soc. Rev.* **40** (2011) 3941–3994.
41. E.D. Cranston and D.G. Gray, Formation of cellulose-based electrostatic layer-by-layer films in a magnetic field, *Sci. Technol. Ad. Mater.* **7** (2006) 319–321.
42. W. Gindl, G. Emsenhuber, G. Maier and J. Keckes, Cellulose in never-dried gel oriented by an AC electric field, *Biomacromol.* **10** (2009) 1315–1318.
43. Y. Habibi, T. Heim and R. Douillard, AC electric field-assisted assembly and alignment of cellulose nanocrystals, *J. Polym. Sci. Part B* **46** (2008) 1430–1436.
44. Y. Nishiyama, K. Shigenori, W. Masahisa and O. Takeshi, Cellulose microcrystal film of high uniaxial orientation, *Macromol.* **30** (1997) 6395–6397.
45. I. Hoeger, O.J. Rojas, K. Efimenko, O.D. Velev and S.S. Kelley, Ultrathin film coatings of aligned cellulose nanocrystals from a convective-shear assembly system and their surface mechanical properties, *Soft Matter* **7** (2011) 1957–1967.

46. L. Csoka, I.C. Hoeger, P. Peralta, I. Peszlen and O.J. Rojas, Dielectrophoresis of cellulose nanocrystals and alignment in ultrathin films by electric field-assisted shear assembly, *J. Coll. Int. Sci.* **363** (2011) 206–212.
47. A. Bohn, H.P. Fink, J. Ganster and M. Pinnow, X-ray texture investigations of bacterial cellulose, *Macromolecular Chem. Phys.* **201** (2000) 1913–1921.
48. W. Gindl and J. Keckes, Drawing of self-reinforced cellulose films, *J. Appl. Polym. Sci.* **103** (2007) 2703–2708.
49. S. Iwamoto, A. Isogai and T. Iwata, Structure and mechanical properties of wet-spun fibers made from natural cellulose nanofibers, *Biomacromol.* **12** (2011) 831–836.
50. H. Sehaqui, N.E. Mushi, S. Morimune, M. Slakjkova, T. Nishino and L.A. Berglund, Cellulose nanofibers orientation in nanopaper and nanocomposites by cold drawing, *ACS Appl. Mater. Interf.* **4** (2012) 1043–1049.
51. M. Wada, J. Sugiyama and T. Okano, Native cellulose on the basis of two crystalline phase (Ia/Ib) system, *J. Appl. Poly. Sci.* **49** (1993) 1491–1496.
52. A. Thygesen, J. Oddershede, H. Lihot, A.B. Thomsen and K. Stahl, On the determination of crystallinity and cellulose content in plant fibres, *Cellulose* **12** (2005) 563–576.
53. J. Sugiyama, R. Vuong and H. Chanzy, Electron diffraction study on the two crystalline phases occurring in native cellulose from an algal cell wall, *Macromol.* **14** (1991) 4168–4175.
54. Y.H. Lee, S.W. Lee, J.R. Youn, K. Chung and T.J. Kang, Characterization of fiber orientation in short fiber reinforced composites with an image processing technique, *Mat. Res. Innovat.* **6** (2002) 65–72.

Chapter 8

Characterization of Cellulose Nanofiber-based Nanocomposite Interfaces

Stephen J. Eichhorn
College of Engineering, Maths & Physical Sciences, University of Exeter, UK

This chapter addresses the challenges and approaches to obtaining detailed and accurate information on the nature and mechanics of interfaces in cellulose nanofiber-based nanocomposites. Covering the nanofiber types of nanocrystals, nanofibrils, and bacterial cellulose, the chapter directs the reader to issues governing the mechanics of nanocomposite materials in general. The chapter also introduces a number of techniques that have been used to follow the micromechanics of cellulose nanocomposites, namely, Raman spectroscopy, X-ray diffraction, and Förster resonance energy transfer imaging.

8.1. Introduction

Cellulose is the most abundant natural resourced polymer on the planet.[1] Industrially, it has been exploited for use in papermaking, textiles and clothing, for ropes, sails, and in pharmaceuticals and some foodstuffs. These traditional uses of cellulose amount to a 10^{12} tonne usage, which when compared to steel is something like 20–25 times its volume. Despite this, the material itself is often overlooked in materials science textbooks.

In recent times, there has been a renewed interest in cellulose biomass as a source of energy and materials.[2] This is not least because of the "fears" over the ever increasing price and diminishing accessible sources of oil.[2] Cellulose, being a renewable resource, and potentially digestible by enzymes into sugars, could form the future for a whole new industry based on this renewable resource. In tandem with this has been the discovery that the very building blocks of plant cell walls, which comprise plant tissues, have dimensions on the nanoscale. These nanofibers are readily extracted from the plant cell walls using a variety of methods: chemical, mechanical, and/or enzymatic.[3]

The nanofibers that have been most reported in the literature are cellulose nanocrystals (CNCs),[4] or cellulose nanofibers (CNFs)[5] and bacterial cellulose (BC). CNCs are produced using chemical hydrolysis of cellulose (mainly from plants, but also bacteria and some animals, e.g., tunicates).[4] CNFs are typically produced using mechanical homogenization and/or grinding of plant cell wall material.[5] In addition to mechanical treatments, chemical and enzymatic pretreatments have been reported in order to reduce energy costs in the processing of CNF.[6,7] Finally, BC is typically produced using specialized culturing conditions by the bacterium *Glucanobacter xylinum*.[3]

All of these nanofiber types have been widely reported as useful materials for the reinforcement of composite materials. Before embarking on a discussion of the ways in which the interfaces in these materials can be better understood, we will first discuss the form and structure of the fibers.

8.2. Form and structure of CNFs

This section will cover the size, shape, and surface chemistry of the different forms of CNFs with particular relevance to their incorporation in composite materials.

8.2.1 *Cellulose nanocrystals*

CNCs typically take the form, post-hydrolysis, of rod-like particles with aspect ratios (length/diameter) in the range 20–100.[4] A typical reaction scheme, showing the final morphology of these rod-like particles, is shown in Fig. 8.1.

The fact that the particles are rod like with small lateral dimensions has many implications in terms of the interfaces present when they are combined with a matrix material. Above a certain concentration of CNCs, and depending on their aspect ratio, they typically form what is called a "percolated network".[8,9] A percolated network in two dimensions is defined as the ability to move from one side of the network to the other without deviating from points in the network. In a fibrous network, this would mean the percolation is achieved if one can pass along fiber

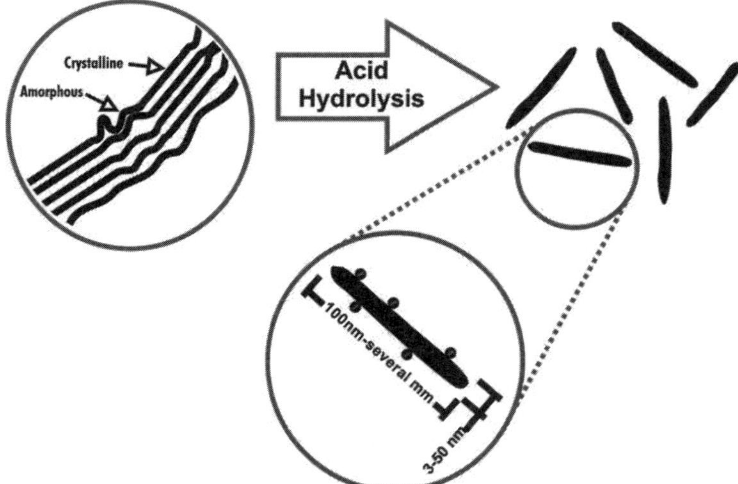

Fig. 8.1. Schematic of reaction to produce CNCs from semicrystalline cellulose source (top left) through hydrolysis to yield rod-like particles (top right) with dimensions on the nanometer to micron scale, with negative surface charges (bottom).

lengths from one side of the network to the other without having to move into empty space, or in the case of a composite, through the matrix. It is beyond the scope of this chapter to delve into the details of percolation theory, but it is important to state that in the percolated network state, there are two potential interfaces present, namely, fiber–fiber and matrix–fiber. Discriminating these interfaces and interrogating their influence in a composite is difficult, and will be discussed later on in this chapter.

The other important influence on the interface is the presence of surface charge. If the CNCs are produced using hydrochloric acid hydrolysis, then they typically have a near to neutral surface charge.[10] The fact that sulfuric acid-hydrolyzed whiskers typically possess surface charges means that this will also have an influence on the interfacial adhesion with polymeric materials. This influence is unknown to a large extent. The other important implication of surface charge is that their effective diameter is increased due to its presence. This is often neglected in calculations of percolation thresholds.

The small diameter of nanofibers in general has implications on their interactions with polymer molecules in a composite.[11] This is most easy to understand by considering their surface to volume ratios (SVRs). The SVR of a cylindrical particle (neglecting ends) is given by the equation[11]

$$SV_{\mathrm{R}} = \frac{2}{r}, \tag{8.1}$$

where r is the radius of the cylinder. It is easy to see that as r decreases, then the particle is more and more dominated by its surface. For a given volume fraction of particles, V_{f} the total available surface area (S_{a}) is therefore

$$S_{\mathrm{a}} = \frac{2V_{\mathrm{f}}}{r}. \tag{8.2}$$

If the surrounding polymer comprises molecular chains with radii of gyration R_{g}, then the volume fraction of particles within one radius of gyration (V_{g}) of the polymer will be

$$V_{\mathrm{g}} = \frac{2V_{\mathrm{f}}R_{\mathrm{g}}}{r}. \tag{8.3}$$

This means that particles with a small r are intimately connected to many polymer chains within the nanocomposite.

Given the large surface area of these nanofibers, they are expected to not only be in intimate contact with a resin material, but also each other. Given that van der Waals forces scale as r^{-6} means that at small distances for objects with intimate contact, the forces of attraction are very large. Overcoming these forces of attraction is one of the major issues of nanocomposites, and why so often people have issues with dispersion of nanoscaled fillers.

8.2.2 *Cellulose nanofibers*

CNFs are typically in the form of a reticulated network of fibers, with strong interconnections, either in a suspension or as a dried sheet of material resembling paper.[5] The term nanofibers can be something of a misnomer as often the fibrils can have a wide range of diameters, and aggregation of fibers in the networks effectively increases diameter into the micron range. Some of the interfacial influences discussed in Sec. 8.2.1 regarding CNCs apply also to CNF. One important distinction, however, is the fact that nanofibers within the network are typically densely packed. It can be shown, using theoretical approaches to fiber network formation, that the mean pore diameter is related to the fiber width (ω; or coarseness in paper physics terms) by the equation[12,13]

$$\bar{d} = \left(\frac{2\bar{\omega}}{\log(1/\varepsilon)}\right), \tag{8.4}$$

where ε is the porosity. It is easy to see that either low porosity or a small fiber width will decrease the mean pore size of the network significantly. This has been shown to effect the ingress of cells into porous electrospun fibrous networks, with larger fiber diameters allowing increased cell ingress and vice versa.[14] The same has been demonstrated for nanofibrillated cellulose networks and resin infiltration.[13] In the latter case, the implications are that unless the fibers are dispersed away from a network structure in the resin, then the properties of the network will have significant implications for the mechanical properties of the composite.[13,15] Thus, in a dense network system, fiber–fiber interactions in the network may dominate the mechanics of the composite. This particular effect will be discussed later on.

8.2.3 *Bacterial cellulose*

BC also, like CNF, comes in the form of a reticulated network of fibers.[1] One main difference between the network structures is the presence of bifurcation points in the network, where presumably bacteria divide (although this is not necessarily proven). These bifurcations are not present in CNF networks. Fiber–fiber bonds are also present in BC, and like CNF networks, they are typically highly packed, low porosity networks. As a consequence, the same arguments regarding mean more diameter can be applied as in Sec. 8.2.2. In addition to this, it is often the case that the transverse sections of the networks appear layered, with weak interfaces between the layers.[16] If resin does not penetrate the network, then these weak interfaces can also act as points of failure in the composite.[17–19]

8.3. Modulus of CNFs

In any composite requiring mechanical reinforcement, typically, a fiber with high modulus is incorporated to a lower stiffness matrix. Stress transfers to the higher stiffness component, whereby the overall modulus of the composite E_c is given by the equation[20]

$$E_c = \eta E_f V_f + (1 - V_f)E_m, \tag{8.5}$$

where E_f is the fiber modulus, V_f is the volume fraction of fibers, and E_m is the modulus of the matrix material. The parameter η is called the stress-transfer efficiency factor. This factor is equal to 1 for a fully aligned system, and 3/8 for an in-plane randomly oriented structure of fibers.[21] In addition to this, another factor η_L can be used, which takes into account fibers of low aspect ratio, and low stress-transfer efficiency. This parameter is called the "fiber length efficiency factor".[21] Cox gives this efficiency factor as[22,23]

$$\eta_L = \left[1 - \frac{\tanh(\beta L/2)}{\beta L/2}\right], \tag{8.6}$$

where β is additionally defined by the equation

$$\beta = \frac{2}{D}\left[\frac{2G_m}{E_f \ln(\sqrt{\pi}/\Psi V_f)}\right]^{1/2}, \tag{8.7}$$

where G_m is the shear modulus of the matrix and Ψ is a parameter related to the packing geometry of the composite. Equation (8.5) can then be written

$$E_c = \eta\eta_L E_f V_f + (1 - V_f)E_m. \tag{8.8}$$

If long fibers are used, then η_L tends toward 1. For short fibers, the stiffness of the composite then becomes dependent on the length of the fibers and their modulus. Below percolation, this is the situation for CNF composites. Above percolation, the modulus of the reinforcing phase still plays a significant role.

The crystal modulus and the stiffnesses of CNFs have been theoretically predicted and experimentally determined. The most notable of these studies are summarized in Table 8.1, with their respective moduli.

Table 8.1. Examples of theoretical and experimental determinations of the crystal and fibril modulus of cellulose.

Method	Modulus (GPa)	Reference
Modeling	167.5	24
Modeling	145	25
Modeling	124–155	26
Modeling	66–168	27
Experimental (Ramie fibers, X-ray)	137	28
Experimental (Ramie fibers, X-ray)	138	29
Experimental (Tunicate CNCs, Raman)	143	30
Experimental (Tunicate fibrils, AFM bending)	145.2–150.7	31
Experimental (Raman, CNF/BC)	29–36/79–88	32
Experimental (BC, AFM)	78	33

The moduli of CNFs are therefore reported to be high (~70–150 GPa), with the exception of CNF, although this requires further experimental confirmation.

8.4. Quantifying stress-transfer in cellulose nanocomposites

8.4.1 *Raman spectroscopy*

Raman spectroscopy relies on the inelastic scattering of light from a molecule and is used to study the vibrational, rotational, and low frequency behavior of a wide range of materials, liquids, and gases. With reference to this chapter, the technique has been widely applied to the study of polymers. The effect was first reported by C.V. Raman in 1928, for which he won the Nobel Prize in 1930. Raman used sunlight to excite vibrations but typically laser sources are used in modern spectrometers. The process by which a Raman spectrum is recorded will now be given.

Theoretically, the effect can be described by first considering Planck's equation to describe the energy of electromagnetic radiation (ΔE) as

$$\Delta E = h\nu_0, \tag{8.9}$$

where h is Planck's constant (=6.63×10^{-34} Js). This incoming radiation is scattered by the molecule. The three components of the scattered radiation can be summarized as follows:

- *Rayleigh scattering*: This is the most intense component of the scattered radiation, and has the same frequency as the incident radiation.
- *Stokes–Raman scattering*: This is a very weak component of the scattered radiation in comparison to Rayleigh scattering. The incident photons excite the molecule to a higher energy state. The loss in energy of the photon is equal to the gain in energy by the molecule. The scattered photon gain in energy by the molecule. The scattered photon has a frequency $\nu_0 - \nu_\mathrm{m}$ where ν_m is the vibrational frequency of the molecule.
- *Anti-Stokes–Raman scattering*: This component is very weak compared to Rayleigh scattering, but it is opposite to Stokes–Raman since the molecules lose energy and the scattered photons have higher frequencies $\nu_0 + \nu_\mathrm{m}$ than the incident photon.

These transitions are schematically represented in Fig. 8.2.

If we consider in a more formal sense the interaction of light with a material, then we have to take into account the electric field component of the electromagnetic field, and how this interacts with the electronic cloud of the molecule. Mathematically, the incident electromagnetic field, which is oscillating sinusoidally in space, can be described by the equation

$$E_\mathrm{s} = E_{\mathrm{s}0} \cos(2\pi\nu_0 t), \tag{8.10}$$

where $E_{\mathrm{s}0}$ is the amplitude of the oscillating wave over time t. This electric field interacts with a molecule inducing an electric dipole moment P, given by the

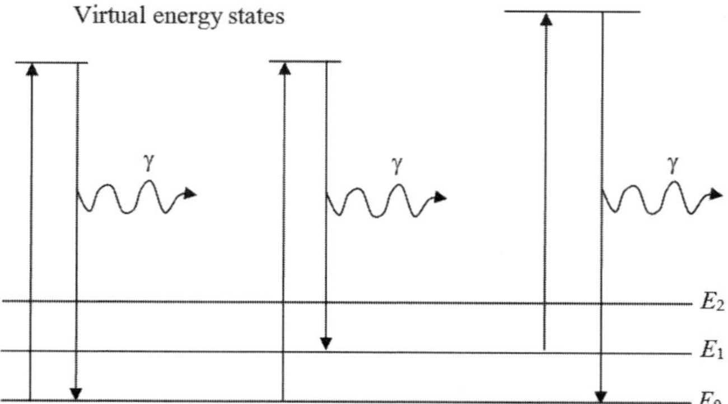

Fig. 8.2. Schematic of Rayleigh, Stoke's–Raman, and anti-Stoke's–Raman scattering, where E_0, E_1, and E3 are the ground, 1st and 2nd energy states of an arbitrary molecule and γ is a photon. (Reproduced from Ref. 34 with permission.)

equation

$$P = \alpha(q)E, \tag{8.11}$$

where $\alpha(q)$ is the polarizability; this is itself a function of the nuclear displacement q. The molecule also vibrates, due to the interaction of the electric field of the electromagnetic wave, with its own frequency ν_m, and therefore the nuclear displacement q can be expressed using the equation

$$q = q_0 \cos(2\pi\nu_\mathrm{m}t), \tag{8.12}$$

where q_0 is the amplitude of the vibration over time t. If the amplitudes are small, then $\alpha(q)$ is best approximated from the first two terms of a power series, as

$$\alpha(q) = \alpha_0 + \left(\frac{\partial\alpha}{\partial q}\right)_0 q_0 + \cdots, \tag{8.13}$$

where α_0 is the polarizability at the equilibrium position. A combination of Eqs. (8.10–8.13) gives the equation

$$P = \alpha_0 E_0 \cos(2\pi\nu_0 t) + \left(\frac{\partial\alpha}{\partial q}\right)_0 q_0 E_0 \cos(2\pi\nu_0 t)\cos(2\pi\nu_\mathrm{m}t), \tag{8.14}$$

which can be re-written, using a trigonometric identity, as

$$P = \alpha_0 E_0 \cos(2\pi\nu_0 t) + \frac{1}{2}\left(\frac{\partial\alpha}{\partial q}\right)_0 q_0 E_0[\cos(2\pi(\nu_0 + \nu_\mathrm{m})t) + \cos(2\pi(\nu_0 - \nu_\mathrm{m})t)] \tag{8.15}$$

This equation shows the contributions to the electric dipole moment from the Rayleigh scattering $\alpha_0 E_0 \cos(2\pi\nu_0 t)$ and the Raman scattering (with two independent frequencies; $\nu_0 + \nu_\mathrm{m}$ (anti-Stokes) and $\nu_0 - \nu_\mathrm{m}$ (Stokes). A Raman active vibration must satisfy the condition that it has a non-zero rate of change of polarizability with respect to the nuclear displacement (i.e., $\partial\alpha/\partial q > 0$). The intensity of a Raman vibration is found to be proportional to the rate of change of polarizability.

The measurement of local stress states in polymer fibers using Raman spectroscopy relies on the fact that when a polymer molecule is deformed, the vibrational state of the bonds is within the polymer change. This change comes about due to an alteration of the force constant for the vibrational mode of the particular bond species. In stressed highly oriented polymers, this deformation is mostly translated along the backbone of the chain — so typically through symmetric bonded species, e.g., C–C, C–O, etc. These symmetric moieties are typically most Raman active, i.e., giving the highest intensity Raman signals. Conversely, non-symmetric species, e.g., C–H, O–H, N–H are more infrared active as they possess a dipole moment. Therefore, molecular stress-states in polymers are best obtained using Raman spectroscopy if axial deformation is the information that is most required.

The change in the force constant can be shown to be related to the vibrational frequency of the Raman mode most simply using an expression of an anharmonic oscillator with force constant F_2' as

$$\nu_i = (4\pi^2 c^2 \mu)^{-1/2}(F_2')^{1/2}, \tag{8.16}$$

where c is the speed of light, μ is the reduced mass of the oscillator. So when there is a change in the force constant, there is a shift in the vibrational frequency. The effect was originally experimentally observed by Baughman *et al.*[34] in 1977 for polydiacetylene single crystals, but also re-confirmed experimentally and further explained theoretically by Batchelder and Bloor.[35]

The change in vibrational frequency is typified by a shift in the position of a characteristic Raman band toward a lower wavenumber position (wavenumber being the inverse of wavelength). Given this, it is easy to show that this shift corresponds to an increase in energy of the vibrational state of the molecule. A wide range of deformed polymer fibers have been shown to exhibit such shifts in vibrational frequencies.[36] The effect was first observed for cellulose in 1997 by Hamad and Eichhorn[37] who showed that deformed regenerated cellulose fibers exhibited shifts in the position of peaks initially located at $\sim$1095 cm^{-1} and $\sim$895 cm^{-1}. Subsequently, the approach has been used to follow the deformation micromechanics of cellulose fibers embedded in a polymer matrix.[38–40] This approach is only possible if the resin is transparent and Raman peaks from the resin do not interfere with peaks from the fibers.

The technique was first applied to cellulose nanocomposites in 2005 by Sturcova *et al.*[30] who dispersed tunicate CNCs in an epoxy resin. They deformed this sample using a customized four-point bending rig under the microscope of the Raman spectroscopy. Incremental deformation was applied to the sample and spectra recorded at each increment. It was shown that the Raman peak initially located at $\sim$1095 cm^{-1} shifted toward a lower wavenumber position, enabling a calculation of the effective stiffness of the CNCs; this gave a value of 143 GPa. Subsequent work on cotton-derived CNCs gave values in the range ~50–100 GPa, depending on whether a 2D in-plane or 3D random arrangement of whiskers is present in the composite.[41] Subsequently, a number of studies have looked at a range of CNCs embedded in different matrices; namely PVAc[42] and in an all-cellulose composite where the matrix itself comprises dissolved cellulose.[43,44]

The approach is not just effective in determining local stress states in the composite. It can also be used to probe local orientation and link this to the local mechanics of the composite. Rusli *et al.*[42] showed that in a heat pressed composite comprising CNCs embedded in PVAc, local orientation of the nanocrystals occurred. This local orientation was mapped using the fact that the intensity of the peak located at ~1095 cm^{-1} is polarization sensitive. Initially, polarized light was used to identify light and dark domains in the sample. The light domains became dark when the sample was rotated, and the dark domains remained dark. This showed that there was optical anisotropy (or birefringence) in the sample, which related to local orientation of the CNCs; the light domains contained oriented CNCs and the dark domains isotropically and randomly oriented CNCs. The polarization axes were fixed parallel to the tensile axis of the sample for the incoming radiation, and parallel for the scattered radiation. Then the samples were rotated, focusing the laser on the light and dark domains, respectively. This showed that local orientation of the CNCs were present via large changes in the intensity of the band located at ~1095 cm^{-1} with rotation for the light domains, whereas no change occurred for the dark domains. Furthermore, when the samples were deformed in tension, it was shown that the shift rate with respect to strain for the band located at ~1095 cm^{-1} was much greater when data were collected from the light domains, with the tensile axis and whisker orientation parallel to each other, and greater than the shift rate obtained from the dark domains. This showed the effectiveness of orienting CNCs in a nanocomposite for the first time, by direct evidence of the whisker deformation. Rusli *et al.*[42] also showed how the presence of moisture could "switch-off" the interface in the composite. This switching-off of the interface exhibited itself in a lack of shift in the position of the band located at ~1095 cm^{-1} when the PVAc–CNC composite samples were wet.

Another way in which the Raman spectroscopic approach has enabled stress transfer in cellulose nanocomposites to be elucidated is discriminating the effects of the reinforcing phase and matrix. This has been particularly effective for all-cellulose nanocomposites where Raman peaks located at ~1095 cm^{-1} and ~895 cm^{-1} have been shown to give information on both the molecular deformation and orientation of the reinforcing CNC phase and the matrix phase, respectively.[43,44] It should be noted that strictly speaking, the peak located at ~1095 cm^{-1} is a composite band, containing information on *both* the matrix and the CNCs.[43] It was shown that the CNC phase of the composite bears much of the load in the all-cellulose composite.[43] It was also demonstrated that orientation of local domains due to the presence of a magnetic field during curing of the composite could be mapped in the material and related to the local mechanics.[44]

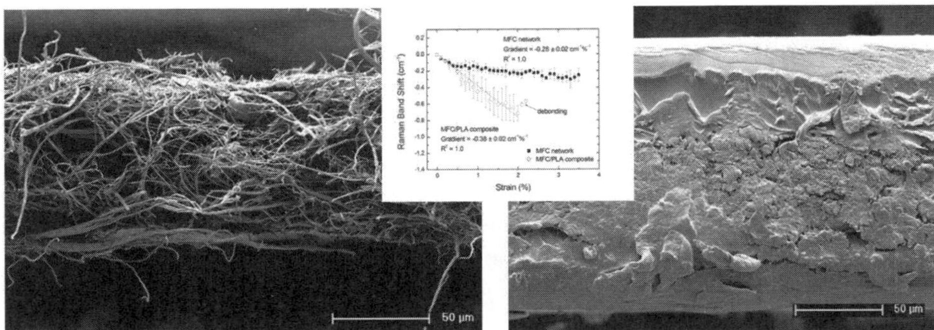

Fig. 8.3. Scanning electron microscope images of an open network of CNFs fibrils made from lyocell (left) impregnated with PLA resin (right). Inset shows shifts in the Raman band located at $\sim 1095\,\text{cm}^{-1}$ for the network and composite, respectively.

The use of Raman spectroscopy has also been applied to better understand the influence of network geometry on the interfacial micromechanics of both CNF and BC-based nanocomposite. In the former system, CNF from a lyocell fiber source has been impregnated with poly(lactic acid) (PLA)[13] and also 2,2,6,6-tetramethylpiperidine-1-oxyl radical (TEMPO)-oxidized CNF with PLA.[15] It has been shown that lateral fiber dimensions are critical for resin impregnation into the composite.[13] Micron-scaled fibrils enable better ingress of resin due to the proportional relationship between mean pore size and mean fiber diameter in a stochastic network of fibers[13] — see Fig. 8.3. It was also shown that the properties of the network dominate the mechanics of TEMPO-oxidized CNF/PLA composites.[15] For BC-based nanocomposites, it has been shown that lower density networks enhance the relative mechanical properties of a composite compared to a higher density network.[19] This was found to be due to the fact that enhanced stress-transfer occurs between the resin and the BC networks if resin impregnation can take place. A dense network does not allow this to happen, and so stress transfer is compromised. The stress-transfer efficiency of the composite system was followed using the Raman spectroscopy approach. Finally, it was also shown that cross-linking using glyoxal could enhance stress transfer between weakly bound layers in a BC network.[18] These cross-linked networks were further incorporated into a composite material, with a PLA resin, and it was shown how the interfaces between the matrix and the BC fibrillar network, and between the fibrils themselves within the network can be discriminated.[17]

8.4.2 *X-ray diffraction*

X-ray diffraction can be used to measure changes in the crystal strain of semicrystalline polymer fibers. This approach was first used by Sakurada *et al.*[28] to measure the crystal strain, and ultimately calculate the crystal modulus of a wide variety of polymers. Subsequently, there have been a number of reports on the measurement of crystal strain in polymer fibers and composite systems, which are too numerous and beyond the scope of this chapter to summarize (see review by Young and Eichhorn[36]). X-ray diffraction has been used to follow the micromechanics of interfaces in cellulose nanocomposites. Gindl *et al.*[45] showed that it is possible to independently follow the orientation of the matrix and reinforcing phases in an all-cellulose nanocomposite. It was shown that orientation of the reinforcing phase is obtained from the diffraction pattern of the cellulose, being distinct for cellulose-I (reinforcement) and cellulose-II (matrix). So far, no other reports have been made on the use of this technique for cellulose nanocomposites, and so remains a topic for further study.

8.4.3 *Förster resonance energy transfer*

Förster resonance energy transfer (FRET) is an effect of energy transfer between two chromophores; a donor and an acceptor. The donor chromophore, being initially in an excited state, may transfer energy to the acceptor chromophore through a non-radiative dipole–dipole coupling. This exchange is particularly sensitive at small distances, and so can be useful for imaging the formation of an interface in a nanocomposite. So far, there has only been one report on the use of this technique to image interfaces in a CNF-based nanocomposite; CNF doped with 5-(4,6-dichlorotriazinyl)aminofluorescein (FL) dispersed in PE doped with Coumarin 30 (C30).[46] The C30 acted as a donor and the FL the acceptor, with the exchange only occurring when there is intimate contact between the two components, hence demonstrating its use in characterizing the interface between the two components. The interface therefore shows up as a bright "fringe" around the nanoparticle. This enables further analysis to assess the thickness of the interface region and could enable better assessment of the "quality of the interface".

8.5. Conclusions

The field of cellulose nanocomposites is still really only emerging, perhaps lagging behind its immediate predecessor where intact plant fibers are combined with resin materials. The interfaces in cellulose nanocomposites are known to be complex, given the interfacial effects between the nanofibers and the resin material surrounding them. In addition to this, there are effects of the networks that these nanofibers form that are not fully understood. There have been some advances in these areas using Raman spectroscopy and other techniques, but many assumptions have to be made in order to fully quantify interfacial adhesion in cellulose nanocomposites. Techniques such as single fiber pull out tests, that have been applied to understand better carbon nanotube–polymer interactions, have not been so successfully applied to cellulose nanocomposites. This is likely to take place as better approaches are developed to deal with and manipulate CNFs in the same context. The combination of techniques (FRET/Raman, X-ray/Raman) may also prove useful in assisting our understanding of CNF–polymer interfaces.

References

1. D. Klemm, B. Philipp, T. Heinze, U. Heinze and W. Wagenknecht, *Comprehensive cellulose chemistry: general principles and analytical methods.* Weinheim: Wiley-VCH (1998).
2. S.J. Eichhorn and A. Gandini, Materials from renewable resources, *MRS Bull.* **35**(3) (2010) 187–190.
3. D. Klemm, F. Kramer, S. Moritz, T. Lindstrom, M. Ankerfors, D. Gray and A. Dorris, Nanocelluloses: a new family of nature-based materials, *Angew. Chemie Int. Edn.* **50**(24) (2011) 5438–5466.
4. S.J. Eichhorn, Cellulose nanocrystals: promising materials for advanced applications, *Soft Matter* **7**(2) (2011) 303–315.
5. I. Siro and D. Plackett, Microfibrillated cellulose and new nanocomposite materials: a review, *Cellulose* **17**(3) (2010) 459–494.
6. M. Paakko, M. Ankerfors, H. Kosonen, A. Nykanen, S. Ahola, M. Osterburg, J. Ruokolainen, J. Laine, P.T. Larsson, O. Ikkala and T. Lindstrom, Enzymatic hydrolysis combined with mechanical shearing and high-pressure homogenization for nanoscale cellulose fibrils and strong gels. *Biomacromol.* **8**(6) (2007) 1934–1941.
7. H. Kamitakahara, A. Koschella, Y. Mikawa, F. Nakatsubo, T. Heinze and D. Klemm, Syntheses and comparison of 2,6-di-O-methyl celluloses from natural and synthetic celluloses, *Macromol. Biosci.* **8**(7) (2008) 690–700.
8. V. Favier, H. Chanzy and J.Y. Cavaille, Polymer nanocomposites reinforced by cellulose whiskers. *Macromolecules* **28**(18) (1995) 6365–6367.
9. J.R. Capadona, K. Shanmuganathan, D.J. Tyler, S.J. Rowan and C. Weder, Stimuli-responsive polymer nanocomposites inspired by the sea cucumber dermis, *Science* **319**(5868) (2008) 1370–1374.
10. R. Rusli, K. Shanmuganathan, S.J. Rowan, C. Weder and S.J. Eichhorn, Stress transfer in cellulose nanocrystal composites-influence of whisker aspect ratio and surface charge, *Biomacromol.* **12**(4) (2011) 1363–1369.
11. A.H. Windle, Two defining moments: a personal view by Prof. Alan H. Windle. *Compos. Sci. Technol.* **67**(5) (2007) 929–930.
12. S.J. Eichhorn and W.W. Sampson, Statistical geometry of pores and statistics of porous nanofibrous assemblies, *J. Roy. Soc. Interface* **2**(4) (2005) 309–318.
13. S. Tanpichai, W.W. Sampson and S.J. Eichhorn, Stress-transfer in microfibrillated cellulose reinforced poly(lactic acid) composites using Raman spectroscopy, *Compos. Part A* **43**(7) (2012) 1145–1152.
14. Q.P. Pham, U. Sharma and A.G. Mikos, Electrospun poly(epsilon-caprolactone) microfiber and multilayer nanofiber/microfiber scaffolds: characterization of scaffolds and measurement of cellular infiltration, *Biomacromol.* **7**(10) (2006) 2796–2805.
15. M.S. Bulota, Tanpichai, M. Hughes and S.J. Eichhorn, Micromechanics of TEMPO-oxidized fibrillated cellulose composites, *ACS Appl. Mater. Interf.* **4**(1) (2012) 331–337.
16. M. Nogi and H. Yano, Transparent nanocomposites based on cellulose produced by bacteria offer potential innovation in the electronics device industry, *Adv. Mater.* **20**(10) (2008) 1849–1852.
17. F. Quero, S.J. Eichhorn, M. Nogi, H. Yano, K.Y. Lee and A. Bismarck, Interfaces in cross-linked and grafted bacterial cellulose/poly(lactic acid) resin composites, *J Polym. Environ.* **20**(4) (2012) 916–925.
18. F. Quero, M. Nogi, K.Y. Lee, G.V. Poel, A. Bismarck, A. Mantalaris, H. Yano and S.J. Eichhorn, Cross-linked bacterial cellulose networks using glyoxalization, *ACS Appl. Mater. Interf.* **3**(2) (2011) 490–499.
19. F. Quero, M. Nogi, H. Yano, K. Abdulsalami, S.M. Holmes, B.H. Sakakini and S.J. Eichhorn, Optimization of the mechanical performance of bacterial cellulose/poly (L-lactic) acid composites, *ACS Appl. Mater. Interf.* **2**(1) (2010) 321–330.

20. D. Hull and T.W. Clyne, *An introduction to composite materials.* Cambridge: Cambridge University Press (1996).
21. M.R. Piggott, *Load bearing fibre composites.* Oxford: Pergamon Press (1980).
22. H.L. Cox, The elasticity and strength of paper and other fibrous materials, *Br. J. Appl. Phys.* **3**(MAR) (1952) 72–79.
23. B. Harris, *Engineering composite materials.* London: IOM Communications Ltd (1999).
24. K. Tashiro and M. Kobayashi, Theoretical evaluation of 3-dimensional elastic-constants of native and regenerated celluloses — role of hydrogen-bonds, *Polymer* **32**(8) (1991) 1516–1530.
25. M. Bergenstrahle, L.A. Berglund and K. Mazeau, Thermal response in crystalline I beta cellulose: A molecular dynamics study, *J. Phys. Chem. B* **111**(30) (2007) 9138–9145.
26. F. Tanaka and T. Iwata, Estimation of the elastic modulus of cellulose crystal by molecular mechanics simulation, *Cellulose* **13**(5) (2006) 509–517.
27. S.J. Eichhorn and G.R. Davies, Modelling the crystalline deformation of native and regenerated cellulose, *Cellulose* **13**(3) (2006) 291–307.
28. I. Sakurada, Y. Nukushina and T. Ito, Experimental determination of elastic modulus of crystalline regions in oriented polymers, *J. Polym. Sci.* **57**(165) (1962) 651–660.
29. T. Nishino, K. Takano and K. Nakamae, Elastic-modulus of the crystalline regions of cellulose polymorphs, *J. Polym. Sci. Part B* **33**(11) (1995) 1647–1651.
30. A. Sturcova, G.R. Davies and S.J. Eichhorn, Elastic modulus and stress-transfer properties of tunicate cellulose whiskers, *Biomacromol.* **6**(2) (2005) 1055–1061.
31. S. Iwamoto, W.H. Kai, A. Isogai and T. Iwata, Elastic modulus of single cellulose microfibrils from tunicate measured by atomic force microscopy, *Biomacromol.* **10**(9) (2009) 2571–2576.
32. S. Tanpichai, F. Quero, M. Nogi, H. Yano, R.J. Young, T. Lindstrom, W.W. Sampson and S.J. Eichhorn, Effective Young's modulus of bacterial and microfibrillated cellulose fibrils in fibrous networks, *Biomacromol.* **13**(5) (2012) 1340–1349.
33. G. Guhados, W.K. Wan and J.L. Hutter, Measurement of the elastic modulus of single bacterial cellulose fibers using atomic force microscopy, *Langmuir* **21**(14) (2005) 6642–6646.
34. V.K. Mitra, W.M. Risen and R.H. Baughman, Laser Raman study of stress dependence of vibrational frequencies of a monocrystalline polydiacetylene, *J. Chem. Phys.* **66**(6) (1977) 2731–2736.
35. D.N. Batchelder and D. Bloor, Strain dependence of the vibrational-modes of a diacetylene crystal, *J. Pol. Sci. Part B* **17**(4) (1979) 569–581.
36. R.J. Young and S.J. Eichhorn, Deformation mechanisms in polymer fibres and nanocomposites, *Polymer* **48**(1) (2007) 2–18.
37. W.Y. Hamad and S. Eichhorn, Deformation micromechanics of regenerated cellulose fibers using Raman spectroscopy, *ASME J. Eng. Mater.* **119**(3) (1997) 309–313.
38. S.J Eichhorn and R.J. Young, Deformation micromechanics of natural cellulose fibre networks and composites, *Compos. Sci. Technol.* **63**(9) (2003) 1225–1230.
39. S.J. Eichhorn and R.J. Young, Composite micromechanics of hemp fibres and epoxy resin microdroplets, *Compos. Sci. Technol.* **64**(5) (2004) 767–772.
40. B. Mottershead and S.J. Eichhorn, Deformation micromechanics of model regenerated cellulose fibre-epoxy/polyester composites, *Compos. Sci. Technol.* **67**(10) (2007) 2150–2159.
41. R. Rusli and S.J. Eichhorn, Determination of the stiffness of cellulose nanocrystals and the fiber-matrix interface in a nanocomposite using Raman spectroscopy, *Appl. Phys. Lett.* **93**(3) (2008) 033111.
42. R. Rusli, K. Shanmuganathan, S.J. Rowan, C. Weder and S.J. Eichhorn, Stress-transfer in anisotropic and environmentally adaptive cellulose whisker nanocomposites, *Biomacromol.* **11**(3) (2010) 762–768.

43. T. Pullawan, A.N. Wilkinson and S.J. Eichhorn, Discrimination of matrix-fibre interactions in all-cellulose nanocomposites, *Compos. Sci. Technol.* **70**(16) (2010) 2325–2330.
44. T. Pullawan, A.N. Wilkinson and S.J. Eichhorn, Influence of magnetic field alignment of cellulose whiskers on the mechanics of all-cellulose nanocomposites, *Biomacromol.* **13**(8) (2012) 2528–2536.
45. W. Gindl, K.J. Martinschitz, P. Boesecke and J. Keckes, Structural changes during tensile testing of an all-cellulose composite by in situ synchrotron X-ray diffraction, *Compos. Sci. Technol.* **66**(15) (2006) 2639–2647.
46. M. Zammarano, P.H. Maupin, L.P. Sung, J.W. Gilman, E.D. McCarthy, Y.S. Kim and D.M. Fox, Revealing the interface in polymer nanocomposites, *ACS Nano* **5**(4) (2011) 3391–3399.

Chapter 9

Toughness and Strength of Wood Cellulose-based Nanopaper and Nanocomposites

Lars Berglund
Wallenberg Wood Science Centre,
Royal Institute of Technology,
Stockholm, Sweden

Cellulose nanopaper in the form of nanofiber networks show superior mechanical performance and new functional characteristics compared with the brittle paper and fiberboard materials and thermoplastic biocomposites, which are commercially available. The chapter analyzes the potential to combine toughness and strength in polymer matrix nanocomposites based on cellulose nanofiber networks.

9.1. Cellulose and CNF

The term *cellulose* has been used in many different ways. For instance, it is often be used for chemical wood pulp fibers, which are really porous wood fibers where lignin has been removed but with substantial amounts of hemicellulose still present. The term can also be used to describe the cellulose molecule itself. In plant structures, a basic cellulose unit is in the physical form of *microfibrils* (plant physiology term), which provide reinforcement to the plant cell wall and carry tensile loads. The microfibrils are long, slender nanofibrous entities 3–10 nm in diameter with a length up to several micrometers. There are many advantages to be gained from using a nanofibrous form of cellulose in new materials.

In a technical sense, cellulose nanofibers (CNFs) also called nanofibrillated cellulose (NFC), disintegrated from the wood fiber cell wall is particularly interesting, since it can be available in large quantities at low cost. The basic unit of the NFC is wood microfibrils, although NFC can be larger in diameter and less well defined. The use of CNFs in polymer matrix biocomposites, foams, and aerogels has been reviewed.[1] This chapter focuses on the specific advantage of high volume fraction wood nanofiber composites in that substantial modulus and strength can be combined with high ductility (strain-to-failure). This is unusual for biocomposites, which tend to be brittle, and is caused by the small NFC dimensions and high mechanical properties of NFC and NFC networks. This chapter mainly focuses on NFC networks, which can be combined with ductile polymer matrices by the use of processing concepts related to papermaking.

9.2. Characteristics of cellulose nanopaper networks

CNFs can be prepared by mechanical disintegration of wood pulp,[2] but disintegration is facilitated by enzymatic[3] or chemical pretreatment.[4] Hydrocolloidal suspensions show high stability and provide the starting form for CNF after disintegration from wood pulp.

CNF networks or nanopaper structures can be made by vacuum filtration of colloidal suspensions of CNF in water.[2,5] A CNF network is formed analogous to wood fiber networks in paper and fiberboard structures. The main difference is in scale. The CNF from wood typically has diameters in the 5–20 nm range whereas plant fiber networks have fiber diameters of about 30 μm. The CNF aspect ratio (ratio of length to diameter) can typically be around 100. The orientation distribution of CNF is normally random-in-the-plane with some limited out-of-plane CNF orientation. The porosity of the network will depend on the drying conditions of the film, but is often 10–20% and much lower than for typical paper structures.

9.2.1 *Stress–strain curves in tension — cellulose nanopaper*

A straightforward way to characterize the mechanical properties of CNF network films is by conventional tensile tests. The early data[6] showed brittle behavior in this type of test. It is likely that the main reason was poor degree of dispersion of CNF, and/or that the CNF fibrils were not fully disintegrated to diameters below 20 nm. This results in agglomerates in the materials, which act as regions of stress concentrations so that premature failure occurs at low strain.

The work by Henriksson *et al.*[5] demonstrated the potential of nanopaper to provide high tensile strength and high strain-to-failure (see Fig. 9.1).

The strain-to-failure is around 10% and the strength is well above 200 MPa for the strongest material. Note that the material shows yielding at a stress level of around 100 MPa, which is much lower than ultimate strength. This is followed

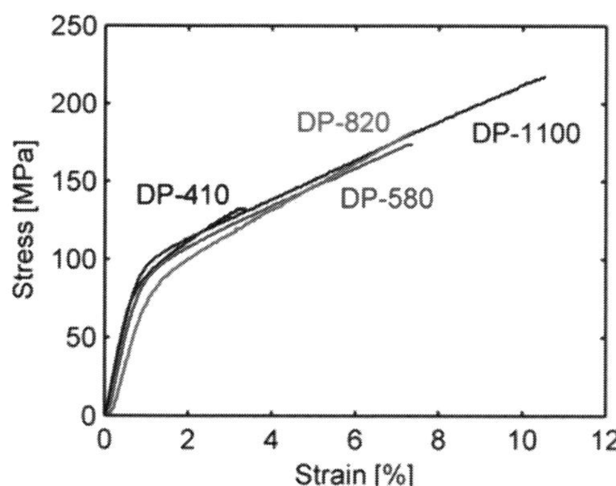

Fig. 9.1. Stress–strain curves of cellulose nanopaper in tension. The designations refer to different degrees of polymerization of the cellulose. The DP-1100 material is topochemically modified so that the CNF surface has carboxylate functionality. (After Henriksson *et al.*,[5] reprinted with permission from Ref. 5. Copyright 2008 American Chemical Society.)

by a plastic deformation region with considerable strain hardening (stress increases steeply with strain). The structure of the material is in the form of a nanofibrous network. Yielding is associated with failure at the CNF–CNF interface, although the specific nature of this mechanism is unknown. In the plastic deformation region, loading–unloading experiments show that the modulus of the network is increased with increasing strain. It is likely that during plastic deformation, the individual NFC fibrils become straightened and better oriented in the direction of loading. Fibrils slide with respect to each other so that the fibril network structure is altered during loading. This deformation mechanism is significant in analysis of CNF nanopaper toughness.

9.2.2 *Parameters of importance to mechanical behavior of the CNF network*

Fiber networks are important in many technical areas, and are used in the form of paper and paperboard products, non-woven textiles, and electrospun membranes. Recently, Silberstein *et al.*[7] published a study where the mechanical behavior of non-woven polymer fiber mats was studied both experimentally and by a theoretical elastic–plastic model. The model is micromechanical and connects the details of the fiber network structure to its global elastic–plastic behavior in tensile loading. This is very helpful for the design of fibril network structures of high toughness. Although the details of their study are outside the scope of this chapter, it is of interest to list some important parameters. The plasticity of the fiber network can have its origin in either the fiber itself, in the deformation of the fiber network, or both. For NFC networks, plastic deformation is dominated by deformation of the fibril network. The NFC fibrils consist of highly ordered cellulose molecules in extended chain conformation with little capacity for plastic deformation in tension.

The following are examples of material parameters of importance for the mechanics of fibril networks:

- Intrinsic mechanical properties of fibril (axial properties, properties in bending).
- Geometry of fibrils (length, diameter, curvature).
- Network mat structure.
- Volume fraction of fibrils, or corresponding porosity.
- Orientation distribution of fibrils.
- Density and characteristics of fibril–fibril bonds.

In the work by Silberstein *et al.*,[7] a representative volume element modeling approach is presented where all the material parameters above are taken into consideration. The strain-hardening behavior and its mechanisms are a challenge to determine and model for CNF networks. From a physical mechanism point of view, density of fibril–fibril bonds and corresponding debonding behavior are important. Also, the frictional behavior between fibrils sliding with respect to each other is important for the strain-hardening behavior of the real material. In this chapter, several studies report high strain-to-failure of "nanopaper" network structures (see Table 9.1).

Table 9.1. Comparison of mechanical properties from tensile tests measured for different types of materials based on CNFs or BC.

Composite	CNF content (% by volume)	E-modulus (GPa)	Ultimate tensile strength (MPa)	Strain-to-failure (%)	Reference
CNF nanopaper	80 20% pores	13	220	10	5
CNF nanopaper (supercritical CO_2-drying)	44 56% pores	1.4	84	17	8
Oriented nanopaper	100	25–33	400–430	2.5	10
CNF/MF	87/13 wt% 8% pores	16	108	0.8	6
CNF/50-50 starch-glycerol	10–70 wt%	0.2–6.2	5–160	8–25	12
BC/HEC	80/20	12.5	290	6	13
CNF/HEC	45/41 14% pores	7.5	180	26	14
CNF-HEC non-woven	20/46 44% pores	1.3	94	55	15

MF, melamine-formaldehyde resin; HEC, hydroxyethyl cellulose polymer.

High strain-to-failure correlates with high specific surface area (SSA) of the CNF network,[8] which possibly correlates with low density of weak fibril–fibril bonds.

The test conditions and specimen geometry will influence the mechanical behavior of CNF nanopaper. Most data in the literature are from thin films of thickness around 100 μm, or sometimes even as thin as 10 μm. Thick specimens may show a different behavior. For instance, the extent of out-of-plane CNF orientation is likely to be higher in thicker materials due to the differences in drying conditions. Under conditions, which favor plastic behavior (adsorbed moisture in the material and low strain rate), the length and width of the typical tensile test specimen should not strongly influence the stress–strain curve. However, in dry conditions or at high strain rates, CNF nanopaper can show brittle behavior and smaller specimens may then show higher strength. This is a well-known brittle material behavior where fracture is controlled by the largest defect in the material. For larger specimens, the probability is higher that a large defect, i.e., residual non-fibrillated wood fibers, is present and the measured average strength becomes lower. In order to have tougher CNF network materials, plastic deformation mechanisms should be promoted.

9.3. Ductile CNF nanopaper networks of high SSA

In Sec. 9.2.1, it was pointed out that ductile nanopaper structures can be prepared. In the study by Henriksson *et al.*,[5] the strongest nanopaper structures are also the most ductile. They have high cellulose molar mass, which should have a positive effect on tensile strength. The best CNF also has carboxylate groups on the topochemically modified cellulose nanofibrils. Testing is carried out at conditions of 50% relative humidity (RH). One may speculate that fibrils with carboxylate functionality in the suspension are hydrated on the surface in the nanopaper form. This could favor plastic deformation since the hydrated region may serve as a lubricant and thus facilitate the sliding of fibrils with respect to each other.

Recently, new drying methods were used for CNF aerogels and nanopaper structures.[8,9] Very porous aerogels were prepared by solvent exchange from water to tert-butanol followed by drying.[9] This resulted in much higher SSA of the aerogels (150–280 m^2/g) compared with freeze-dried aerogels prepared from water. The resulting aerogels showed a fine CNF network structure and much lower modulus and yield strength compared with freeze-dried foams with cellular structure. Fitting of modulus data to a theoretical fiber network model resulted in estimated CNF segment lengths of 300–480 nm between CNF–CNF joints, in agreement with structures observed in field emission scanning electron microscopy (FE-SEM).

The aerogel study inspired work on CNF nanopaper structures dried by different routes.[8] Stress–strain curves for such CNF nanopaper structures of high SSA show high ductility. Drying from supercritical carbon dioxide results in an SSA of 480 m^2/g. The strain-to-failure of this "SC-CO_2" nanopaper is 17% at 56% porosity and 50% RH. The modulus is 1.4 GPa (quite low) and the strength in tension is 84 MPa. In a fiber network model context, we may relate higher SSA to increased fiber segment length between CNF–CNF joints. This explains lowered modulus and yield strength. The increased strain-to-failure indicates that CNF sliding and reorientation with respect to neighboring fibrils is facilitated.

9.4. Oriented CNF films

The properties of oriented CNF nanopaper structures are interesting, both for potential applications and for comparison with randomly oriented CNF structures. Nanopaper films with preferred orientation of CNF have been prepared by cold-drawing.[10] The preparation route is papermaking-like and includes vacuum filtration of a CNF water suspension, drawing in wet state and drying. At high draw-ratio, the degree of CNF orientation is as high as 82%, and the Young's modulus is 33 GPa. The highest average strength reported is 430 MPa. This is much higher than mechanical properties of isotropic nanopaper from the same CNF where the typical modulus is 15 GPa and the strength is around 220 MPa. However, the strain-to-failure of the oriented sample is only 2%, since fibrils do not reorient and slide when they are in parallel orientation. The cold drawing method can also be applied to CNF nanocomposites. This has been demonstrated for CNF/hydroxyethyl cellulose (CNF/HEC) nanocomposites.[10] The soft HEC matrix allows further tailoring of the mechanical properties. The strength is somewhat reduced compared with neat CNF, but strain-to-failure increases when 44% by volume of HEC matrix is used.

9.5. Tough polymer matrix nanocomposites and non-wovens — high CNF content

The network characteristics of CNF nanopaper are unique and should be utilized also in polymer matrix nanocomposites. The early thermoset composites prepared by impregnation of CNF networks with water solutions of phenol formaldehyde[2,11] or melamine formaldehyde (MF)[6] showed high modulus but were very brittle in tension.[6,11] This reflects the typically brittle characteristics of cured polyphenol formaldehyde (PF) and MF thermoset resins. Based on the discussions in Secs. 9.1 and 9.2, it is clear that the mechanisms for plastic deformation in the CNF network (nanofiber straightening, reorientation, sliding, etc.) are unlikely in high T_g resin composites of high yield stress. Furthermore, a likely consequence of the curing reaction mechanisms for PF and MF is that the hydroxyls at CNF surfaces are covalently linked to the polymer matrix. This probably reduces the extent of toughening by CNF pull-out mechanisms.

In order to study CNF composites based on a very soft matrix of low yield stress, Svagan *et al.* combined CNF with a thermoplastic starch matrix.[12] The composition was a highly plasticized 50/50 mixture of amylopectin-rich potato starch and glycerol. This polymer is almost viscous at room temperature due to the plasticizing effect of the glycerol. The highly hydrated primary plant cell wall was an inspiration since it has good mechanical function despite the high water content. The cellulose fibril network provides the major mechanical function in the plant cell wall. In the bioinspired CNF/starch materials, a modulus of 6.2 GPa, a tensile strength as high as 160 MPa, and a strain-to-failure of 8.1% were observed at 70 wt% of CNF reinforcement. The work of fracture (area under stress–strain curve) was as high as 9.4 MJ/m^3. An important reason for this unique combination of strength and strain-to-failure in a viscous matrix composite is the nanoscale network structure of the CNF.

Although thermoplastic starch is of industrial importance, cellulose derivatives may be used to prepare CNF nanocomposites with more favorable combinations of properties. Cellulose derivatives are synthesized by dissolution of cellulose followed by chemical reactions. The glucose rings in the main chain are given new functionalities and are usually unable to crystallize so that cast films are amorphous. The cellulose backbone and high molar mass provide attractive film properties and yet the polymers are water soluble. HEC was first combined with bacterial cellulose (BC) during BC biosynthesis in a CNF/HEC weight ratio of 80/20.[13] The resulting composite showed a modulus of 12.5 GPa, yield strength of around 100 MPa, ultimate tensile strength of 290 MPa, strain-to-failure of about 6%, and work of fracture 11 MJ/m^3. These data are high values, and the HEC was distributed as a coating around the individual BC fibrils. This highly controlled way of HEC distribution is supported by nuclear magnetic resonance (NMR) and transmission electron microscopy (TEM) data in the study. The data for strain-to-failure (6%) indicate that BC fibril sliding can take place in this HEC matrix material, most likely due to the low yield stress of HEC and its nanoscale distribution as a fibril coating. Later, HEC was combined with CNF from wood in a papermaking type of process.[14] HEC adsorbs to CNF and very little HEC is lost during filtration.[14] Since the HEC content is fairly high, some HEC is adsorbed to CNF but there are also HEC-rich regions between CNF-rich "flocs". Results from tensile tests are presented in Fig. 9.2. The 45/41 material has 45 vol.% CNF, 41% HEC, and an estimated porosity of 14%. The mechanical properties are high, see also Table 9.1.

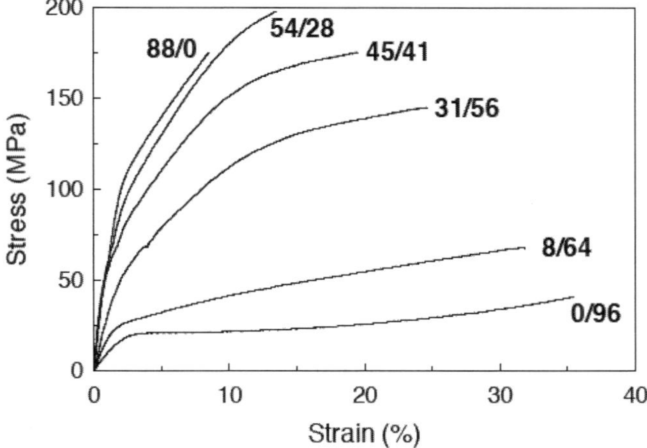

Fig. 9.2. Stress–strain curves for CNF/HEC nanocomposites prepared by vacuum filtering from CNF suspension containing dissolved HEC, which largely adsorbs to CNF. 45/41 means 45 vol.% CNF and 41% HEC estimated from the nanocomposite density, component weight fractions, and component densities. 100 minus volume fraction of CNF and HEC is the estimated porosity (14% for 45/41 CNF/HEC). (Image from Ref. 14, reproduced by permission of the Royal Society of Chemistry.)

No other biocomposite material is known to us with this favorable combination of strength, modulus, and strain-to-failure.

The concept of HEC-coated cellulose fibrils was also used to prepare non-woven nanopaper networks.[15] For non-wovens with a porosity of 44%, the nominal strain-to-failure was 55% (see Table 9.1). The fibril–fibril interaction is expected to be very low in this high SSA material. Again, this example shows the potential to tailor modulus, strength, and toughness of non-woven nanopaper structures. Not only the porosity, but also the density and strength of fibril–fibril bonds are important structural parameters to control in the development of such materials.

9.6. Summary

Thermoplastic wood biocomposites used in molded components are often quite brittle and show limited modulus and strength. Biocomposites with improved performance are therefore of great interest. Low strain-to-failure (<3–4%) is caused by fracture initiation at agglomerates or large wood particles. Limitations in modulus (<7 GPa) and strength (<100 MPa) are related to short aspect ratio of the wood reinforcement phase and low content of wood reinforcement (<30% by volume).

Some data for cellulose nanopaper and CNF polymer matrix nanocomposites are therefore summarized in Table 9.1. Data are discussed from top to bottom. Porous cellulose nanopaper has a modulus of 13 GPa, a tensile strength of 220 MPa, and a strain-to-failure approaching 10%. With lower porosity, the strength can be even higher. The nanopaper shows plastic yielding and then reaches its ultimate strength through substantial strain hardening. The porous nanopaper network can be used as reinforcement for polymer matrix nanocomposites.

The second cellulose nanopaper material in Table 9.1 is more porous and prepared by drying from supercritical CO_2. This nanopaper has higher porosity, weaker fibril–fibril bonds, and therefore deforms to a higher strain-to-failure (17%).

If CNF in wet nanopaper is oriented by slow stretching and dried, then the strength can be as high as 430 MPa (see Table 9.1). The highest modulus obtained is 33 GPa, which can also serve as a conservative lower bound value for NFC modulus.

The next material in Table 9.1 is a MF thermoset matrix composite. Since MF thermosets typically have high yield stress, the plastic deformation characteristics of the NFC/MF biocomposites are unfavorable and strain-to-failure is low. As an extreme contrast, highly plasticized thermoplastic starch was therefore used and a favorable effect of this low yield stress matrix was observed on strain-to-failure. NFC could reorient and slide so that a strain-to-failure of 8–25% was obtained, depending on NFC content. However, the high glycerol content limits the tensile strength.

The three last materials in Table 9.1 all have HEC matrix. This is a linear, water-soluble high molar mass polymer (cellulose derivative) with high modulus but low yield stress in film form. In mixtures with cellulose fibrils in water, individual HEC molecules adsorb to cellulose surfaces, and the fibrils become coated by a thin HEC layer.[13] Good nanostructural control of the polymer matrix distribution is obtained. Each fibril is compartmentalized by the polymer coating. For the BC, we obtain very high tensile strength (290 MPa). The approach is then extended to CNF and the volume fraction of HEC is increased. The strain-to-failure then becomes as high as 26% with a good combination of modulus, strength, and toughness. A non-woven material of HEC-coated CNF reaches 55% strain. Again, the reason is that the weakly connected CNF fibrils can reorient and slide to allow large deformation. The "new", more oriented, CNF structure also results in reasonably high strength (94 MPa).

In summary, CNFs can be combined with low yield stress polymers such as HEC to form cellulose biocomposites with a unique combination of high strength, modulus, and strain-to-failure. The plastic deformation mechanisms in such materials have been discussed. Papermaking type of processes can be used and materials with high volume fraction of the reinforcement phase can be formed. Recently, clay platelets have also been added to the hydrocolloidal mixture and high-performance nanocomposites with good flame and fire-retardant function was obtained.[16]

References

1. L.A. Berglund and T. Peijs, Cellulose biocomposites-from bulk moldings to nanostructured systems, *MRS Bull.* **35** (2010) 201–207.
2. A.N. Nakagaito and H. Yano, Novel high-strength biocomposites based on microfibrillated cellulose having nano-order-unit web-like network structure, *Appl. Phys. A* **80** (2005) 155–159.
3. M. Henriksson, G. Henriksson, L.A. Berglund and T. Lindström, An environmentally friendly method for enzyme-assisted preparation of microfibrillated cellulose (MFC) nanofibers, *Eur. Polym. J.* **43** (2007) 3434–3441.

4. T. Saito, S. Kimura, Y. Nishiyama and A. Isogai, Cellulose nanofibers prepared by TEMPO-mediated oxidation of native cellulose, *Biomacromolecules* **8** (2007) 2485–2491.
5. M. Henriksson, L.A. Berglund, P. Isaksson, T. Lindstrom and T. Nishino, Cellulose nanopaper structures of high toughness, *Biomacromolecules* **9** (2008) 1579–1585.
6. M. Henriksson and L.A. Berglund, Structure and properties of cellulose nanocomposite films containing melamine formaldehyde, *J. Appl. Pol. Sci.* **106** (2007) 2817–2824.
7. M.N. Silberstein, C.L. Pai, G.C. Rutledge and M.C. Boyce, Elastic–plastic behavior of non-woven fibrous mats, *J. Mech. Phys. Solids* **60** (2012) 295–318.
8. H. Sehaqui, Q. Zhou, O. Ikkala and L.A. Berglund, Strong and tough cellulose nanopaper with high specific surface area and porosity, *Biomacromolecules* **12** (2011) 3638–3644.
9. H. Sehaqui, Q. Zhou and L.A. Berglund, High-porosity aerogels of high specific surface area prepared from nanofibrillated cellulose (NFC), *Comp. Sci. Technol.* **71** (2011) 1593–1599.
10. H. Sehaqui, N. Mushi, S. Morimune, M. Salajkova, T. Nishino and L. Berglund, Cellulose nanofiber orientation in nanopaper and nanocomposites by cold drawing, *ACS Appl. Mater. Interfaces* **4** (2012) 1043–1049.
11. A.N. Nakagaito and H. Yano, The effect of fiber content on the mechanical and thermal expansion properties of biocomposites based on microfibrillated cellulose, *Cellulose* **15** (2008) 555–559.
12. A.J. Svagan, M.A.S.A. Samir and L.A. Berglund, Biomimetic polysaccharide nanocomposites of high cellulose content and high toughness, *Biomacromolecules* **8** (2007) 2556–2563.
13. Q. Zhou, E. Malm, H. Nilsson, P.T. Larsson, T. Iversen, L.A. Berglund and V. Bulone, Nanostructured biocomposites based on bacterial cellulosic nanofibers compartmentalized by a soft hydroxyethylcellulose matrix coating, *Soft Matter* **5** (2009) 4124–4130.
14. H. Sehaqui, Q. Zhou and L.A. Berglund, Nanostructured biocomposites of high toughness-a wood cellulose nanofiber network in ductile hydroxyethylcellulose matrix, *Soft Matter* **7** (2011) 7342–7350
15. H. Sehaqui, S. Morimune, T. Nishino and L.A. Berglund, Stretchable and strong cellulose nanopaper structures based on polymer-coated nanofiber networks: an alternative to nonwoven porous membranes from electrospinning, *Biomacromolecules* **13** (2012) 3661–3667.
16. A. Liu, A. Walther, O. Ikkala, L. Belova and L.A. Berglund, Clay nanopaper with tough cellulose nanofiber matrix for fire retardancy and gas barrier functions, *Biomacromolecules* **12** (2011) 633–641.

Chapter 10

Reinforcing Efficiency of Nanocelluloses in Polymer Nanocomposites

Yvonne Aitomäki and Kristiina Oksman

Composite Centre Sweden, Division of Materials Science, Luleå University of Technology, Luleå, Sweden

Cellulose nanofibers (CNFs) have been shown to offer good reinforcement in polymer composites. The recent advancements in combined processing techniques have lowered the energy requirements to produce these nanosized celluloses advancing both the availability and commerciality of these biobased materials. Here, we focus on the current status of the mechanical properties of reinforced polymer nanocomposites. A comparison is made of the reinforcing efficiency of CNF composites by back-calculating a reinforcing efficiency parameter using established micromechanical models. Included is a brief review of the factors affecting reinforcing efficiency such as the sources and production methods of CNF as well as the matrices and methods used to manufacture nanocomposites from them. Some comparisons are made to current reinforced polymer composites, such as glass fiber composites. Highlighted are some of the issues that need addressing if these materials are to be used as structural applications.

10.1. Introduction

There is a general acceptance of global warming as well as an increased awareness of the effect greater global industrialization has on the environment and ultimately how it affects our society. As a consequence, interest is high in finding lightweight, sustainable materials that are efficient to produce and manufacture as alternatives to existing fossil-fuel-based plastics and even metals. Focus is also on using the resources we have as efficiently as possible. Since cellulose is the most abundant natural occurring organic compound on the planet and it is produced by plants making it a renewable and carbon dioxide neutral, it is not surprising that solutions based on this compound are the focus of much attention. Publications on cellulose nanomaterials have as a consequence risen rapidly (Fig. 10.1) after studies showed the positive effect of using them as reinforcements.

Most cellulose nanomaterials are found in plant cellulose walls but sources also include algae, animals such as sea cucumber and tunicates, and certain types of bacteria, e.g., *Acetobacter*. Cellulose nanomaterials from all of these sources have been used as reinforcements in polymers in the form of cellulose nanofibers (CNFs) and cellulose nanocrystals (CNCs). The CNCs are the base-reinforcing unit made

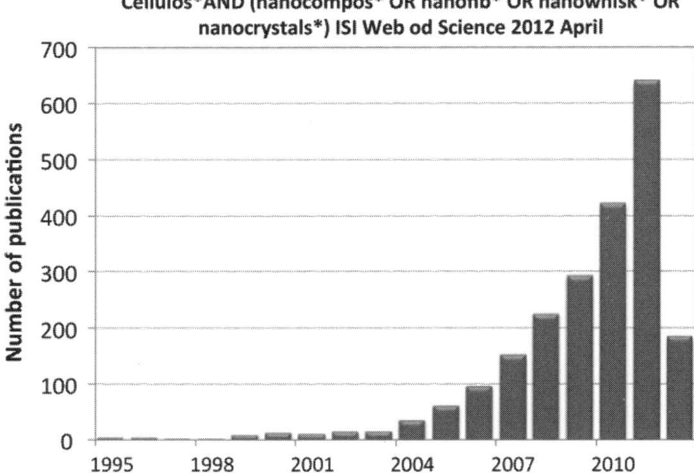

Fig. 10.1. The increased number of publications in the field of nanocellulose research including their nanocomposites.

up of aligned cellulose molecules with diameters of approximately 5 nm and length of approximately 2 μ.[1,2] The stiffness and strength of these crystals are approximately 150 GPa and 10,000 MPa.[3] CNFs are made up of these crystalline structures bounded together by regions of amorphous cellulose. The diameter and length of CNF varies and depends on the source as well as the method by which the CNFs are extracted. For example, CNF isolated from softwood pulp has diameters of 15–60 nm and length estimated at several microns.[4]

Using CNF in materials requires breaking down cellulose pulp to extract the CNF, then reconstructing material from them. A typical reconstructed material is where CNFs are used as reinforcements in polymers. However, if the goal is structural, lightweight composites from sustainable resources, the material should possess strong mechanical properties that would remain good under the conditions of use and the lifetime of the product. Added to this, if these materials are going to be easily accepted, then they also have to be economically competitive. To achieve this requires not just cheap and plentiful raw materials, but efficient processes to isolate them and to manufacture products from them. Addressing all of the issues at an initial stage of development is not possible but as CNFs become a more established material, solutions to increasing efficiency of extraction and the potential for industrial upscaling are being investigated.[5–8]

In this chapter, we focus on the development of composite materials from different nanocelluloses in the form of cellulose nanocomposite. In particular, we focus on the mechanical properties of cellulose nanocomposite as a means of assessing their use as structural materials and compare them to standard glass and flax fiber composites. The objective is to highlight which areas require further work to establish these materials as competitive alternatives to existing materials in structural applications in particular.

10.2. Theoretical efficiency

Interest in the field of using nanocellulose fibers or crystals as reinforcement in polymer matrices is rising rapidly as discussed in the introduction and it is not surprising that a very large number of combinations of nanofibers and matrices have been tested with widely varying concentrations and mechanical properties. This is shown in the review by Hubbe *et al.*[9] that lists 119 articles where the reinforcing effect of nanocellulose is used to motivate their investigations. Of this list, 49 articles reported an increase in the stiffness and strength of the polymer when reinforced with nanocellulose.

A more modest, but detailed list, of 28 articles is presented in a review by Siró *et al.*[10] on the reinforcement effects of CNF from wood pulp. In comparing nanocomposites, we focus on the stiffness achieved by the addition of CNF and compare results to other polymer composites. In structural composites, high fiber fractions (>50%) are of interest since the fibers provide the stiffness and strength. Hence of interest is a comparison between typical structural polymer composites and nanocomposites with high volume fractions. We have chosen to compare these composites by calculating the efficiency of the reinforcement in stiffness of the CNF composite using the efficiency of the reinforcement parameter in the Halpin–Tsai model.[11] The Halpin–Tsai model has frequently been used to model the Young's modulus of nanocomposites[12–14]; however, unlike here, the efficiency parameter, ζ, is usually assumed to be twice the aspect ratio of the fibers.

The stiffness is only one property of the bionanocomposites but is an important factor in selecting a material for use in structural applications. Strength is another important property in structural materials; however, the mechanisms of crack propagation, initiation, and failure of the matrix/fiber interface, all of which affect strength, are more complex than models of stiffness and less accurate even for well-established materials. Due to the lack of strength models for nanocellulose composites, efficiency of the strength of reinforcement is not considered here.

The stiffness efficiency is highest for a composite where there are aligned, continuous fibers. In the case of randomly orientated nanofibers in composites, the fibers are neither aligned nor continuous. To take into account the random alignment of the fibers, classical laminate theory (CLT) is used with a quasi-isotropic composite made of different orientation of layered unidirectional fibers in the calculation of the longitudinal stiffness of the CNF composite. The non-continuous nature of the nanofibers is not taken into account since the aspect ratio of many of the nanocellulose fibers is very high and in such cases, the non-continuous fiber behavior approaches that of the continuous fibers.[15]

The approach to calculate the efficiency of the CNF reinforcement in selected nanocellulose composites is to make an initial estimate of the reinforcement parameter, ζ. From this value, the stiffness of a single unidirectional laminate is calculated and used in the CLT to calculate the stiffness of the randomly orientated CNF composite. This calculated stiffness is then compared to experimentally measured stiffness of the composite and the value of ζ is then iterated until the theoretical stiffness matches the measured stiffness. This then gives a value of the reinforcing efficiency ζ for that CNF composite.

To understand this value of the reinforcing efficiency, it is important to establish the upper and lower boundaries. In a composite, the Voigt model for predicting the Young's modulus is often used to give its upper limit.[15] In the derivation of the Voigt model, the assumptions are that the fibers are continuous, the load is in the direction of the fibers, and the interface between the matrix and the fibers acts as a non-slip boundary hence the matrix is constrained by the fibers giving the condition of constant strain in the composite (see Fig. 10.2).

In the Reuss model, a constant stress is assumed and thus the more pliant constituent of the composite, usually the matrix, dominates the behavior. This model usually underestimates the behavior of the composite and provides a lower bound for the prediction of stiffness. The Halpin–Tsai model can be used to predict different mechanical properties, including Young's modulus. It can be used to fit both of the extremes predicted by the Voigt and Reuss model. It is expressed as

$$P_{\mathrm{c}} = P_{\mathrm{m}} \frac{1 + \zeta\eta\phi}{1 - \eta\phi}, \tag{10.1}$$

where,

$$\eta = \frac{P_{\mathrm{f}}/P_{\mathrm{m}} - 1}{P_{\mathrm{f}}/P_{\mathrm{m}} + \zeta}, \tag{10.2}$$

and where P_{c} is a property of the composite, P_{f} is the property of the fiber, P_{m} is the property of the matrix, ϕ is the volume ratio, and ζ is the efficiency of the reinforcement. In this model, the same result as that of the Voigt model for the modulus, E_f can be achieved by setting ζ to infinitely. Setting this parameter to zero instead results in the Halpin–Tsai model equating to the Reuss model. Values of ζ between 0 and infinity therefore reflect the efficiency of the reinforcing elements in composites[15] and it is this value that we use to compare the efficiency of the reinforcement of the CNF in different polymers.

In using the Halpin–Tsai model with CNF-reinforced composites, we use values of E_{f} from X-ray diffraction measurements of the modulus of crystalline cellulose, i.e., the crystalline elements of the nanofibers. For this to be achievable, the entire length of the nanocellulose fiber would have to be made up of nanocrystals. Whilst not correct, it does provide the upper limit of what stiffness can be expected of the nanocellulose fiber. A review by Eichhorn *et al.* (2001)[16] shows that the modulus values for the nanocrystals have been found in the range of 100–160 GPa. In the determination of the reinforcing efficiency, we use a value of 138 GPa for the modulus of the fibers, E_{f}. This was determined by X-ray diffraction of deformed bleached ramie fiber bundles[17] and is a value that is commonly referred to by many authors.

This value was used despite the fact that a recent study to determine the modulus of CNF from wood pulp using Raman spectroscopy[18] estimated the value to be as low as 33 GPa. However, the results from the study, which estimated the modulus of both bacterial cellulose (BC) and CNF from wood pulp, rely heavily on the

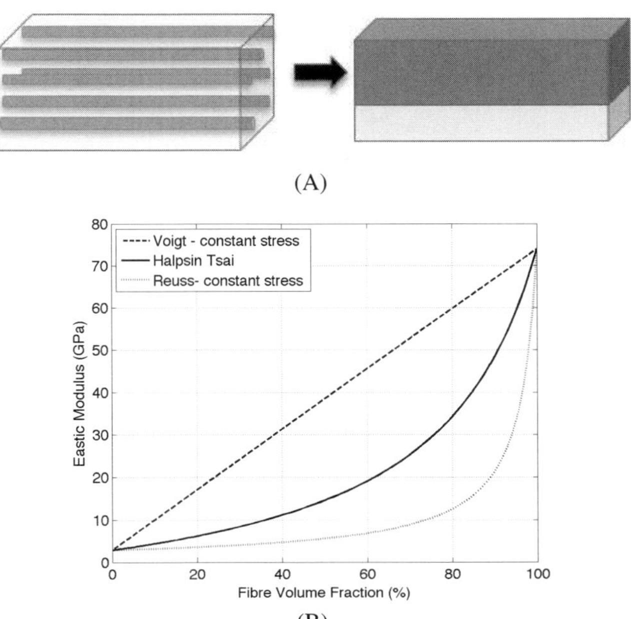

Fig. 10.2. (A) In the models, the fibers in the matrix are considered as a volume of solid fiber material bonded to a volume of solid polymer material. A constant strain assumption is used in Voigt model where the fibers restrict the displacement of the matrix. The constant stress condition is used in the Reuss model and leads to dominance of the matrix properties. The reinforcing parameter in Halpin–Tsai model can be set to fit either the Reuss or Voigt model or values of elastic modulus between these two extremes for a given volume fraction. (B) Example here is given of when this model is applied to a glass fiber/epoxy composite, with the reinforcing efficiency parameter set to 5 for Halpin–Tsai model.

value of the reinforcing efficiency, η_o. In the study, η_o was taken as 3/8. This value is theoretically derived for short fibers randomly orientated in 2D. The relationship between η_o, and the estimated E_f is plotted in Fig. 10.3 for BC. As can be seen, if a 3D random arrangement is assumed ($\eta_o = 0.2$), then the values for the BC reach up to 160 GPa instead of those suggested by the study of a 88 GPa. Whilst a 3D random arrangement is not the case and there is considerable alignment of both BC and CNF in the 2D planes as shown in Fig. 10.4, deviation from the 2D plane (waviness) and other factors that reduce efficiency will lead to significant increases in the estimation of the modulus. This supports the use of the higher value of E_f used here.

For the derivation of the quasi-isotropic composite, a layup of [0 45–45 90] was used as is typical for CLT[19] and has also been used by authors modeling of CNF composites.[13] For this calculation, a total of four independent engineering constants of the composite laminate are required. These are the Young's modulus, E_1, its in-plane Poisson's ratio, v_{12}, the in-plane shear modulus, G_{12}, and the Young's modulus in the transverse direction, E_2. Poisson ratio, v_{21}, is calculated from the other engineering constants. For E_1, Eq. (10.1) is used, substituting P_c for E_1 and letting $P_f = E_f = 138$ GPa. The matrix is assumed to be isotropic and the values are taken

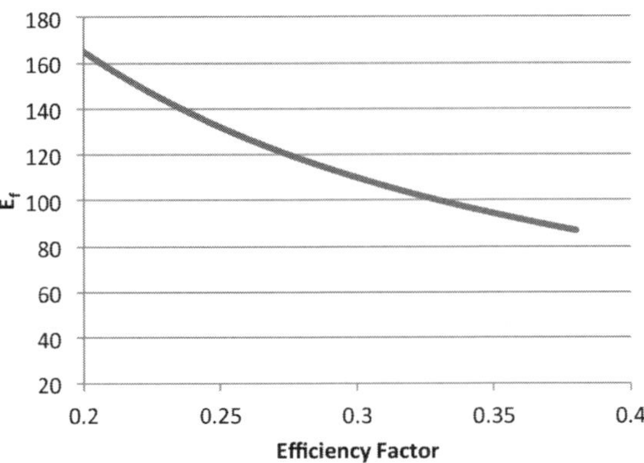

Fig. 10.3. Krenchel model with different values for the efficiency factor ranging from the values for 3D to 2D theory of random fiber arrangements with the elastic model of the network derived from Tanpichai *et al.* (2012) with Raman spectroscopy.

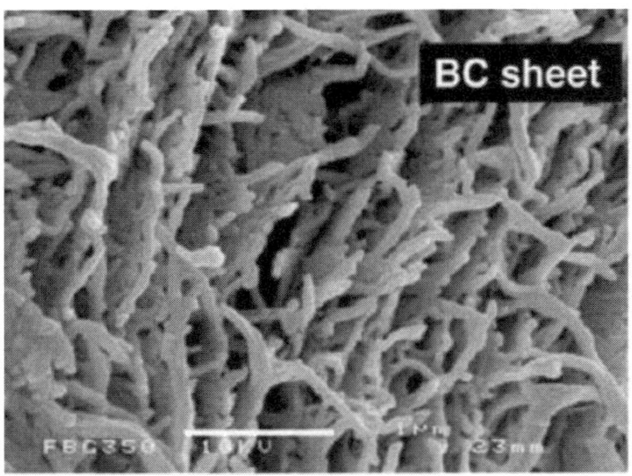

Fig. 10.4. Fracture surface of a BC sheet showing that although there is considerable alignment in the 2D plane, it is not without waviness. (Reprinted (with permission) from "All-cellulose nanocomposites by surface selective dissolution of bacterial cellulose".[43.])

from those given in different articles in which composites based on nanocellulose have been investigated. Equation (10.1) can also be used for v_{12} and Poisson's ratio for the fibers, v_{f} is set to 0.3 as is done by Pakzad *et al.*[20] Since many matrices have values of Poisson's ratio close to 0.3, the value of the v_{m} is assumed to be 0.3. When calculating G_{12}, Capadona *et al.*[13] used Eq. (10.1) with $\zeta = 1$, where ζ in this case is the reinforcing efficiency of the shear modulus, hence referred to as ζ_s for clarity. This value was found to provide good estimation of G_{12}, for composites with circular filaments packed in square arrays.[15] Other values of ζ_{s} can be used if different packing arrangements are considered but since there is a random arrangement of nanofibers, closer packing arrangement such as hexagonal arrangements are likely to overestimate G_{12}. Thus, following Capadona *et al.*,[13] we use $\zeta = 1$ in Eq. (10.1). We also follow Capadona *et al.*[13] and use their value of transverse shear for the fibers of 1.7 GPa. The final value, E_2, is calculated using Eq. (10.1) such that

$$E_2 = E_m \frac{1 + \zeta\eta\phi}{1 - \eta\phi}, \tag{10.3}$$

and

$$\eta = \frac{E_{2f}/E_m - 1}{E_{2f}/E_m + \zeta}, \tag{10.4}$$

where E_{2f} is the transverse fiber Young's modulus. Here $\zeta = 1$ is used as it is typically used for circular filaments packed in square arrays for the same reasons as explained earlier. In the calculation of E_2, we use $E_{2f} = 25$ GPa derived by Pakzad *et al.*[20] for wood-based nanocellulose.

The reinforcing factor is then calculated for each of the composites in Table 10.1. This is done by comparing the reported value of Young's modulus for each composite with a calculated value of the stiffness using CLT and the Halpin–Tsai model as previously described. The values for the parameters in the model are set as described above for the CNFs. The matrix values are set to the matrix values used in the articles, as is the volume fraction. The reinforcing factor is set to an initial value of three unless convergence to a solution was unsuccessful. The initial value is shown in Table 10.1. The value is iterated using a Matlab (The MathWorks, Inc.) search function based on the optimization route described by Forsythe *et al.*[21] When the difference between the model and the experimental value is within one decimal place of each other, the solution for the reinforcing values is assumed as found.

10.3. Efficiency comparison

Table 10.1 gives an overview of the reinforcing effect of CNFs where they have been used in high volume concentrations. A calculation of ζ for each of the nanocomposites reported is done using the model described above based on the information available from the different studies. Very high values ($\zeta \to \infty$) imply that the composites are close to performing like the Voigt model and strain in the fibers and matrix is equal which in turn suggests good load transfer between the fibers and the matrix and that the load is taken by the fiber. It can be seen quite clearly that in the nanocomposites with soft matrices such as starch or polyurethane (PU), the reinforcing effect of the nanofibers is high. These cases of low matrix stiffness and high ζ are particularly valuable where nanocellulose-reinforced biobased matrices are used as the stiffness increases and in such a way, these matrices can become competitive with resin commonly used in structural composites such as epoxies and acrylics. However, though they compete with unreinforced traditional matrices, these nanocomposites with these levels of stiffness, do not, in terms of mechanical properties, take us to the next level of biobased composites in that they are not in the region of being equivalent to glass composites. As a means of taking the stiffness of the matrix into account, and thus its competitiveness as a structural material, the reinforcement factor is weighted to stiffer materials such that $\zeta_w = \zeta E_m$. This weighted parameter, ζ_w, is included in Table 10.1.

Using this indicator, the list is topped by Yano and coworkers who have pioneered much of the work on nanocomposite with high stiffness and strength, and with high volume fractions.[18,22–26] In the study shown in Table 10.1, Yano *et al.* (2005)[26] used a UV-cured epoxy and BC fibers. The study was aimed at producing transparent film with high mechanical properties. The stiffness of the composite was 21 GPa, which the authors state is a third that of glass, i.e., a unidirectional glass fiber composite. In fact, this undervalues the stiffness value of the nanocomposite as this is achieved from an isotropic material. These values are comparable to typical epoxy glass fiber composites (see Table 10.1) and this is a more fair comparison as the nanofibers are randomly orientated in the 2D plane thus nanocomposites will be transversely isotropic whereas the unidirectional glass fibers composites have only high stiffness in the fiber direction. Good impregnation and compatibility between the matrix and the resin is reflected by the high transparency found in these composites since poor impregnation and poor compatibility leads to air gaps between the matrix and the nanofibers, thus reducing the transparency.[27] The impregnation method used in this study uses the capillary action and the application of a buoyancy force on air trapped in the CNF network to cause the air to rise, thus drawing the resin into the network. The force is provided by the application of a vacuum and has been used by this group in particular to give good but slow impregnation with the impregnation usually carried out over a period of 12 to 72 h or more.[22,24–26,35]

In a study by Juntaro *et al.* (2012),[27] PU was reinforced with BC fibers. The impregnation period was extremely short (1 min) and the cross-sectional scanning electron microscopy (SEM) image showed that the resin had formed a layer on either side of the nanofiber network giving rise to a sandwich structure (see Fig. 10.5). Despite this, the optical transparency is high and the increase in stiffness over the dry nanofiber network stiffness suggests synergy between the matrix and the CNFs.

Example of nanocomposites with low stiffness matrices have been investigated by Sehaqui *et al.* who used hydroxyethyl cellulose (HEC) and Svagan *et al.* who used starch. Both of these nanocomposites give high values for ζ. These high values are thought to be due to the network effect of fibers that, in cases where the matrix stiffness is much lower than the fibers, will cause the strain of nanocomposite to be closer to that of the fibers.

The composites using HEC were produced by mixing this matrix with CNF followed by filtering. This produced good impregnation as is seen in the SEM images of the fracture surfaces (see Fig. 10.6). The stiffness of the composite is high considering the low stiffness of the matrix used and the authors also showed increased toughness and strength of the nanocomposite compared to the non-impregnated paper.

The nanocomposites by Svagan *et al.* (2007)[30] with starch had a stiffness of 6.2 GPa, which is comparable with natural fiber composites with non-orientated

Table 10.1. Cellulose nanocomposites with high fiber fraction.

Matrix	Nano cellulose	Matrix stiffness (GPa)	Fiber mat stiffness (GPa)	Composite stiffness (GPa)	Weight fraction (%)	Vol. fraction[a] (%)	Initial value	Reinforcement efficiency (ζ)	Weighted efficiency (ζ_w)	Reference
Epoxy	BC	2.8	20	21	65	43	100	190	532.0	26
PU	BC	0.19	5.1	11.6	51	43	100	541	102.8	27
Epoxy	Glass 0/90	2.9	74	24	—	60	3	28	81.2	28
Hydroxyethyl cellulose	CNF	0.95	9.9	8.2	68	43	3	40	38.0	29
Starch	CNF	0.0016	13	6.2	70	61	3	9530	15.2	30
Polylactic acid (PLA)	CNF	2	—	3	18	15	3	5.6	11.2	31
Epoxy	Glass 45/-45	2.9	74	11	—	60	3	3.4	9.9	28
Phenol formaldehyde	CNF	8.3	10	19	—	73	3	0.64	5.3	24
Epoxy	CNF	2.98	12	8.7	—	60	3	1.7	5.1	32
Chitosan	CNF	1.4	8	2.1	20	17	3	3.5	4.9	33
Melamine formaldehyde	CNF	8.3	14	16.6	87	79	3	<0	—	34

[a]Volume fraction was estimated from the values available in the articles for matrix and fiber density as well as porosity.

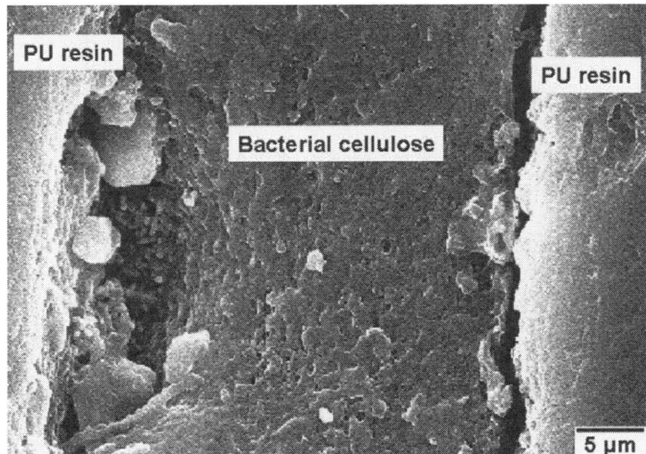

Fig. 10.5. Sandwich structure of the PU/BC nanocomposite. (Reprinted from "Bacterial cellulose reinforced polyurethane-based resin nanocomposite: A study of how ethanol and processing pressure affect physical, mechanical and dielectric properties", Copyright (2012) with permission from Elsevier.[34])

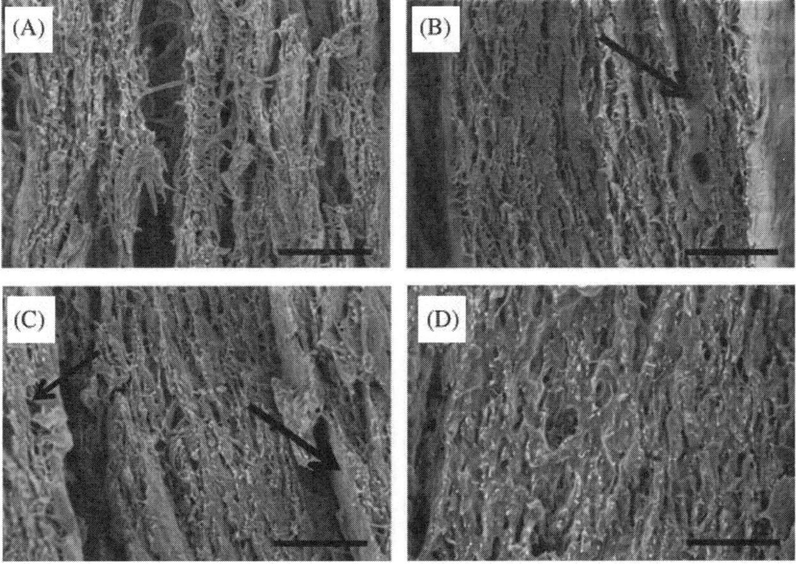

Fig. 10.6. Fractured cross-section SEM images of (A) CNF nanopaper (CNF/HEC 88/0) and CNF/HEC biocomposite samples at CNF/HEC volume fractions of (B) 54/28, (C) 45/41 and (D) 31/56. Scale bar is 500 nm. (Reproduced from Ref. 29 with permission from the Royal Society of Chemistry.)

fibers.[36] The authors point out that they also have a very high work fracture $9.4\,\mathrm{MJm}^{-3}$ as compared to that of steel with $1\,\mathrm{MJm}^{-3}$.

The studies by Tingaut *et al.*,[31] using 18% CNF in PLA, and Hassan *et al.*,[33] using 20% CNF in chitosan, represents the large number of nanocomposites made with volume fractions of less the 20%. The reinforcing efficiency is good with the PLA nanocomposite giving a weighted reinforcing efficiency of 11.2 and the chitosan nanocomposite giving a weight efficiency of 4.9, as shown in Table 10.1. However, these results do not reflect the more spectacular reported results found in the increase of the storage modulus when using very low concentration (5–6 wt%) in low stiffness matrices.[37,38] This occurs once the T_g of the matrix is exceeded and is due in part to the network effect of the nanocellulose in the matrix, the stiffness of the nanocomposite remains at a similar stiffness to that which it had at temperature below the T_g of the matrix. Thus, the increase in the stiffness is remarkable. However, since we are focusing on the use of nanocellulose reinforcement as structural material, we are examining the reinforcement effect in nanocomposites at room temperature and typically below the T_g of the polymer. Although there is an increase in the stiffness, this increase is well represented by the results shown by Tingaut *et al.*[31] and Hassan *et al.*[33] Again, although the reinforcement is efficient, the overall stiffness as a composite material is low.

Nakagaito *et al.*[24] and Henriksson *et al.*[34] are examples of nanocomposites using very stiff matrices. The resulting nanocomposites are some of the stiffest produced despite the fact that the reinforcement efficiency is low for the phenol formaldehyde and below zero for the melamine formaldehyde. This is because the stiffness of the matrix is high and is combined with a high volume fraction of nanofibers. Although the negative value is likely to be caused by overestimating the value of the stiffness of the nanofibers, the relevance of the value is in illustrating that the composites are perhaps not performing as well as they should at these levels of volume fraction. These results suggest that whilst the formation of the network for reinforcing soft matrices is beneficial, in stiffer matrices, this network formation could be limiting the potential of the nanofibers to reinforce polymers. In Henriksson *et al.*,[34] it is pointed out that these composites are brittle with low strain to failure and hence the strength is controlled by local defects. This means that although the average strength is good (142 MPa) and higher than some natural fiber composites, the sensitivity to local defects causes large variation in the strength.

10.4. Accuracy

The reinforcing factor in the nanocomposites in Table 10.1 is difficult to calculate accurately from the different studies for a number of reasons. The main errors are from the estimation of E_f that was assumed to be equal to that of the cellulose crystals. One can argue that all the nanocomposites were approximated in a similar manner, hence the relative value of ζ is accurate. However, different sources and methods of deriving the nanocellulose have been used and this is likely to affect the reinforcing elements. This potential difference in stiffness means that the accuracy of the assumption will be different in the different cases. A second source of error is the calculation of the volume fraction. This is not given in most of the articles hence has to be estimated from the density of the matrix, density of the CNF, and void content of the nanocomposite. The density of the matrix is not given in all cases, for example for the melamine formaldehyde, hence an estimate from data available from other sources was used. Where the porosity was not given, it was assumed to be zero.

A point of issue is the use of the Halpin–Tsai model when percolation models have been shown to fit the data better.[13,38,39] However, percolation models are derived for mixtures of crystalline and non-crystalline polymers. The modulus of the crystalline polymer used is that of a pure crystalline polymer. Although it is mentioned in the derivation in Takayanagi *et al.*[40] that the volume fraction related parameters (λ and φ) are modified for non-spherical particles, account is not taken of possible non-isotropic modulus values of particles with high aspect ratios. Hence as the volume fraction ϕ tends to one in the percolation model, and when the rigid phase is much greater than the soft phase (valid in most cases), then E_c tends to E_f. However, unless the fibers are orientated, the maximum value must be lower than this. This means that the percolation model, at high volume fraction will overestimate the value of the modulus of the nano composite. For that reason, the Halpin–Tsai model with the CLT has been used. The Halpin–Tsai model does not capture the behavior of the network close to or below the volume fraction percolation threshold; however, low volume fractions networks are not the focus of the chapter so this should not affect the accuracy of the reinforcement estimate.

10.5. Summary

A method of comparing the efficiency of reinforcements in different nanocomposites is presented. This efficiency factor is subsequently weighted to more clearly reflect the potential of CNFs composites as structural materials.

When comparing the stiffness of nanocomposite using a weighted reinforcing factor for stiffness, BC and epoxy was one of the most successful nanocomposites. With a very stiff matrix, the efficiency of reinforcement is low, whilst in soft matrices, the efficiency is high. This suggests that in soft matrices, the formation of a CNF network enhances the composite stiffness whereas in stiffer matrices, this network formation could be limiting the potential of the nanofibers. Alignment of the CNF could be a key to achieving greater efficiency of stiffness. Alternatively, the isotropic nature of the CNF material can be exploited.

Only one factor, the stiffness, was examined in the comparison of cellulose nanocomposites. Strength is another key factor in structural applications and the effect of CNF reinforcements on strength is a reoccurring feature in many nanocomposites. Durability and their behavior in different environmental conditions are important factors that depend heavily on the matrix used in combination with the CNF. These effects have been and continue to be the subject of investigation.[41–43]

Alternative applications other than as structural materials are also key development areas for cellulose nanocomposites, with focus on the other properties of nanocomposites such as their transparency, their water and solvent barrier properties as well as possibility for electrical conduction.

Acknowledgments

The authors are grateful to the Bio4Energy Program (Sweden) for the financial support of this research.

References

1. D. Bondeson, A. Mathew and K. Oksman, Optimization of the isolation of nanocrystals from microcrystalline cellulose by acid hydrolysis, *Cellulose* **13** (2006) 171–180.
2. G. Siqueira, A.P. Mathew and K. Oksman, Processing of cellulose nanowhiskers/ cellulose acetate butyrate nanocomposites using sol–gel process to facilitate dispersion, *Comp. Sci. Techno.* **71** (2011) 1886–1892.
3. W. Hamad, *Cellulosic materials: fibers, networks, and composites*, Springer (2002).
4. G. Gong, J. Pyo, A.P. Mathew and K. Oksman, Tensile behavior, morphology and viscoelastic analysis of cellulose nanofiber-reinforced (CNF) polyvinyl acetate (PVAc), *Comp. Part A: Appl. Sci. Manufac.* **42** (2011) 1275.
5. M. Jonoobi, A.P. Mathew and K. Oksman, Producing low-cost cellulose nanofiber from sludge as new source of raw materials, *Ind. Crop. Prod.* **40** (2012) 232.
6. T. Saito, S. Kimura, Y. Nishiyama and A. Isogai, Cellulose nanofibers prepared by TEMPO-mediated oxidation of native cellulose, *Biomacromol.* **8** (2007) 2485–2491.
7. S. Janardhnan and M. Sain, Isolation of cellulose microfilbrils-an enzymatic approach, *BioResources* **1** (2006) 176–188.
8. K.L. Spence, R.A. Venditti, O.J. Rojas, Y. Habibi and J.J. Pawlak, A comparative study of energy consumption and physical properties of microfibrillated cellulose produced by different processing methods, *Cellulose* **18** (2011) 1097–1111.
9. M.A. Hubbe, O.J. Rojas, L.A. Lucia and M. Sain, Cellulosic nanocomposites: a review, *BioResources* **3** (2008) 929–980.
10. I. Siró and D. Plackett, Microfibrillated cellulose and new nanocomposite materials: a review, *Cellulose* **17** (2010) 459–494.
11. J.C. Halpin and J.L. Kardos, The Halpin-Tsai equations: a review, *Polym. Eng. Sci.* **16** (1976) 344–352.
12. V. Favier, G.R. Canova, Cavaillé J.Y., H. Chanzy, A. Dufresne and C. Gauthier, Nanocomposite materials from latex and cellulose whiskers, *Polym. Adv. Technol.* **6** (1995) 351–355.
13. J.R. Capadona, K. Shanmuganathan, D.J. Tyler, S.J. Rowan and C. Weder, Stimuli-responsive polymer nanocomposites inspired by the sea cucumber dermis, *Science* **319** (2008) 1370–1374.
14. P.M. Visakh, S. Thomas, K. Oksman and A.P. Mathew, Crosslinked natural rubber nanocomposites reinforced with cellulose whiskers isolated from bamboo waste: processing and mechanical/thermal properties, *Comp. Part A-Appl. Sci. Manufac.* **43** (2012) 735–741.
15. R.L. McCullough, L.A. Carlsson and J.W. Gillespie (Eds.) Technomic Publishing Co (1990) pp. 49–90.
16. S. Tanpichai, F. Quero, M. Nogi, H. Yano, R.J. Young, T. Lindström, W.W. Sampson and S.J. Eichhorn, Effective Young's modulus of bacterial and microfibrillated cellulose fibrils in fibrous networks, *Biomacromol.* **13** (2012) 1340–1349.
17. I. Sakurada, Y. Nubushina and T. Ito, Experimental determination of the elastic modulus of crystalline regions in oriented polymers, *J. Polym. Sci.* **57** (1962) 651–660.
18. S. Tanpichai, F. Quero, M. Nogi, H. Yano, R.J. Young, T. Lindström, W.W. Sampson and S.J. Eichhorn, Effective Young's modulus of bacterial and microfibrillated cellulose fibrils in fibrous networks, *Biomacromol.* **13** (2012) 1340–1349.

19. J. Varna and L.A. Berglund, *Mechanics of fiber composite materials*, Division of polymer engineering, Luleå University of Technology (1996).
20. A. Pakzad, J. Simonsen, P.A. Heiden and R.S. Yassar, Size effects on the nanomechanical properties of cellulose I nanocrystals, *J. Mater. Res.* **27** (2012) 528–536.
21. G.E. Forsythe, M.A. Malcolm, and C.B. Moler, *Computer methods for mathematical computations*, Prentice-Hall (1976).
22. K. Abe, S. Iwamoto and H. Yano, Obtaining cellulose nanofibers with a uniform width of 15 nm from wood, *Biomacromol.* **8** (2007) 3276–3278.
23. A.N. Nakagaito and H. Yano, Novel high-strength biocomposites based on microfibrillated cellulose having nano-order-unit web-like network structure, *Appl. Phys. A* **80** (2005) 155–159.
24. A.N. Nakagaito, S. Iwamoto and H. Yano, Bacterial cellulose: the ultimate nano-scalar cellulose morphology for the production of high-strength composites, *Appl. Phys. A* **80** (2005) 93–97.
25. Y. Shimazaki, Y. Miyazaki, Y. Takezawa, M. Nogi, K. Abe, S. Ifuku and H. Yano, Excellent thermal conductivity of transparent cellulose nanofiber/epoxy resin nanocomposites, *Biomacromol.* **8** (2007) 2976–2978.
26. H. Yano, J. Sugiyama, A. Nakagaito, M. Nogi, T. Matsuura, M. Hikita and K. Handa, Optically transparent composites reinforced with networks of bacterial nanofibers, *Adv. Mater.* **17** (2005) 153–155.
27. J. Juntaro, S. Ummartyotin, M. Sain and H. Manuspiya, Bacterial cellulose reinforced polyurethane-based resin nanocomposite: a study of how ethanol and processing pressure affect physical, mechanical and dielectric properties, *Carbohydr. Polym.* **87** (2012) 2464.
28. B. Nyström and Y. Aitomäki, Structural biobased composite processed by vacuum infusion using flax fibres and biobased epoxy in 12th international conference on biomaterials: transition to green materials, niagara falls, (2012) May 6–8 Canada.
29. H. Sehaqui, Q. Zhou and L.A. Berglund, Nanostructured biocomposites of high toughness—a wood cellulose nanofiber network in ductile hydroxyethylcellulose matrix, *Soft Matter* **7** (2011) 7342–7350.
30. A.J. Svagan, M.A.S. Azizi Samir and L.A. Berglund, Biomimetic Polysaccharide nanocomposites of high cellulose content and high toughness, *Biomacromol.* **8** (2007) 2556–2563.
31. P. Tingaut, C. Eyholzer and T. Zimmermann, A. Hashim (Ed.) Intech (2011) pp. 319–334.
32. K. Lee, T. Tammelin, K. Schulfter, H. Kiiskinen, J. Samela and A. Bismarck, High performance cellulose nanocomposites: comparing the reinforcing ability of bacterial cellulose and nanofibrillated cellulose, *ACS Appl. Mater. Interf.* **4** (2012) 4078–4086.
33. M.L. Hassan, E.A. Hassan and K.N. Oksman, Effect of pretreatment of bagasse fibers on the properrties of chitosan/microfibrillated cellulose nanocomposites, *J. Mater. Sci.* **46** (2011) 1732–1740.
34. M. Henriksson, L.A. Berglund, P. Isaksson, T. Lindström and T. Nishino, Cellulose nanopaper structures of high toughness, *Biomacromol.* **9** (2008) 1579–1585.
35. S. Iwamoto, A.N. Nakagaito, H. Yano and M. Nogi, Optically transparent composites reinforced with plant fiber-based nanofibers, *Appl. Phys. A* **81** (2005) 1109–1112.
36. K. Oksman, Mechanical properties of natural fibre mat reinforced thermoplastic, *Appl. Comp. Mater.* **7** (2000) 403–414.
37. A. Dufresne and M.R. Vignon, Improvement of starch film performances using cellulose microfibrils, *Macromol.* **31** (1998) 2693–2696.
38. V. Favier, J.Y. Cavaille, G.R. Canova and S.C. Shrivastava, Mechanical percolation in cellulose whisker nanocomposites, *Polym. Eng. Sci.* **37** (1997) 1732–1739.

39. A.P. Chatterjee and D.A. Prokhorova, An effective medium model for the elastic moduli of fiber networks and nanocomposites, *J. Appl. Phys.* **101** (2007) 104301.
40. M. Takayanagi, S. Uemura and S. Minami, Application of equivalent model method to dynamic rheo-optical properties of crystalline polymer, *J. Polym. Sci. Pol. Chem.* **5** (1964) 113–122.
41. A. Dufresne, J. Cavaille and W. Helbert, New nanocomposite materials: Microcrystalline starch reinforced thermoplastic, *Macromol.* **29** (1996) 7624–7626.
42. S. Dammström, L. Salmén and P. Gatenholm, The effect of moisture on the dynamical mechanical properties of bacterial cellulose/glucuronoxylan nanocomposites, *Polymer* **46** (2005) 10364–10371.
43. A.P. Mathew, K. Oksman, D. Pierron and M-F. Harmad, Crosslinked fibrous composites based on cellulose nanofibers and collagen with in situ pH induced fibrillation, *Cellulose* **19** (2012) 139–150.

Chapter 11

Advanced Bacterial Cellulose Composites

Koon-Yang Lee and Alexander Bismarck
*Polymer and Composite Engineering group,
Institute of Materials Chemistry and Research,
Faculty of Chemistry, University of Vienna, Vienna,
Austria and Department of Chemical Engineering,
Imperial College London, London, UK*

Bacterial cellulose (BC) is one of the strongest materials produced by nature, possessing high modulus and strength, estimated to be 114 GPa and in excess of 1500 MPa, respectively. It has been shown to be an effective nano-reinforcement for polymers to produce lightweight and mechanically strong nanocomposites when high loading fractions of BC were used. This chapter discusses the applications of BC in advanced polymeric materials. These materials include optically transparent nanocomposites and BC-reinforced, natural fiber-reinforced hierarchical composites. We also discuss the use of BC in the so-called "all-cellulose nanocomposites" and biomimetic BC-reinforced nanocomposites. The application of BC simultaneously as stabilizer and nano-reinforcement to produce macroporous polymers by mechanical frothing of acrylated epoxidized soybean oil is also discussed in this chapter.

11.1. Introduction

Cellulose is used in the paper,[1] pharmaceutical, and cosmetic industries,[2,3] explored as reinforcement for polymers[4–7] and natural fiber-reinforced polymer nanocomposites.[8–11] Numerous products are also derived from cellulose; technical textile fibers, such as viscose and Lyocell[12] and thermoplastic polymers,[13] such as cellulose acetate. Currently, much research activity and attention focused on the isolation and production of nanoscale cellulose fibers from woody materials. For comprehensive reviews on the production and application of nanocellulose, the readers are referred to publications by Klemm *et al.*[14] and Siró *et al.*[15] Interest in nanocellulose stems from the fact that nanoscale cellulose combines the physical and chemical properties of cellulose, such as hydrophilicity and the ability to be easily modified chemically, with other features such as high specific surface area and aspect ratio.

Nanocellulose can be obtained by two approaches: (i) top-down and (ii) bottom-up. The top-down approach involves the disintegration of (ligno)cellulose biomass, such as wood fibers into nanofibers. This technique was first reported by Wuhrmann *et al.*,[16] whereby ultrasonication was used to prepare nanocellulose from ramie, hemp, and cotton. Nanocellulose can also be produced from wood pulp by feeding the pulp through a high-pressure homogenizer to reduce the size of the fibers down to the nanometer scale.[17,18] A more recent method of producing nanocellulose from plant-based cellulosic fibers involves using grinders;[19] wood pulp is fibrillated by the high shear generated from passing through the slit between a static and rotating grindstone. The bottom-up approach, on the other hand, utilizes the fermentation of low molecular weight sugar using cellulose-producing bacteria such as from the *Acetobacter* species to produce nanocellulose,[2,20–22] also termed as microbial or bacterial cellulose (BC). BC is pure cellulose without the presence of hemicellulose, pectin, or lignin.[23] The cellulose is excreted by bacteria into an aqueous culture medium directly as nanofibers, with diameters ranging from 25 to 100 nm.[21,23] These nanofibers make up the pellicles in the culture medium.[21]

BC was first used as nano-reinforcement in polymer matrices by Gindl and Keckes.[24] The authors reinforced cellulose acetate butyrate (CAB) with different BC loading fractions. The tensile properties of the resulting BC-reinforced CAB nanocomposites improved by as much as five times compared to neat CAB (see Table 11.1). Since then, studies on utilizing BC as nano-reinforcement in polymer matrices have increased significantly over the years. The major driver for this is the potential of exploiting the high stiffness of cellulose crystals. X-ray diffraction, Raman spectroscopy, and numerical simulations estimated the stiffness of a cellulose crystal to be approximately 100–160 GPa,[25–27] which is highly desirable as reinforcing filler for polymer matrices. However, it is not clear what the true crystal modulus of cellulose or its maximum attainable stiffness as reinforcing filler is.[28] Nonetheless, nanocellulose has been shown to improve the mechanical properties of polymers. Table 11.2 summarizes the tensile properties of BC-reinforced polymer nanocomposites at various BC loadings and in various polymer matrices.

For a general overview of the area of nanocomposites, the readers are referred to publications by Khalil *et al.*,[39] Blaker *et al.*,[40] and Eichhorn *et al.*[28] This chapter is not intended to provide a general overview of BC nanocomposites research but the progress to date on the applications of BC in advanced (functional) materials.

Table 11.1. Tensile properties of BC-reinforced CAB nanocomposites. v_f, E, σ, and ε denote fiber volume fraction of BC, tensile modulus, tensile strength, and strain-at-break of the material. (Adapted from Ref. 24)

Samples	v_f (%)	E (GPa)	σ (MPa)	ε (%)
CAB	—	1.2	25.9	3.5
BC-reinforced CAB	10	3.2	52.6	3.5
	32	5.8	128.9	3.6

Table 11.2. The tensile properties of BC-reinforced polymer nanocomposites obtained from selected studies. w, $E_{polymer}$, $E_{composite}$, $\sigma_{polymer}$, and $\sigma_{composite}$ denote the BC loading by weight, tensile modulus of the neat polymer, tensile modulus of the nanocomposites, tensile strength of the neat polymer, and tensile strength of the nanocomposites, respectively.

Matrix	w (%)	$E_{polymer}$ (GPa)	$E_{composite}$ (GPa)	$\sigma_{polymer}$ (MPa)	$\sigma_{composite}$ (MPa)	Reference
Epoxy	70	—	20	—	325	29
Acrylic	5	0.025	0.355	4	21	30
Chitosan	5	1.6	2.9	35	60	31
	10		3.6		70	
TPS	1	0.001	0.004	0.5	1.0	32
	5		0.021		3.1	
PLLA	2	1.34 ± 0.04	1.75 ± 0.05	60.7 ± 0.08	57.5 ± 1.4	33
	5		1.89 ± 0.02		60.9 ± 0.5	
PLLA	4.3	2.0 ± 0.2	2.1 ± 0.2	27.7 ± 2.5	46.9 ± 2.7	34
	18		4.0 ± 0.1		115.2 ± 9.8	
PE	1[a]	0.771 ± 0.09	0.962 ± 0.068	25.5 ± 1.9	20.6 ± 1.3	35
	1[b]		1.011 ± 0.087		21.1 ± 0.8	
PLLA	11.5	0.687 ± 0.100	2.444 ± 0.285	16.05 ± 1.10	23.7 ± 1.9	36
Acrylic	1	0.16	0.14	7	7.2	37
	2.5		0.19		8.8	
	5.0		0.23		9.2	
	10.0		0.36		11.6	
ESO	25	0.45 ± 0.1	2.8 ± 0.4	5.5 ± 0.4	25 ± 0.4	38
	75		5.9 ± 0.5		81 ± 0.7	

PLLA, TPS, PE, and ESO denote poly(L-lactic acid), thermoplastic starch, polyethylene, and epoxidized soybean oil, respectively.
[a]BC used in freeze-dried form.
[b]BC used in pellicle form.

11.2. Optically transparent BC nanocomposites

Optically transparent and flexible polymers are of significant interest due to the rapidly expanding electronic industries. However, flexible polymers exhibit large coefficients of thermal expansion (CTE), in the order of 200 (ppm K^{-1}).[28] As a result, any functional materials deposited onto this substrate could be damaged due to the mismatch in CTE. Hence, it is essential to develop novel flexible and lightweight materials with a CTE close to glass ($\sim$8 ppm K^{-1}), which are optically transparent. In this context, reinforcing polymers with BC to produce nanocomposites is an ideal way forward. BC possesses low CTE[31] in the order of 0.1 ppm K^{-1} and high Young's modulus, estimated to be 114 GPa.[27] Yano *et al.*[29] first demonstrated that optically transparent nanocomposites can be produced by impregnating BC with an acrylic resin (see Fig. 11.1). The authors have managed to achieve BC loading fractions of up to 70 wt%. The optical transparency of the resulting materials arises from the fact that (i) BC possesses dimensions that are less than one-tenth of the wavelength of light and (ii) the refractive indices between the two materials match ensuring that light is not scattered by these nanocomposites.[41] More importantly, the resulting material possesses a very low CTE of only 3–6 ppm K^{-1}. The resulting nanocomposites had a Young's modulus and tensile strength of 21 GPa

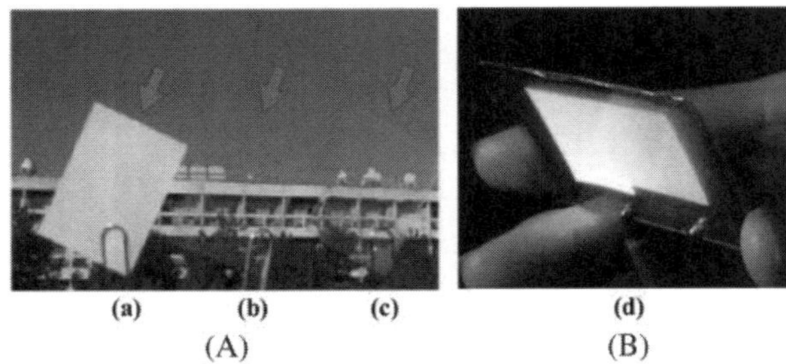

Fig. 11.1. The appearance of (A) 65 μm thick BC sheet, (B) BC sheet-reinforced acrylic resin, (C) BC sheet-reinforced epoxy resin, and (D) luminescence of an organic light-emitting diode (OLED) deposited onto a transparent BC nanocomposites. (Images A and B are reproduced from Ref. 29 and Ref. 30, respectively with kind permission from Wiley.)

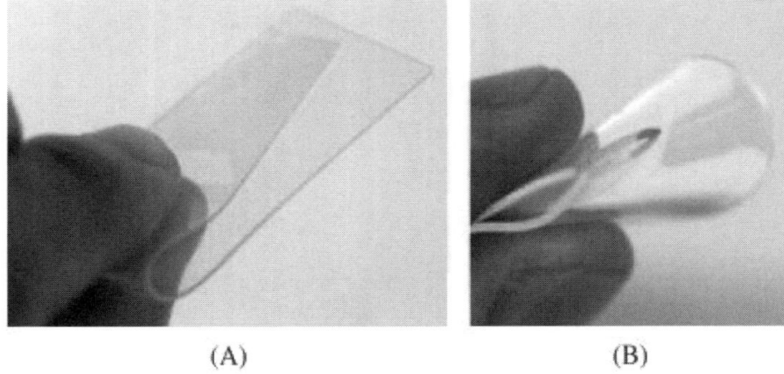

Fig. 11.2. Foldable transparent nanocomposites reinforced with BC (A) and the more fragile acrylic resin with the same thickness (B). (Obtained from Ref. 30 with kind permission from Wiley.)

and 325 MPa, respectively.[29] Moreover, the authors have also successfully deposited an electroluminescent layer onto the transparent BC nanocomposites (Fig. 11.1). Researchers from the same group,[42] also reported that by reducing the BC loading fraction in acrylic resin to 7.4 wt%, the light transmittance through the material reduced only by 2.4% compared to neat resin. The CTE of the resulting nanocomposites also reduced from 86 ppm K^{-1} (neat resin) to 38 ppm K^{-1}. A subsequent study by the same authors showed that it is to produce BC-reinforced acrylic nanocomposites with 5 wt% BC loading that possesses a CTE of only 4 ppm K^{-1}, similar to nanocomposites at 70 wt% BC loading.[30] This difference in CTE of the low BC loading-reinforced acrylic resin is attributed to the Young's modulus of the acrylic resin used. When acrylic resin with low Young's modulus was used, the induced thermal stresses in the resin are negligible and the CTE of the resulting nanocomposites is completely determined by the rigid BC network within the material. The resulting BC nanocomposites are also highly flexible (see Fig. 11.2) and possess Young's modulus and tensile strength of 355 MPa and 25 MPa, respectively.

The BC nanocomposites also exhibited a ductile behavior, with a measured strain-to-failure of up to 15%. Subsequent studies showed that acetylation of BC could reduce the moisture uptake of the resulting nanocomposites while the optical transparency of the resulting BC nanocomposites was retained.[43,44]

11.3. Hierarchical composites

11.3.1 *Creating hierarchical structures within composites*

Natural fibers have been considered for numerous composite applications[45] but they suffer from drawbacks such as dimensional inconsistency and variability in mechanical properties, even when harvested from a single cultivation.[46] Moreover, using natural fibers as reinforcement does not always improve the composites' properties.[47,48] One major reason is the anisotropicity of natural fibers themselves. In addition to this, the failure of natural fibers to deliver the desired performance in composites could be attributed to the high linear thermal coefficient of expansion (LTCE) of natural fibers, which results in a poor fiber–matrix interface.[49] Coating of natural fibers with BC could be a potential solution to this problem during thermal processing of the composites. The presence of BC on the natural fiber surface could potentially bridge the gap that forms between the fibers and the matrix due to the high LCTE of natural fibers. By culturing cellulose-producing bacteria in the presence of natural fibers in an appropriate culture medium, BC is preferentially deposited in culture onto the surface of natural fibers (see Fig. 11.3)[8,9,50,51]

Simple weight gain measurements showed that approximately 5 to 6 wt% of BC was deposited onto the surface of these natural fibers.[9] However, the mechanical properties of the natural fibers after BC modification depend on the type of natural fibers used (see Table 11.3). The modification process did not affect the mechanical properties of sisal fibers but the properties of hemp fibers were severely affected. This is attributed to the separation of technical bast fibers into smaller individual fibers as a result of the intrinsically non-cohesive structure of bast fibers during culturing. These BC-modified sisal and hemp fibers have also been used to produce unidirectional natural fiber-reinforced CAB and polylactide (PLLA) model composites.[8,51]

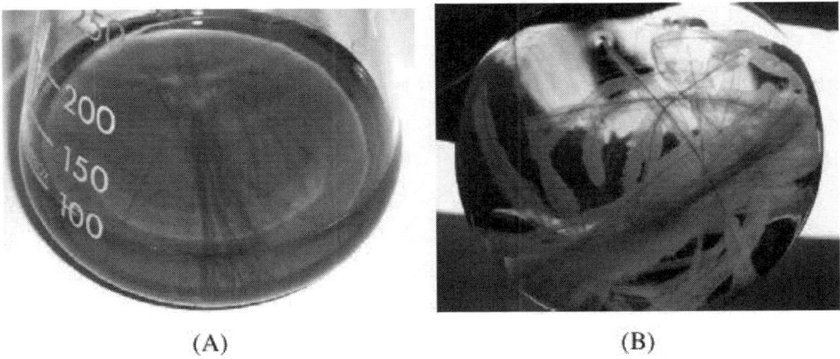

Fig. 11.3. Images showing (A) natural fibers immersed in a culture medium of *Gluconacetobacter xylinum* before bacteria culturing (B) the culture medium after 2 days. (Reprinted from Ref. 9 with kind permission from ACS publication.)

Table 11.3. The mechanical properties of natural fibers modified with BC nanofibrils. (Adapted from Ref. 9.)

Natural fibers	Young's modulus (GPa)	Tensile strength (MPa)	Elongation at break (%)
Neat sisal fiber	15.0 ± 1.2	342 ± 33	2.9 ± 0.1
BC-modified sisal fiber	12.5 ± 1.0	324 ± 33	4.5 ± 0.4
BC-modified sisal fiber with purification[a]	12.0 ± 0.9	310 ± 32	4.1 ± 0.5
Neat hemp fiber	21.4 ± 2.0	286 ± 31	2.0 ± 0.2
BC-modified hemp fiber	8.8 ± 0.7	171 ± 11	2.9 ± 0.2
BC-modified hemp fiber with purification[a]	8.0 ± 0.6	130 ± 12	2.9 ± 0.2

[a]Purification indicates the extraction of post-BC-modified sisal fibers with NaOH at 80°C.

Table 11.4. Mechanical properties of BC-modified hemp and sisal fibers-reinforced CAB and PLLA composites.

Composites	Neat fiber		Modified fiber		Improvements	
	σ (MPa)	E (GPa)	σ (MPa)	E (GPa)	σ (%)	E (%)
CAB/Hemp[a]	98.1 ± 12.7	8.5 ± 1.3	86.7 ± 13.6	5.8 ± 0.5	−12	−35
PLLA/Hemp[a]	110.5 ± 27.2	11.8 ± 4.2	104.8 ± 9.1	7.9 ± 1.2	−5	−33
CAB/Sisal[a]	92.9 ± 9.3	5.5 ± 0.5	100.4 ± 7.0	8.8 ± 1.4	8	59
PLLA/Sisal[a]	78.9 ± 14.7	7.9 ± 1.3	113.8 ± 14.0	11.2 ± 1.2	44	42
CAB/Hemp[b]	15.8 ± 2.2	1.9 ± 0.1	13.4 ± 1.4	0.6 ± 0.2	−15	−69
PLLA/Hemp[b]	13.4 ± 3.6	3.2 ± 0.2	13.3 ± 2.5	2.3 ± 0.3	−1	−28
CAB/Sisal[b]	10.9 ± 1.7	1.6 ± 0.1	14.4 ± 3.7	1.8 ± 0.3	32	15
PLLA/Sisal[b]	10.0 ± 3.1	2.1 ± 0.1	16.8 ± 4.1	3.1 ± 0.2	68	49

[a]The loading direction is parallel (0°) to the fibers.
[b]The loading direction is perpendicular (90°) to the fibers.

The mechanical properties of BC-coated sisal fiber-reinforced polymers showed significant improvements over neat polymers (Table 11.4). The tensile strength and modulus for sisal/PLLA composites improved by as much as 68% and 49%, respectively. However, improvements were not observed for composites containing BC-coated hemp fibers. The tensile strength and modulus decreased by as much as 15% and 69%, respectively for hemp/CAB composites. One should note that the fibers were damaged during bacteria culture and their tensile strength were only one-third of those of the original fibers.

Inspired by the potential of "hairy" natural fibers, a novel, simple, and cost-effective method based on slurry dipping to coat sisal fibers with nanosized BC without the need of bioreactors was developed.[52] This technique relies on the hydrophilic nature of natural fibers, which when immersed into dispersions of BC in water the fibers will draw in the water and BC nanofibrils from the medium. BC nanofibrils are filtered against the surface of sisal fibers, resulting in BC-coated sisal fibers (BC loading fraction on sisal fibers is approximately 10 wt%). The fast drying rate of the coated fibers under vacuum resulted in the collapse of BC nanofibrils onto the surface of sisal fibers. By pressing the wet BC-coated sisal fibers between filter

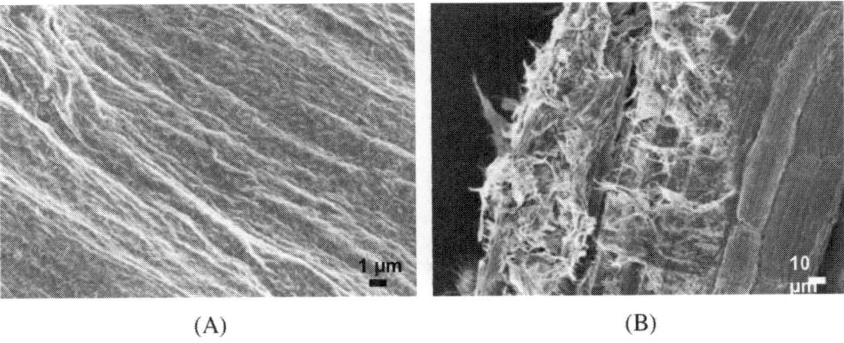

(A) (B)

Fig. 11.4. Scanning electron micrographs showing (A) sisal fibers coated with a dense layer of BC and (B) "hairy" sisal fibers produced using a novel slurry dipping method. A dense layer of BC on sisal fibers was obtained by drying the slurry-dipped fibers under vacuum 80°C. "Hairy" sisal fibers were obtained by partially drying the slurry-dipped fibers between filter papers, followed by drying in an air oven held at 40°C. (Obtained with kind permission from Elsevier.)

papers, the fibers were partially dried. The combination of capillary action with the slow drying of the coated fibers, which prevents the collapse of the nanofibrils, results in a BC coating in which BC nanofibrils were oriented perpendicular to the sisal surface. The morphology resembles true "hairy fibers" (Fig. 11.4).

The tensile properties of randomly oriented short (BC-coated) sisal fiber-reinforced PLLA composites have also been studied.[52] Two different types of hierarchical composites were prepared; (i) BC-coated sisal fiber-reinforced PLLA and (ii) BC-coated sisal fiber-reinforced PLLA–BC nanocomposites. The former composites contained BC on the surface of sisal fibers only and the latter composites contained BC both on the fiber surfaces and dispersed within the PLLA matrix. It can be seen from Fig. 11.5 that with (BC-coated) sisal fibers as reinforcement, the tensile moduli for all composites increased.

The increase in the tensile modulus of the hierarchical composites was enhanced when BC was additionally dispersed in the matrix. This is thought to be due to the stiffening of the matrix by BC.[33] With BC dispersed in the matrix and attached to the fibers, both the matrix and the fiber–matrix interface could be reinforced (or stiffened). The tensile strength of the hierarchical composites showed a slightly different trend compared to tensile modulus. A decrease in tensile strength was observed when PLLA is reinforced with (BC-coated) sisal fibers, with no BC dispersed in the matrix. When the hierarchical composites were reinforced with BC dispersed in the PLLA matrix, the tensile strength improved by as much as 11% when compared to neat PLLA and 21% when compared to BC-coated sisal fiber-reinforced PLLA composites without BC dispersed in the matrix. This could be due to enhanced interfacial adhesion between BC-coated fibers and BC-reinforced PLLA matrix. With BC dispersed in the matrix, the matrix is stiffened.

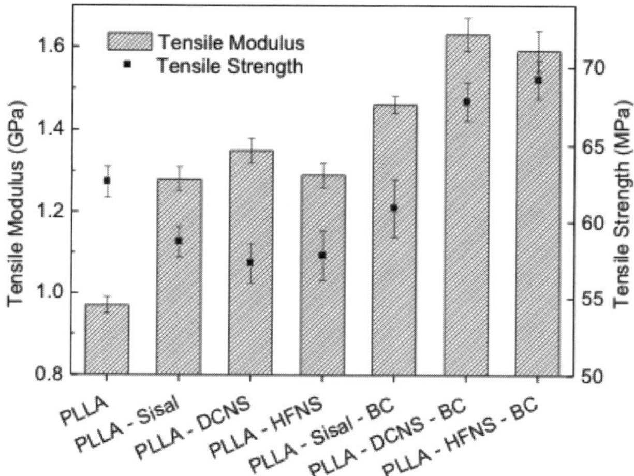

Fig. 11.5. Tensile properties of (hierarchical) sisal fiber-reinforced PLLA (nano)composites. PLLA–sisal, PLLA–DCNS, and PLLA–HFNS denote PLLA (nano)composites reinforced with 20 wt% neat sisal fibers, densely-coated neat sisal fibers, and "hairy" fibers of neat sisal, respectively. PLLA–sisal–BC, PLLA–DCNS–BC, and PLLA–HFNS–BC represent PLLA nanocomposites reinforced with 15 wt% neat sisal fibers, densely-coated neat sisal fibers and "hairy" fibers of neat sisal, respectively, with 5 wt% BC dispersed in the matrix. (Adapted from Ref. 54.)

11.3.2 *Utilizing BC as binder to produce robust fiber preforms*

Hornification is a physical phenomenon observed when making paper,[53] whereby irreversible formation of hydrogen bonds occur during the drying of wood pulp, which consist mainly of cellulose. When this hornified cellulose is re-suspended in water, the cellulose will not regain its original well-dispersed state in water. Similarly, BC, which consists of pure cellulose, will also undergo hornification and this will produce high modulus and high strength papers.[54] This effect of hornification can be combined with the previously described slurry dipping method to create non-woven natural fiber preforms using a papermaking process.[55] However, instead of just dipping the sisal fibers into a water dispersion of BC, the dispersion of sisal fibers–BC can be vacuum filtered, wet pressed, and dried to produce rigid fiber preforms using BC as binder for the production of composite materials. With BC as the binder for the sisal fibers, a tensile strength (defined as the maximum load required to break the sample per unit width of the specimen as the cross-sectional area of the fiber mat) of 13.1 kN m^{-1} was achieved for a 10 wt% BC loading in sisal fibers. However, the tensile strength of the neat sisal fiber preforms without BC binder was not measureable. This is due to the fact that these sisal fibers are loose and held together only by friction between the fibers even after the wet pressing step to consolidate them into fiber preforms. The improved mechanical performance of BC–sisal fiber preforms can be attributed to the use of BC as the binder, which also promotes fiber–fiber stress transfer. The nanosized BC holds the otherwise loose sisal fibers together due to hornification (irreversible hydrogen bonding between the nanocellulose).[53] The high tensile strength of the BC network, which formed in

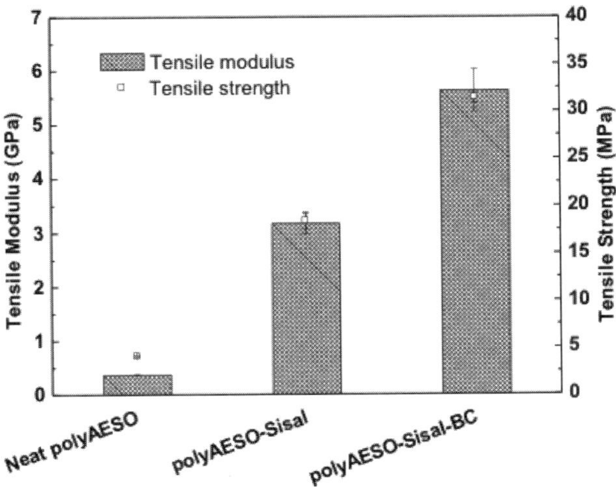

Fig. 11.6. Tensile properties of neat polyAESO and sisal fiber preform-reinforced polyAESO (hierarchical) composites. polyAESO–sisal and polyAESO–sisal–BC indicate composites reinforced with neat sisal fiber preform and sisal fiber preform using BC as binder, respectively. (Adapted from Ref. 57.)

between the sisal fibers, provided the mechanical performance of the manufactured BC–sisal fiber preforms.

The natural fiber preforms were infused with acrylated epoxidized soybean oil (AESO) and polymerized to produce sisal fiber-reinforced hierarchical composites.[55] The tensile properties of neat polyAESO (polymerized AESO) and the (hierarchical) composites reinforced with sisal fiber preforms are shown in Fig. 11.6.

The fiber volume fractions of sisal–polyAESO and BC–sisal–polyAESO is 4.0%. When sisal fibers were used as reinforcement for polyAESO, the tensile modulus improved from 0.4 GPa for neat polyAESO to 3.2 GPa for 4.0% sisal fiber-reinforced polyAESO composites. A further improvement of the tensile modulus of the composites from 3.2 GPa to 5.6 GPa was achieved when BC was used as the binder for the natural fiber preform. This is thought to be due to the stiffening of the polymer matrix when the fiber preform contained a hornified network of BC. A similar trend was observed for the tensile strength of the (hierarchical) composites. This significant improvement when BC–sisal fiber preforms were used to create hierarchical composites can be attributed to (i) the enhanced fiber–matrix interaction and (ii) enhanced fiber–fiber stress transfer. The use of BC as binder for the fibers resulted in the formation of continuous but hornified BC network, encasing sisal fibers bonding them together. It is postulated that this enhances the fiber–fiber stress transfer compared to sisal fiber-only preforms, where the fibers are mostly isolated. In addition to this, it has been shown that using BC as binder enhances the tensile properties of the BC–sisal fiber preforms compared to sisal fiber preforms. This translates to the improved tensile strength of the manufactured BC–sisal–polyAESO.

11.4. All-cellulose nanocomposites

Recent advances in polymer processing technologies have focused on mono-material designs, leading to the production of "all-polymer composites" or "self-reinforced polymers".[56–59] The essence of "all-polymer composites" or "self-reinforced polymers" lies in the compaction process, whereby a fraction of the surface of each fiber is melted under low contact pressure and subsequently consolidated at a higher pressure for a short period of time to consolidate the material.[60] Upon cooling of the material, the polymer recrystallizes and binds the fibers together, very much like the resin in a fiber-reinforced composites system. All-polyethylene[56–60] and all-polypropylene[60] composites have been produced with great commercial success. By following a similar concept, self-reinforced cellulose composites can also be produced. However, it should be noted that unlike thermoplastics, cellulose does not possess melting temperatures (T_{m}) as it degrades prior to reaching its T_{gm}. Two approaches can be taken: (i) impregnating cellulose fibers with a cellulose matrix, followed by its subsequent regeneration and (ii) selective dissolution of the cellulose fiber surface to bond the fibers together.[28] The approach mentioned in (i) was first demonstrated by Nishino *et al.*,[61] whereby Kraft pulp was dissolved in a solution containing 8 wt% LiCl in dimethylacetamide (DMAc). This solution was then infused into ramie fibers and the solvent was removed by solvent exchange with methanol and subsequently air-dried at room temperature. This creates a composite material whereby both the fiber and the matrix are cellulose. Approach (ii) has also been demonstrated by Nishino *et al.*;[62] filter papers were first immersed in distilled water, followed by acetone and DMAc, respectively and then immersed in the same cellulose dissolving solvent (8 wt% LiCl in DMAc) to selectively dissolve the fiber. Methanol was then used to extract the solvent and the all-cellulose composites were dried. This concept resembles the previously described self-reinforced polymer. The selective dissolution of the cellulose fiber surface followed by cellulose regeneration forms the matrix whilst the fiber core retains its original structure and acts as reinforcement for the regenerated cellulose matrix (Fig. 11.7). All-cellulose nanocomposites using BC[63,64] and regenerated cellulose fibers[65–67] as the starting cellulose can also be prepared using similar manner.

The tensile properties of the resulting all-cellulose nanocomposites are tabulated in Table 11.5. These results showed that a short dissolution time of 10 min is the optimum, giving rise to the highest tensile strength (411 MPa) and good Young's

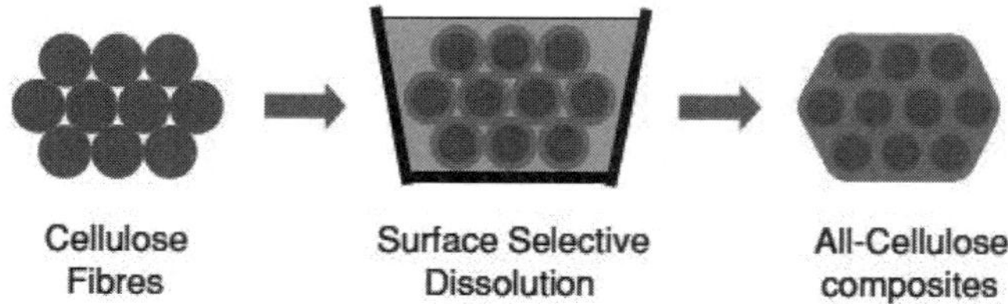

Fig. 11.7. The process of selective dissolution of the cellulose fiber surface to produce all-cellulose composites. (Image obtained from Ref. 65 with kind permission from Springer.)

Table 11.5. The tensile properties of all-cellulose nanocomposites. E and σ denote the tensile modulus and strength, respectively, of the materials. (Data obtained from Ref. 63.)

Materials	E (GPa)	σ (MPa)	Work of fraction ($MJ\,m^{-3}$)
Neat BC	20 ± 1.7	395 ± 36	7.3 ± 1.87
BC-5 min	16 ± 1.0	392 ± 23	8.2 ± 1.1
BC-10 min	18 ± 0.8	411 ± 22	9.0 ± 1.1
BC-15 min	16 ± 0.6	309 ± 25	5.5 ± 1.2
BC-20 min	11 ± 0.6	188 ± 19	3.5 ± 0.6
BC-40 min	4.4 ± 0.3	97 ± 15	6.0 ± 1.4
BC-60 min	2.6 ± 0.4	75 ± 15	15.8 ± 4.1

modulus (18 GPa) value. This optimum point corresponds to adequate dissolution of BC to bond the fibers together whilst not disrupting the fibrous network and properties within the composites.[63] At shorter dissolution period, the BC surface has not been dissolved enough to form the matrix. At longer dissolution period on the other hand, the crystals of BC dissolves away, which then lead to poorer mechanical performance of the nanocomposites.

Comparing this all-cellulose nanocomposites to the all-cellulose composites made with filter papers, the nanocomposites possess higher tensile modulus at short dissolution time compared to the composites. This is attributed to the presence of stiff BC nanofibrils in the nanocomposites. The trend in the tensile strength, on the other hand, differs significantly between the all-cellulose nanocomposites and all-cellulose composites. Table 11.5 shows that the tensile strength of the nanocomposites decreases with increasing dissolution time but the tensile strength of the composites increases from 50 MPa to nearly 250 MPa with increasing dissolution time. Filter papers are made of loosely bound fibrous network with large voids. It was postulated that the dissolution process led to the filling of these voids and improve the tensile strength of the composites.

11.5. Biomimetic BC-reinforced nanocomposites

Inspiration to manufacture high-performance cellulose nanocomposites can be obtained from nature. Wood, for example, is a composite material consisting of cellulose, hemicellulose, lignin, and pectin.[46] Cellulose is the reinforcing agent and lignin serves as the matrix. Hemicellulose coats cellulose within plant cell wall and function as "Velcro hook" between cellulose and lignin.[68] This configuration gives wood its mechanical rigidity. However, despite our understanding of a woody structure, translating this into engineering materials still remains a challenge. The biomimetic coating of cellulose often fails to deliver the required mechanical performance as a result of the tendency of cellulose to form aggregates. Moreover, the cost involved in the isolation of nanocellulose from woody materials (fibrillation or homogenization processes) is often very energy consuming,[14] which further limits the development of biomimetic coating of nanocellulose. It was recently shown that nanocellulose obtained from woody materials does not differ much in its reinforcing ability in a composite structure compared to BC.[54] This comes as a surprise given the vastly different crystallinity and anticipated mechanical properties of the two types of nanocellulose. Compared to nanocellulose obtained from woody materials, the ribbons of BC can be modified during biosynthesis[69–72] and this approach can be taken to produce truly bio-inspired high-performance engineering materials.

Hydroxyethyl cellulose (HEC) can be introduced into the culture medium during the biosynthesis of cellulose by *Acetobacter aceti*.[73] After culturing for 7 days, the BC-containing HEC (BCHEC) was removed, purified with dilute alkaline solution, and pressed to obtain BC-reinforced HEC composite. This biomimetic nanocomposite contains 80 wt% BC loading. X-ray diffraction shows the broadening of 110, $1\bar{1}0$, and 200, indicating a decrease in crystallite size. These results suggest an incomplete aggregation of cellulose when HEC is introduced into the culture medium. The introduction of HEC is postulated to reduce the extent of co-crystallization between the elementary cellulose fibrils. When comparing this biomimetic composites to conventional BC-reinforced HEC composites with 80 wt% BC loading not prepared *in situ* (BC/HEC), both the tensile modulus and strength of BCHEC are more superior than BC/HEC (see Table 11.6). The remarkable tensile properties of BCHEC can be attributed to the improved degree of fibril dispersion within the nanocomposites, which is induced by coating individual BC nanofibrils with HEC.

Another water-soluble polymer, polyvinyl alcohol (PVA), has also been explored. PVA has attracted a lot of interest due to its solubility in water, good mechanical properties, and biocompatibility. Gea *et al.*[74] have cultured BC in the presence of PVA *in situ* (BC/PVA) and compared that to BC impregnated with PVA (BCPVA). After purification, the BC loading was estimated to be 96.3 wt% for BC/PVA whilst the BCPVA possessed BC loading of 98.6 wt%. It can be seen from Table 11.6 that the tensile modulus of BC/PVA is less than that of BCPVA. This could be attributed to the lower BC loading fraction in the nanocomposites prepared *in situ*. The tensile strength, on the other hand showed a different trend. BC/PVA possesses higher tensile strength of ~210 MPa, closer to neat BC sheet, compared to ~110 MPa of BCPVA. This is attributed to the aforementioned

Table 11.6. The tensile properties of BC reinforced HEC biomimetic nanocomposites. (Data obtained from Ref. 73 and Ref. 74. E and σ denote the Young's modulus and tensile strength, respectively.)

Sample	E (GPa)	σ (MPa)	Work of fracture ($\mathrm{MJ\,m^{-3}}$)
BC sheet	12.5 ± 0.3	225.6 ± 3.7	10.7 ± 0.5
BCHEC	12.5 ± 0.7	289.4 ± 13.87	11.0 ± 1.0
BC/HEC[a]	8.25 ± 0.3	178.0 ± 5.0	8.1 ± 0.4
BC-PVA[b]	9.1	110	
BC-PVA[c]	6.4	210	

[a]80 wt% BC-reinforced HEC, not prepared *in situ* in the culture medium.
[b]BC-reinforced PVA prepared *in situ*. This composite consists of 96.3 wt% BC loading.
[c]BC-reinforced PVA prepared by impregnation method. This composite consists of 98.6 wt% BC loading.

reasons. More importantly, the authors found that BCPVA prepared *in situ* possesses more characteristics of PVA, with a plastic deformation process with necking, followed by a strain-hardening region with a modulus of 14.7 GPa. It is postulated that the co-culture of BC with PVA leads to better coating of individual nanofibers and to a certain degree. This disruption of hydrogen bonding network between BC promotes slippage within the BC network.

11.6. BC in macroporous polymer nanocomposites

In addition to the application in advanced fiber composites, BC can also be used as thickener and stabilizer for the production of polymer foams. Blomfeldt *et al.*[75] introduced 1 wt% BC into wheat gluten foams. The foams were prepared by dispersing gluten in water, freezing the suspension at −25°C for 6 h and subsequently freeze-dry. The BC is introduced to increase the viscosity of the suspension and reduce the variation in the foam density.[76] In addition to this, the BC also acts as a stiffening agent, enhancing the mechanical performance of the resulting nanocomposite foams. BC increased the compression modulus of the foams by as much as 50% (2.1 MPa for neat foam to 3.1 MPa for 1wt% BC in the foam).[75]

BC was also used simultaneously as a thickening and reinforcing agent for soybean oil-based macroporous polymers.[77] Oil foams were produced by mechanically frothing the viscous AESO to introduce air into the system. The foam template was subsequently polymerized using microwave irradiation. When BC was introduced into AESO at a concentration of as low as 0.5 wt%, the stability indices of the oil foam increased significantly; no observable phase separation occurred for 2 months after AESO was frothed (Fig. 11.8).

Neat AESO without BC, on the other hand, phase separated almost completely 1 month after preparation. BC in the liquid foam is postulated to aggregate in the

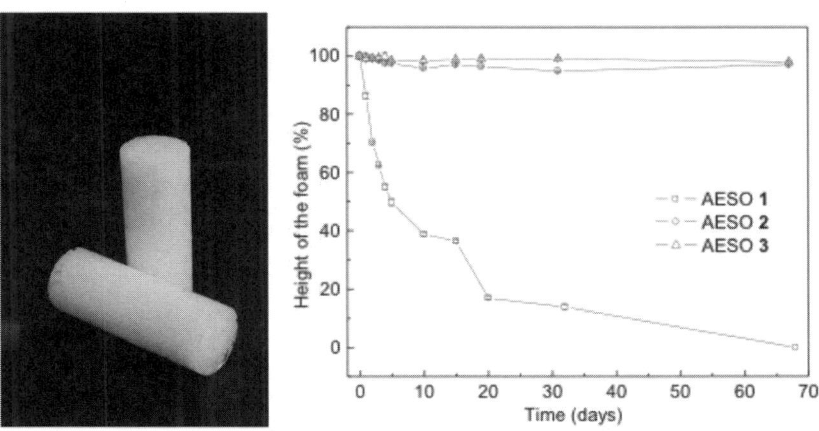

Fig. 11.8. The image on the left shows the air-templated polyAESO and the graph on the right shows the stability indices of the foams. AESO 1, AESO 2, and AESO 3 indicate 0 wt% BC, 0.5 wt% BC, and 1 wt% BC, respectively. (Obtained with kind permission from RSC publishing.)

plateau borders, obstructing the flow of liquid from the lamella. This ultimately hinders the coalescence of the air bubbles in the liquid foam. The specific compression modulus of the foam increased from 253 MPa $kg^{-1} m^3$ to 338 MPa $kg^{-1} m^3$ (0.5 wt% BC) but decreased to 90 MPa $kg^{-1} m^3$ (1 wt% BC). The decrease in the modulus is attributed to the poor pore structure of the resulting macroporous polymer containing high BC loading, which is a direct result of difficulties in frothing the highly viscous AESO.

11.7. Conclusions

Recent interests in nanomaterials have sparked the use of BC in composite applications. It has been shown that BC is able to reinforce polymers to produce nanocomposites with improved properties. The applications of BC in advanced polymer materials are discussed in this chapter. BC can be used to fabricate optically transparent high modulus and strength nanocomposites. Electroluminescent materials can be deposited onto these composites to manufacture the next generation OLED. BC has also been shown to improve the mechanical performance of conventional natural fiber-reinforced composites by mechanical interlocking and matrix stiffening. It can also be used as a binder to bind the otherwise loose fibers into a rigid fiber preform, which has improved downstream handling strength for composites manufacturing.

Lessons can be learned from the area of polymer materials to produce "all-cellulose nanocomposites". These nanocomposites exhibit excellent mechanical performance as the full strength and stiffness of the BC sheet can be exploited. Inspired by nature, BC-based biomimetic nanocomposites can also be manufactured. These composites are produced by introducing water-soluble polymers into the culture medium during the biosynthesis of BC. This produces exfoliated BC within the nanocomposites, similar to a natural composite — wood. In addition to these, BC can also be used as a stabilizer and thickener for the manufacturing of macroporous polymers, moving from the conventional use in advanced fiber composites. The potential of BC is endless and only limited by a researcher's imagination. It is believed that BC is one of the most promising materials for the manufacturing of the next generation of advanced fiber-reinforced renewable composites that are lightweight yet mechanically stable materials.

Acknowledgments

The authors greatly appreciate funding provided by the University of Vienna for KYL.

References

1. A. Sczostak, Cotton linters: an alternative cellulosic raw material, *Macromol. Symp.* **280**(2009) 45–53.
2. D. Klemm, B. Heublein, H.P. Fink and A. Bohn, Cellulose: fascinating biopolymer and sustainable raw material, *Angew. Chem.Int. Ed.* **44**(22) (2005) 3358–3393.

3. S. Kamel, N. Ali, K. Jahangir, S.M. Shah and A.A. El-Gendy, Pharmaceutical significance of cellulose: a review, *eXPRESS Polym. Lett.* **2**(11) (2008) 758–778.
4. J.K. Pandey, S.H. Ahn, C.S. Lee, A.K. Mohanty and M. Misra, Recent advances in the application of natural fiber based composites, *Macromol. Mater. Eng.* **295**(11) (2010) 975–989.
5. K.G. Satyanarayana, G.G.C. Arizaga and F. Wypych, Biodegradable composites based on lignocellulosic fibers-an overview, *Prog. Polym. Sci.* **34**(9) (2009) 982–1021.
6. J.J. Blaker, K.Y. Lee and A. Bismarck, Hierarchical composites made entirely from renewable resources, *J. Biobased Mater. Bioenergy* **5**(1) (2011) 1–16.
7. K.Y. Lee, J.J. Blaker and A. Bismarck, Surface functionalisation of bacterial cellulose as the route to produce green polylactide nanocomposites with improved properties, *Compos. Sci. Technol.* **69**(15–16) (2009) 2724–2733.
8. J. Juntaro, M. Pommet, G. Kalinka, A. Mantalaris, M.S.P. Shaffer and A. Bismarck, Creating hierarchical structures in renewable composites by attaching bacterial cellulose onto sisal fibers, *Adv. Mater.* **20**(16) (2008) 3122–3126.
9. M. Pommet, J. Juntaro, J.Y.Y. Heng, A. Mantalaris, A.F. Lee, K. Wilson, G. Kalinka, M.S.P. Shaffer and A. Bismarck, Surface modification of natural fibers using bacteria: depositing bacterial cellulose onto natural fibers to create hierarchical fiber reinforced nanocomposites, *Biomacromolecules* **9**(6) (2008) 1643–1651.
10. K.Y. Lee, P. Bharadia, J.J. Blaker and A. Bismarck, Short sisal fibre reinforced bacterial cellulose polylactide nanocomposites using hairy sisal fibres as reinforcement, *Compos. Pt. A-Appl. Sci. Manuf.* **43**(11) (2012) 2065–2074.
11. K.Y. Lee, K.K.C. Ho, K. Schlufter and A. Bismarck, Hierarchical composites reinforced with robust short sisal fibre preforms utilising bacterial cellulose as binder, *Compos. Sci. Technol.* **72**(13) (2012) 1479–1486.
12. H. Harms, Lenzing LYOCELL — chances of a new generation of manmade fibres, *Materialwiss. Werkstofftech.* **34**(3) (2003) 267–271.
13. K.J. Edgar, C.M. Buchanan, J.S. Debenham, P.A. Rundquist, B.D. Seiler, M.C. Shelton and D. Tindall, Advances in cellulose ester performance and application, *Prog. Polym. Sci.* **26**(9) (2001) 1605–1688.
14. D. Klemm, F. Kramer, S. Moritz, T. Lindstrom, M. Ankerfors, D. Gray and A. Dorris, Nanocelluloses: a new family of nature-based materials, *Angew. Chem. Int. Ed.* **50**(24) (2011) 5438–5466.
15. I. Siro and D. Plackett, Microfibrillated cellulose and new nanocomposite materials: a review, *Cellulose* **17**(3) (2010) 459–494.
16. K. Wuhrmann, A. Heuberger and K. Mühlethaler, Elektronenmikroskopische untersuchungen an zellulosefasern nach behandlung mit ultraschall, *Experientia* **2**(3) (1946) 105–107.
17. F.W. Herrick, R.L. Casebier, J.K. Hamilton and K.R. Sandberg, *Microfibrillated cellulose: morphology and accessibility*, New York (1983).
18. A.F. Turbak, F.W. Snyder and K.R. Sandberg, Microfibrillated cellulose, a new cellulose product: properties, uses and commercial potential, *J. Appl. Polym. Sci. Appl. Polym. Symp.* **37**(1983) 459–494.
19. S. Iwamoto, A.N. Nakagaito, H. Yano and M. Nogi, Optically transparent composites reinforced with plant fiber-based nanofibers, *Appl. Phys. A-Mater. Sci. Process.* **81**(6) (2005) CP8–1112.
20. A.J. Brown, The chemical action of pure cultivations of bacterium aceti, *J. Chem. Soc. Trans.* **49**(1886) 172–187.
21. R. Jonas and L.F. Farah, Production and application of microbial cellulose, *Polym. Degrad. Stabil.* **59**(1–3) (1998) 101–106.
22. P. Ross, R. Mayer and M. Benziman, Cellulose biosynthesis and function in bacteria, *Microbiol. Rev.* **55**(1) (1991) 35–58.

23. M. Iguchi, S. Yamanaka and A. Budhiono, Bacterial cellulose — a masterpiece of nature's arts, *J. Mater. Sci.* **35**(2) (2000) 261–270.
24. W. Gindl and J. Keckes, Tensile properties of cellulose acetate butyrate composites reinforced with bacterial cellulose, *Compos. Sci. Tech.* **64**(15) (2004) 2407–2413.
25. S.J. Eichhorn and G.R. Davies, Modelling the crystalline deformation of native and regenerated cellulose, *Cellulose* **13**(3) (2006) 291–307.
26. M. Matsuo, C. Sawatari, Y. Iwai and F. Ozaki, Effect of orientation distribution and crystallinity on the measurement by X-ray-diffraction of the crystal-lattice moduli of cellulose-I and cellulose-II, *Macromolecules* **23**(13) (1990) 3266–3275.
27. Y.C. Hsieh, H. Yano, M. Nogi and S.J. Eichhorn, An estimation of the Young's modulus of bacterial cellulose filaments, *Cellulose* **15**(4) (2008) 507–513.
28. S.J. Eichhorn, A. Dufresne, M. Aranguren, N.E. Marcovich, J.R. Capadona, S.J. Rowan, C. Weder, W. Thielemans, M. Roman, S. Renneckar, W. Gindl, S. Veigel, J. Keckes, H. Yano, K. Abe, M. Nogi, A.N. Nakagaito, A. Mangalam, J. Simonsen, A.S. Benight, A. Bismarck, L.A. Berglund and T. Peijs, Review: current international research into cellulose nanofibres and nanocomposites, *J. Mater. Sci.* **45**(1) (2010) 1–33.
29. H. Yano, J. Sugiyama, A.N. Nakagaito, M. Nogi, T. Matsuura, M. Hikita and K. Handa, Optically transparent composites reinforced with networks of bacterial nanofibers, *Adv. Mater.* **17**(2) (2005) 153–155.
30. M. Nogi and H. Yano, Transparent nanocomposites based on cellulose produced by bacteria offer potential innovation in the electronics device industry, *Adv. Mater.* **20**(10) (2008) 1849–1852.
31. S.C.M. Fernandes, L. Oliveira, C.S.R. Freire, A.J.D. Silvestre, C.P. Neto, A. Gandini and J. Desbrieres, Novel transparent nanocomposite films based on chitosan and bacterial cellulose, *Green Chem.* **11**(12) (2009) 2023–2029.
32. I.M.G. Martins, S.P. Magina, L. Oliveira, C.S.R. Freire, A.J.D. Silvestre, C.P. Neto and A. Gandini, New biocomposites based on thermoplastic starch and bacterial cellulose, *Compos. Sci Tech.* **69**(13) (2009) 2163–2168.
33. K.-Y. Lee, J.J. Blaker and A. Bismarck, Surface functionalisation of bacterial cellulose as the route to produce green polylactide nanocomposites with improved properties, *Compos. Sci. Tech.* **69**(15–16) (2009) 2724–2733.
34. F. Quero, M. Nogi, H. Yano, K. Abdulsalami, S.M. Holmes, B.H. Sakakini and S.J. Eichhorn, Optimization of the mechanical performance of bacterial cellulose/poly(L-lactic) acid composites, *ACS Appl. Mater. Interfaces* **2**(1) (2010) 321–330.
35. R.J. Gu, B.V. Kokta, K. Frankenfeld and K. Schlufter, Bacterial cellulose reinforced thermoplastic composites: preliminary evaluation of fabrication and performance, *BioResources* **5**(4) (2010) 2195–2207.
36. Z.Q. Li, X.D. Zhou and C.H. Pei, Preparation and characterization of bacterial cellulose/polylactide nanocomposites, *Polym.Plast. Technol. Eng.* **49**(2) (2010) 141–146.
37. E. Trovatti, L. Oliveira, C.S.R. Freire, A.J.D. Silvestre, C.P. Neto, J. Pinto and A. Gandini, Novel bacterial cellulose-acrylic resin nanocomposites, *Compos. Sci. Tech.* **70**(7) (2010) 1148–1153.
38. A. Retegi, I. Algar, L. Martin, F. Altuna, P. Stefani, R. Zuluaga, P. Ganan and I. Mondragon, Sustainable optically transparent composites based on epoxidized soybean oil (ESO) matrix and high contents of bacterial cellulose (BC), *Cellulose* **19**(1) (2012) 103–109.
39. H. Khalil, A.H. Bhat and A.F.I. Yusra, Green composites from sustainable cellulose nanofibrils: a review, *Carbohydr. Polym.* **87**(2) (2012) 963–979.

40. J.J. Blaker, K.Y. Lee and A. Bismarck, Hierarchical composites made entirely from renewable resources, *J. Biobased Mater. Bioenergy* **5** (2011) 1–16.
41. B.M. Novak, Hybrid nanocomposite materials – between inorganic glasses and organic polymers, *Adv. Mater.* **5**(6) (1993) 422–433.
42. M. Nogi, S. Ifuku, K. Abe, K. Handa, A.N. Nakagaito and H. Yano, Fiber-content dependency of the optical transparency and thermal expansion of bacterial nanofiber reinforced composites, *Appl. Phys. Lett.* **88**(13) (2006).
43. S. Ifuku, M. Nogi, A. Kentaro, H. Keishin, F. Nakatsubo and H. Yano, Surface modification of bacterial cellulose nanofibers for property enhancement of optical transparent composites: dependence on acetyl-group DS, *Biomacromolecules* **8**(6) (2007) 1937–1978.
44. M. Nogi, K. Abe, K. Handa, F. Nakatsubo, S. Ifuku and H. Yano, Property enhancement of optical transparent bionanofiber composites by acetylation, *Appl. Phys. Lett.* **89**(23) (2006) 233123.
45. A.K. Mohanty, M. Misra and G. Hinrichsen, Biofibres, biodegradable polymers and biocomposites: an overview, *Macromol. Mater. Eng.* **276**(3–4) (2000) 1–24.
46. A. Bismarck, S. Mishra and T. Lampke, In *Natural Fibers, Biopolymers and Biocomposites*, A. K. Mohanty, M. Misra and L.T. Drzal (Eds.) Boca Raton: CRC Press (2005) 1st edn. Ch. 2.
47. F.R. Cichocki and J.L. Thomason, Thermoelastic anisotropy of a natural fiber, *Compos. Sci. Tech.* **62**(5) (2002) 669–678.
48. C. Baley, Y. Perrot, F. Busnel, H. Guezenoc and P. Davies, Transverse tensile behaviour of unidirectional plies reinforced with flax fibres, *Mater. Lett.* **60**(24) (2006) 2984–2987.
49. J.L. Thomason, *Why are natural fibres failing to deliver on composite performance?* Edinburgh (2009).
50. J. Juntaro, PhD thesis, Imperial College London, 2009.
51. J. Juntaro, M. Pommet, A. Mantalaris, M. Shaffer and A. Bismarck, Nanocellulose enhanced interfaces in truly green unidirectional fibre reinforced composites, *Compos. Interface* **14**(7–9) (2007) 753–762.
52. K.-Y. Lee, P. Bharadia, J.J. Blaker and A. Bismarck, Short sisal fibre reinforced bacterial cellulose polylactide nanocomposites using hairy sisal fibres as reinforcement, *Compos. Pt. A-Appl. Sci. Manuf* **43**(11) (2012) 2065–2074.
53. J. Diniz, M.H. Gil and J. Castro, Hornification — its origin and interpretation in wood pulps, *Wood Sci. Technol.* **37**(6) (2004) 489–494.
54. K.Y. Lee, T. Tammelin, K. Schulfter, H. Kiiskinen, J. Samela and A. Bismarck, High performance cellulose nanocomposites: comparing the reinforcing ability of bacterial cellulose and nanofibrillated cellulose, *ACS Appl. Mater. Interfaces* **4**(8) (2012) 4078–4086.
55. K.-Y. Lee, K.K.C. Ho, K. Schlufter and A. Bismarck, Hierarchical composites reinforced with robust short sisal fibre preforms utilising bacterial cellulose as binder, *Compos. Sci. Tech.* **72**(13) (2012) 1479–1486.
56. P.J. Hine, I.M. Ward, R.H. Olley and D.C. Bassett, The hot compaction of high modulus melt-spun polyethylene fibers, *J. Mater. Sci.* **28**(2) (1993) 316–324.
57. M.A. Kabeel, D.C. Bassett, R.H. Olley, P.J. Hine and I.M. Ward, Compaction of high-modulus melt-spun polyethylene fibers at temperatures above and below the optimum, *J. Mater. Sci.* **29**(18) (1994) 4694–4699.
58. M.A. Kabeel, D.C. Bassett, R.H. Olley, P.J. Hine and I.M. Ward, Differential melting in compacted high-modulus melt-spun polyethylene fibers, *J. Mater. Sci.* **30**(3) (1995) 601–606.

59. R.H. Olley, D.C. Bassett, P.J. Hine and I.M. Ward, Morphology of compacted polyethylene fibers, *J. Mater. Sci.* **28**(4) (1993) 1107–1112.
60. I.M. Ward and P.J. Hine, Novel composites by hot compaction of fibers, *Polym. Eng. Sci.* **37**(11) (1997) 1809–1814.
61. T. Nishino, I. Matsuda and K. Hirao, All-cellulose composite, *Macromolecules* **37**(20) (2004) 7683–7687.
62. T. Nishino and N. Arimoto, All-cellulose composite prepared by selective dissolving of fiber surface, *Biomacromolecules* **8**(9) (2007) 2712–2716.
63. N. Soykeabkaew, C. Sian, S. Gea, T. Nishino and T. Peijs, All-cellulose nanocomposites by surface selective dissolution of bacterial cellulose, *Cellulose* **16**(3) (2009) 435–444.
64. A. Abbott and A. Bismarck, Self-reinforced cellulose nanocomposites, *Cellulose* **17**(4) (2010) 779–791.
65. T. Huber, S. Pang and M.P. Staiger, All-cellulose composite laminates, *Compos. Pt. A-Appl. Sci. Manuf.* **43**(10) (2012) 1738–1745.
66. T. Huber, J. Muessig, O. Curnow, S. Pang, S. Bickerton and M.P. Staiger, A critical review of all-cellulose composites, *J. Mater. Sci.* **47**(3) (2012) 1171–1186.
67. T. Huber, S. Bickerton, J. Muessig, S. Pang and M.P. Staiger, Solvent infusion processing of all-cellulose composite materials, *Carbohydr. Polym.* **90**(1) (2012) 730–733.
68. D. Kretschmann, Natural materials — Velcro mechanics in wood, *Nat. Mater.* **2** (12) (2003) 775–776.
69. M. Seifert, S. Hesse, V. Kabrelian and D. Klemm, Controlling the water content of never dried and reswollen bacterial cellulose by the addition of water-soluble polymers to the culture medium, *J. Polym. Sci. Pol. Chem.* **42**(3) (2004) 463–470.
70. C. Tokoh, K. Takabe, J. Sugiyama and M. Fujita, Cellulose synthesized by acetobacter xylinum in the presence of plant cell wall polysaccharides, *Cellulose* **9**(1) (2002) 65–74.
71. H. Yamamoto, F. Horii and A. Hirai, In situ crystallization of bacterial cellulose .2. Influences of different polymeric additives on the formation of celluloses I-alpha and I-beta at the early stage of incubation, *Cellulose* **3**(4) (1996) 229–242.
72. C.H. Haigler, A.R. White, R.M. Brown and K.M. Cooper, Alteration of *in vivo* cellulose ribbon assembly by carboxymethylcellulose derivatives, *J. Cell Biol.* **94**(1) (1982) 64–69.
73. Q. Zhou, E. Malm, H. Nilsson, P.T. Larsson, T. Iversen, L.A. Berglund and V. Bulone, Nanostructured biocomposites based on bacterial cellulosic nanofibers compartmentalized by a soft hydroxyethylcellulose matrix coating, *Soft Matter* **5**(21) (2009) 4124–4130.
74. S. Gea, E. Bilotti, C.T. Reynolds, N. Soykeabkeaw and T. Peijs, Bacterial cellulose-poly(vinyl alcohol) nanocomposites prepared by an *in-situ* process, *Mater. Lett.* **64**(8) (2010) 901–904.
75. T.O.J. Blomfeldt, R. Kuktaite, E. Johansson and M.S. Hedenqvist, Mechanical properties and network structure of wheat gluten foams, *Biomacromolecules* **12**(5) (2011) 1707–1715.
76. T.O.J. Blomfeldt, R.T. Olsson, M. Menon, D. Plackett, E. Johansson and M.S. Hedenqvist, Novel foams based on freeze-dried renewable vital wheat gluten, *Macromol. Mater. Eng.* **295**(9) (2010) 796–801.
77. K.Y. Lee, L.L.C. Wong, J.J. Blaker, J.M. Hodgkinson and A. Bismarck, bio-based macroporous polymer nanocomposites made by mechanical frothing of acrylated epoxidised soybean oil, *Green Chem.* **13**(11) (2011) 3117–3123.

Chapter 12

Optically Transparent Nanocomposites

Antonio Norio Nakagaito[1] and Hiroyuki Yano[2]
[1] *Institute of Technology and Science, The University of Tokushima, Tokushima, Japan*
[2] *Research Institute for Sustainable Humanosphere, Kyoto University, Kyoto Japan*

Flexible circuit boards are currently ubiquitous in electronic products, and eventually the substrate of image displays will soon be made of flexible materials as well. Plastics are prospective candidates due to their inherent flexibility and optical qualities but they also present large thermal expansion. The expansion of the substrate has to be compatible with those of the active layers deposited on it, to avoid damages during the thermal cycles involved in the display manufacture. One way to reduce the thermal expansion of plastics is to reinforce them with nanofibers, without losing transparency. Nanofibers are available in large amounts in the form of cellulose and chitin in nature, being produced by plants and animals. Here, some of the researches to produce optically transparent composites based on natural nanofibers for use in flexible displays are presented and discussed.

12.1. Introduction

The next generation of large display technology is the so-called OLED, an acronym for organic light-emitting diode. It offers enhanced brightness, vivid colors, and high contrast ratio, along with fast response time and wide viewing angles. The screen panels can be made extremely thin because they do not require backlights and color filters typical of the current liquid crystal display (LCD) panels, with the added advantage of low power consumption. LCDs have the backlight always lit and images are formed by controlling the transmission of light at each pixel. Meanwhile, OLEDs have the individual pixels switched on only when needed, therefore consuming less power. Each pixel consists of a light-emitting diode made of a polymer film about 100 nm in thickness sandwiched between a metallic cathode and an optically transparent anode, all deposited on a transparent substrate. By the application of an electrical signal to the electrodes, charge carriers, namely holes from the anode and electrons from the cathode are recombined, resulting in light emission by radiative decay known as electroluminescence.[1] As a way to switch these elements on and off, for each pixel, there is a thin film transistor (TFT) circuit that provides a controlled current source for the pixel, in a type of panel called active matrix

OLED (AMOLED). As a consequence, AMOLED display requires an additional TFT backplane deposited on the substrate, which means it is subjected to thermal stress cycles during panel manufacture, hence requiring extremely low coefficient of thermal expansion (CTE) for the substrate material. Nonetheless, AMOLED is the technology needed to realize the fabrication of large high-definition panels.

The all solid-state structure of OLED panels with active layers less than 1 μm in thickness makes them suitable for the fabrication of flexible displays. The use of flexible plastic substrates instead of the traditional rigid glass brings the possibility to make tough, thin, and lightweight panels that flexes, bends, and conforms to the point of being wearable or being rolled for portability. Cost effective roll-to-roll processing becomes a possibility in display manufacture as well. However, plastic substrates still have to offer the high transparency, surface smoothness, barrier capacity, solvent resistance, thermal and dimensional stability, and reduced thermal expansion of glass. Bottom-emissive displays in which images are viewed through the substrate, requires a total light transmittance greater than 85% over the visible spectrum (400 to 800 nm), with a haze of less than 0.7%.[1] The barrier and solvent resistance can be achieved by applying a hard coating, whereas the surface smoothness is obtained through a scratch-resistant coating. The thermal and dimensional stabilities are properties intrinsic to the material, so a proper selection of the type of plastics is needed in order to resist the high temperatures and thermal cycles during the deposition of coatings and active layers in the display manufacture. However, the required CTE should be typically below $2.0 \times 10^{-5}\,K^{-1}$ to avoid any mismatch in thermal expansion relative to the device layers deposited on the substrate, otherwise they would be readily damaged during a thermal cycle.[1]

From an atomistic view of materials, thermal expansion and Young's modulus are both correlated to the depth of the atomic bond energy function. However, the relationship between the two properties is inversely proportional so that flexible plastics having low modulus exhibit intrinsically high CTEs. In order to reduce the CTE of plastics while preserving its flexibility, one way is to reinforce it with fibers possessing low CTE. However, the inclusion of elements even at micrometer sizes reduces the transparency, unless the refractive indexes of fibers and plastic material are exactly matched. A better solution is to use tinier nanofibers instead of microfibers, with lateral dimensions much smaller than the wavelength of the visible light. This required dimension should be an order of magnitude smaller than the wavelength of visible light. The visible spectrum being in the range of 400 to 800 nm, the reinforcing elements should have sizes in the order of 40 to 80 nm. Such nanofibers are already available in nature as renewable resources and in large quantities. Typical examples of these materials are cellulose and chitin found primarily as the framework of wood and the exoskeleton of arthropods, respectively.

This chapter will give an overview of the studies on transparent nanocomposites based on cellulose and chitin nanofibers carried at the Research Institute of Sustainable Humanosphere, Kyoto University, and at the Department of Chemistry and Biotechnology, Tottori University, both located in Japan.

12.2. Cellulose nanofiber-reinforced transparent materials

Cellulose nanofibers consist of semicrystalline cellulose molecular chains parallel to their axes. In the crystalline domains, the cellulose chains are arranged in a way that each long molecule is connected by hydrogen bonds to the neighboring chains forming a highly ordered crystalline form. Each one of these chains is made of glucose rings joined together without foldings, just as the benzene rings are joined in aramid. Even the density and the modulus of the two materials are very similar.[2] The cellulose nanofiber possesses a Young's modulus close to that of a perfect cellulose crystal, 138 GPa,[3] and the estimated tensile strength is well above 2 GPa. The first report of a transparent nanocomposite reinforced by cellulose nanofibers is due to Yano *et al.* in 2005.[4] Cellulose nanofibers spun by bacteria, known as bacterial cellulose (BC), were used as the reinforcing phase of a transparent resin matrix. BC consists of pellicles made of a random assembly of ribbon-shaped fibrils less than 100 nm in width, biosynthesized and secreted by some bacterial species. As the BC pellicles contain about 99 wt% of water, they were compressed to remove the excess water and completely dried, impregnated with acrylic, epoxy, and phenol formaldehyde resins, and cured by UV light or hot pressed depending on the matrix. The obtained composites had fiber contents between 60 and 70 wt%.

Despite the high nanofiber content, which delivered mechanical strength about five times that of engineered plastics and CTE similar to that of silicon crystal, the total light transmittance in the visible wavelength between 500 and 800 nm was greater than 80%. The reduction of light transmittance relative to the neat matrix resin due to the reinforcement was less than 10 percentage points. In conventional microcomposites, the light transmittance is primarily controlled by the matching of the refractive indexes of the fibers and the matrix as transparency is severely reduced if there is a slight mismatch. However, nanofibers are able to avoid light scattering due to their lateral sizes much smaller than the wavelength of visible light. The refractive index of cellulose is 1.618 along the fiber and 1.544 in the transverse direction whereas that of the impregnated phenol-formaldehyde, for instance, is 1.483 at 587.6 nm and 23°C. Similar differences hold for the other types of resin. The high transparency is due primarily to the lateral size of the reinforcing elements, because the scattering of light is prevented by nanofibers much smaller than the wavelength of light.

In terms of mechanical properties, the CTE of the BC/epoxy composite was largely reduced from the $120 \times 10^{-6}\,\mathrm{K}^{-1}$ of the epoxy matrix to just $6 \times 10^{-6}\,\mathrm{K}^{-1}$. The CTE of BC/phenol-formaldehyde was even lower at $3 \times 10^{-6}\,\mathrm{K}^{-1}$, a figure as low as that of silicon crystal. The tensile strength reached 325 MPa with a modulus around 20 to 21 GPa. BC nanocomposites are light, easy to mold, and flexible despite the high modulus. In more detailed studies of the effects caused by the mismatch between the refractive indices of matrix and nanofibers, Nogi *et al.* evaluated the changes in transparency of BC nanocomposites in relation to the refractive index of the matrix acrylic resin.[5] At a wavelength of 590 nm, the total and regular transmittances were above 85% and 75%, respectively, when the refractive index of the

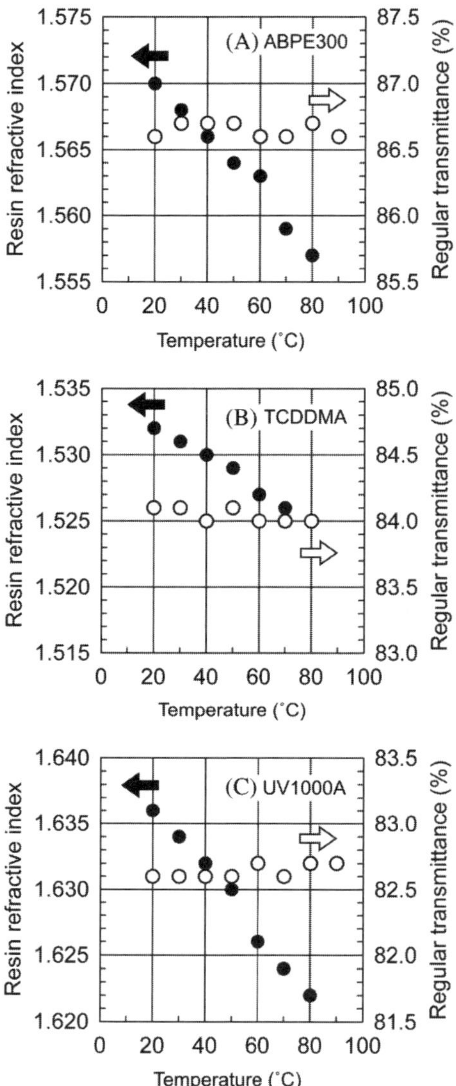

Fig. 12.1. The stability of the optical transmittance of BC nanofiber composites compared to the variations in refractive index of matrix resins (three types shown) due to temperature change.[5]

resin varied from 1.492 to 1.636. Similarly, the regular transmittance of composites did not vary when the matrix refractive index varied by 0.014 due to a temperature variation from room temperature to 80°C, as shown in Fig. 12.1. The evaluation of the transmittance of BC-reinforced acrylic resin[6] revealed that although the transmittance is linearly reduced with increasing fiber contents, the reduction is just 13.7 percentage points from the neat resin to the composite at a fiber content of 66 wt%. Meanwhile, the CTE is suppressed drastically with the addition of BC nanofibers, from $86 \times 10^{-6}\,K^{-1}$ of the acrylic resin to just $15 \times 10^{-6}\,K^{-1}$ at around 30 wt% fiber content, further declining to $10 \times 10^{-6}\,K^{-1}$ at a 50 wt% fiber content.

A typical and intrinsic drawback of cellulose is the high hygroscopicity due to the presence of hydroxyl groups in their molecular chains, making cellulose-based materials highly hydrophilic. In order to reduce the susceptibility of BC nanocomposites to moisture, Nogi *et al.*[7] and Ifuku *et al.*[8] acetylated the BC nanofibers replacing the hydroxyl groups on the surface of the fibers with acetyl groups. When exposed to a 55% relative humidity at room temperature, the untreated BC composite at 60 wt% fiber content had a 3.12% moisture content whereas the acetylated BC composite had the moisture content reduced to 1.33% at a fiber load of 66 wt%. Besides, the CTE of the BC sheet improved from $3 \times 10^{-6}\,\mathrm{K}^{-1}$ of untreated to just $0.8 \times 10^{-6}\,\mathrm{K}^{-1}$ of the acetylated sheet, even though the Young's modulus was reduced from 23.1 to 17.3 GPa. The regular transmittance in one of the BC/acrylic resin combinations revealed that the acetylated BC had increased transparency in the wavelength around 400 nm, keeping unchanged the transparency in the rest of the spectrum. The scanning electron microscopy (SEM) images showed that the untreated BC sheet is densely packed whereas the acetylated one shows nanofibers separated from each other, implying that the presence of acetyl groups in the surface of fibrils prevented their aggregation, keeping the uniformity of the individualized nanosized lateral dimensions of the nanofibers. The resistance to thermal degradation in terms of optical transparency was also improved as the acetylated BC nanocomposite showed a lower drop in transmittance than the untreated BC nanocomposite when both were exposed to a temperature of 200°C for several hours.

As another approach to decrease the thermal expansion of composites, especially at a lower fiber content of just 5 wt%, BC was combined with a low Young's modulus transparent resin.[9] As the modulus of the resin was only 25 MPa, the resulting composite had a modulus of 355 MPa, in the order of high-density polyethylene. This low modulus made the composite not only flexible but even foldable without causing damage (Fig. 12.2). Despite the low modulus, the in-plane CTE was unusually low at just $4 \times 10^{-6}\,\mathrm{K}^{-1}$. From an atomistic point of view, in general higher Young's moduli translate into lower CTEs. However, in these nanocomposites, the

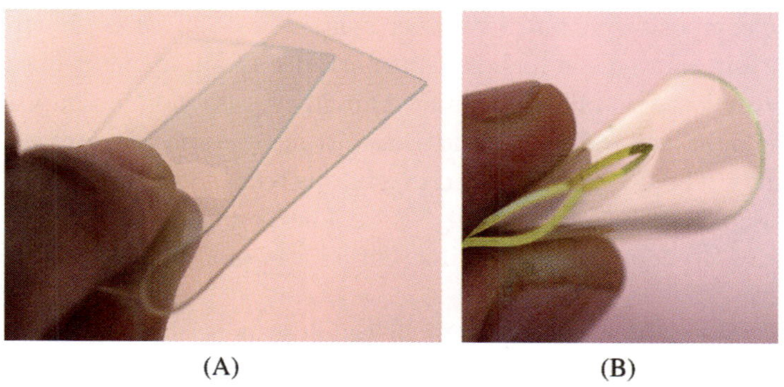

Fig. 12.2. Transparent BC nanofiber composite being folded with no damage (A) compared to the fragile corresponding neat matrix resin (B).[9]

unique anisotropic structure of BC pellicles, consisting of multiple layers of nanofiber networks which are weakly connected along the thickness direction of the pellicles makes them behave differently. Even at low contents, fibers are able to constrain the small stresses generated by the expansion of the low modulus matrix in the in-plane direction while in the thickness direction most of the thermal expansion takes place. This anisotropic behavior results in ultra-low in-plane CTE, but also allows enhanced flexibility for the material to be easily folded.

BC is a very suitable source of nanofibers to produce transparent composites, but the fermentation processes employed to produce BC are extremely costly and time consuming. An alternative resource would be the more abundant wood or plant fibers, even if the nanofibers have to be extracted before use. Cellulose nanofibers are found embedded in a matrix of hemicelluloses and lignin in the fiber cell wall of plants. In order to fibrillate wood pulp fibers and reduce the size of fibril bundles to a greater extent, Iwamoto *et al.* started with a commercially available cellulose known as microfibrillated cellulose (MFC), a morphology obtained by refining and high-pressure homogenization processes. He complemented the refining/high-pressure homogenization steps with an additional grinding treatment.[10] MFC obtained by 14 passes through the homogenizer has a wide distribution of fibrils width, from some tens of nanometers to some micrometers, that produces composites significantly less transparent than BC-based composites. The reduced transparency is attributed to light scattering caused by the larger fibril elements. The 14 passes through homogenizer MFC was supplemented with a grinding treatment, in a total of 10 passes through the grinder. This additional treatment resulted in fibril bundles with dimensions 50 to 100 nm in width. This grinder-fibrillated MFC was dispersed in water and the suspension was vacuum filtered, producing a thin sheet. Sheets were oven dried, impregnated with acrylic resin and cured by UV light.

The light transmittance of grinder-fibrillated fiber/acrylic resin composite, pure acrylic resin, BC/acrylic resin, and homogenizer-fibrillated fiber/acrylic resin were measured. At the wavelength of 600 nm, grinder fibrillated fiber/acrylic composites with 70 wt% fiber content showed a total transmittance of 70%, meaning a transmission reduction of just 20 percentage points from the transmittance of the pure acrylic resin. In contrast, the light transmittance of homogenizer fibrillated fiber/acrylic composites at a 62 wt% fiber content was only 30%. Even though the BC/acrylic resin showed the highest transmittance of all, as plant fibers consist of elementary cellulose microfibrils 4 nm by 4 nm in cross section, an improved fibrillation process might lead to higher light transmittances, eventually exceeding those of BC composites. The CTE of the matrix resin was reduced to one-fifth of its original value, from $86 \times 10^{-6}\,K^{-1}$ to $17 \times 10^{-6}\,K^{-1}$. The Young's modulus was lower than the modulus of BC composites at 7 GPa, but a lower modulus means a more flexible material.

It was found that another benefit of cellulose nanofiber reinforcement is the enhanced thermal conductivity of the composites. Shimazaki *et al.*[11] reported that epoxy resin reinforced with MFC increased the in-plane thermal conductivity to over $1.0\,Wm^{-1}K^{-1}$, a value three to five times higher than that of conventional transparent resins. As the amount of nanofibers was related to the improvement of thermal conductivity and there was no improvement in the direction normal to the in-plane surface, they concluded that the percolated nanofibers were acting as heat pathways to realize high thermal conductivity.

In order to extract higher quality nanofibers by grinding, Abe *et al.*[12] developed a new protocol that produced uniform nanofibers with 15 nm in diameter by only one pass through the grinder, while minimizing the damage to the nanofibers. In previous studies by Iwamoto *et al.*, the severity of mechanical shear applied by the grinder to fibrillate was evaluated based on the mechanical properties of the final nanocomposites.[13] Although the transparency of composites increased up to five passes through the grinder and above it, neither transparency nor the morphology of the nanofibers had changed. On the other hand, the tensile and thermal expansion properties were significantly reduced, following the decrease of crystallinity and degree of polymerization of cellulose. The role of hemicelluloses on nanofibrillation[14] was also clarified. According to this study, when the pulp fibers are dried, the presence of hemicelluloses impedes the formation of irreversible hydrogen bonds between microfibrils, known as hornification. And by rewetting, hemicelluloses are plasticized by absorbing water, favoring the ease of posterior nanofibrillation. Abe chose to keep the fiber in a water-swollen state after chemical removal of hemicelluloses and lignin, skipping the drying process in a typical pulp production that causes hornification. The new process delivers perhaps the most undamaged cellulose nanofibers extracted by only mechanical processes.

Later, Okahisa *et al.*[15] reported the fabrication of an optically transparent film made of a low Young's modulus resin reinforced with cellulose nanofibers extracted from wood by the new method, using it as a substrate on which an OLED layer was deposited. This was a proof of concept of flexible displays based on plastic substrates reinforced with low-cost plant derived nanofibers.

The optically transparent composites evolved later to a completely new material, without the use of matrix resins. It was called by Nogi *et al.* as "cellulose nanofiber paper",[16] and it is literally made of the same chemical constituents of the ordinary paper, but with a physical difference in the size of the fibers, the nanofibers obtained by the protocol developed by Abe *et al.*[12] Hence, the only difference from ordinary paper resides in the morphology of cellulose. A nanofiber aqueous suspension was slowly filtered and carefully dried. The density of the nanofiber paper was $1.53\,g/cm^3$, quite close to $1.59\,g/cm^3$ of the cellulose crystallite,[17] indicating that the small interfibrillar cavities were practically eliminated, as also confirmed by SEM observations. This paper was initially just translucent, but once the roughness of its surfaces was minimized by careful polishing, the transparency at 600 nm wavelength reached 71.6%. The CTE of $8.5 \times 10^{-6}\,K^{-1}$ is comparable to glass, along the tensile modulus of 13 GPa and the strength of 223 MPa. Further studies consisted in coating the cellulose nanofiber paper with transparent resins to smooth out the surfaces,[18] and demonstrated the potential to produce transparent film substrates by simple and cost-effective processes, suggesting the possibility of continuous roll-to-roll manufacturing methods.

12.3. Chitin nanofiber-based transparent materials

Chitin is another abundant polysaccharide and also present as nanofibers in the hard composite material forming the exoskeleton of marine crustaceans, insects, and in the cell walls of many filamentous fungi. The molecular structure of chitin is identical to that of cellulose except for a hydroxyl group replaced by an acetamido group,[19] as seen in Fig. 12.3. Although these nanofibers are originally embedded in hierarchically complex composites, the extraction is relatively less difficult compared to cellulose nanofibers from wood. This might be related to their functional roles in nature, since trees are structures with heights of several meters capable of resisting numerous stresses caused by winds or by their own weight, whereas crustaceans are much smaller beings relying on exoskeletons to give support and protection to their bodies.

Chitin consists of crystalline arrangements of 18–25 chitin molecular chains wrapped by proteins, forming nanofibrils of about 2 to 5 nm in diameter and 300 nm in length. These chitin–protein fibrils cluster to form 50 to 300 nm wide nanofibers. These in turn form a planar woven branched network, in which the space between the nanofibers is filled with proteins and minerals, mostly $CaCO_3$ crystallites (see Fig. 12.4). These planes are stacked as a twisted plywood known as Bouligand pattern.[20–23] Among several different methods to produce chitin nanofibers, Ifuku *et al.*[24] extracted chitin nanofibers by a mechanical process based on a grinding

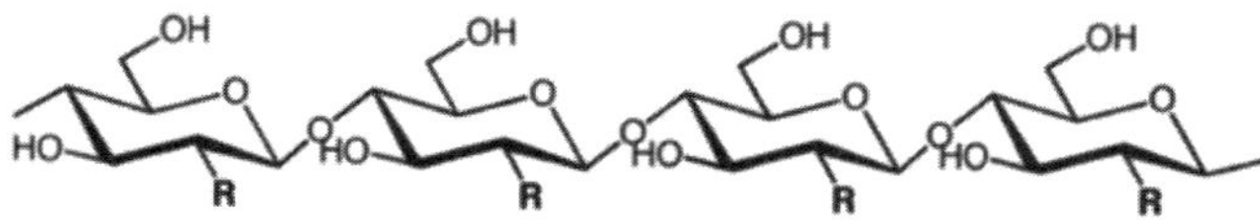

Fig. 12.3. Structural formula of cellulose or chitin. For cellulose, R is hydroxyl group (–OH), whereas for chitin, R is an acetamido group ($-NHCOCH_3$).

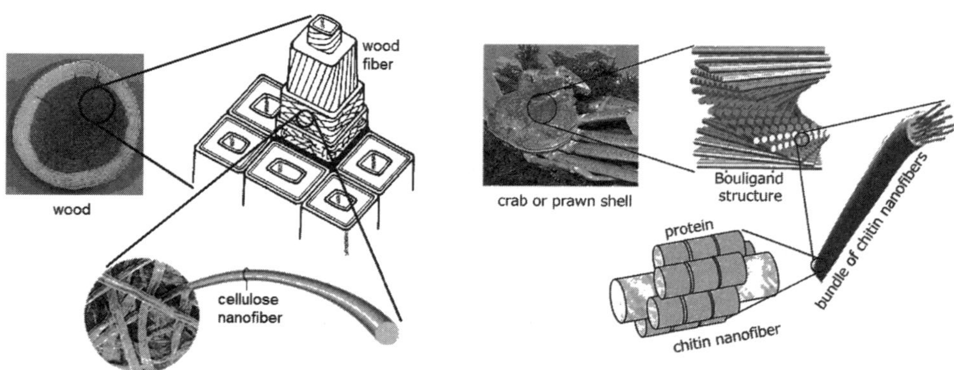

Fig. 12.4. Ultrastructures of wood and crustacean shell.

device. The raw material consisted of commercial crab shell powder sold as fertilizer, which had proteins removed and demineralized by chemical means following a standard method.[25,26] The purified chitin was kept always in a wet state to avoid the formation of hydrogen bonds that hinders individualization of nanofibers. After grinding, a close look by SEM revealed that the widths of fibrils obtained had a wide distribution, ranging from 10 to 100 nm. The authors attributed this to the fact that in crab shells, there are two distinct main layers, the exocuticle and the endocuticle. The exocuticle is composed of a very fine twisted plywood structure, whereas the endocuticle, comprising about 90% of the volume of the crab shell,[21] is made of a much coarser structure[23] being difficult to fibrillate, resulting in thicker bundles. Hence, based on previous studies by Fan *et al.*,[27] the pH of the slurry was adjusted to about 3 by addition of acetic acid, so that the cationization of the amino groups on the surface of chitin nanofibers would cause electrostatic repulsion and facilitate individualization. The grinded material under acidic condition formed a gel after only a single pass, due to the exposure of a greatly expanded surface area of nanoscalar individualized elements and SEM observations confirmed that nanofibers 10 to 20 nm in diameter were obtained. These dimensions were in accordance to the nanofibers observed in the original chitin structure after purification and before grinding, and elemental analysis revealed that deacetylation did not occur during the mechanical treatment. The X-ray diffraction measurements showed that the α-chitin crystalline structure was also preserved. The chitin nanofibers were extracted without damage from the crab shell, keeping the original molecular and crystalline structures intact. The new protocol proved to be so powerful that it delivered successful fibrillation of purified chitin once dried prior to grinding,[28] producing nanofibers of the same quality as those obtained from never-dried purified chitin. This finding brings a highly versatile nanofiber extraction process in terms of logistics of transportation and storage since weight and volume of semiprocessed materials can be reduced by drying, and fibrillation can be accomplished after redispersion in water. From the extraction perspective, it also positions chitin nanofibers at an advantage over cellulose nanofibers.

Transparent composites were produced by impregnating chitin nanofiber sheets with an acrylic resin with a refractive index 1.532, and cured by UV radiation (Fig. 12.5). The nanocomposites had a thickness of 70 μm with a nanofiber load of 60 wt%. The regular transmittances in the visible wavelength of 400 to 800 nm were measured for neat acrylic resin and acrylic resin reinforced with nanofibers from never-dried chitin, once-dried chitin, and dried chitin powder. At the middle of the spectrum, 600 nm, the transmittances of nanocomposites were 89.8% for the never-dried chitin, 88.7% for once-dried chitin, and 87.9% for dried powder chitin. Drying chitin before nanofiber extraction resulted in nanocomposites with transmittance loss less than 2 percentage points. The grinding under acidic medium was the determining factor to the successful fibrillation from a dried state. This is yet not possible with cellulose nanofibers as shown by Iwamoto *et al.*[14]

The fibrillation of chitin in acidic medium is so effective, that the mechanical treatment requirement can be fulfilled by accessible devices as simple as household

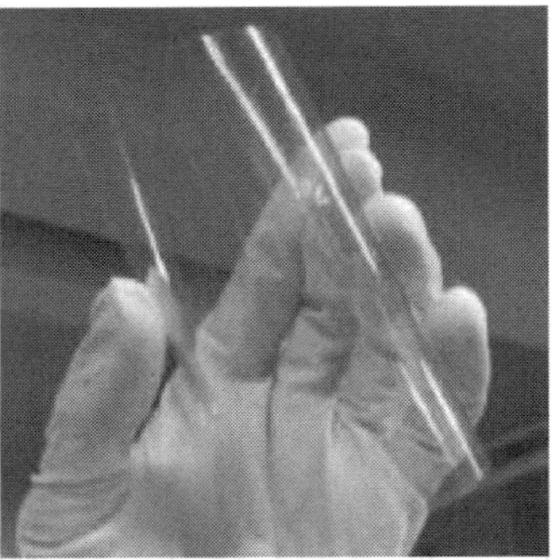

Fig. 12.5. Chitin nanofiber-based transparent composite.

blenders as reported by Shams *et al.*[29] Crab shell powder was chemically purified and an aqueous suspension was prepared adjusting the pH to 3–4 by addition of acetic acid. Chitin nanofibers 20 to 30 nm in width were obtained just by agitating the never-dried chitin slurry for 10 min by a high-speed blender originally intended for food processing. The authors observed that the filtration process to produce thin sheets from the nanofiber suspension was nine times faster for chitin nanofibers if compared to cellulose nanofibers, demonstrating the less hydrophilic character of chitin. These sheets, when impregnated with acrylic resin to produce nanocomposites, delivered optical transparency superior to cellulose nanofiber composites, showing about 10 and 20 percentage points higher regular transmittance at 600 and 400 nm, respectively. The phenomenon was attributed to the better affinity of less hydrophilic chitin with the hydrophobic matrix resin.

Concerning the thermal expansion restriction, the reinforcement by 40 wt% chitin nanofibers reduced the CTE of the acrylic resin from $213 \times 10^{-6}\,K^{-1}$ down to $15.6 \times 10^{-6}\,K^{-1}$, a value as low as that obtained by cellulose nanofiber reinforcement. The reasoning was based on the high axial chitin crystal Young's modulus of 80 GPa,[30] which explains the large thermal expansion reduction, but the final CTE achieved being similar to that of cellulose reinforcement is an open point since crystalline cellulose possess a modulus almost twice as high, at 138 GPa.[3] Probably, it is somehow connected to the reduced hydrophilicity of chitin resulting in enhanced fiber–matrix interactions that better counteract the resin expansion.

Using the same chitin nanofibers extracted from dry chitin powder under acidic condition, Ifuku *et al.*[31] reinforced silsesquioxane (SSQ) resins. These compounds have molecules formed by silica cores connected to organo-functional groups that give them properties of inorganic–organic hybrid materials. The inorganic character confers higher rigidity and thermal stability than usual polymers, while the organic part allows better processability than conventional ceramics, giving SSQs enhanced flexibility, optical transparency, and thermal stability. Due to the excessive brittleness of SSQs, a bifunctional urethaneacrylate (UA) oligomer was mixed at different fractions and copolymerized. With a nanofiber content of about 50 wt%, all composites showed high transparency, with regular transmittances at 600 nm wavelength, in the range between 80 and 85%, compared to about 90% for the pure SSQ resin. Among the nanocomposites, the pure SSQ matrix delivered the highest transparency, while the 1:4 SSQ to UA ratio matrix had it reduced by about 10 percentage points. The CTEs of SSQ-UA blends were in the range of 96 to $164 \times 10^{-6}\,K^{-1}$, but were reduced to about $30 \times 10^{-6}\,K^{-1}$ when reinforced with chitin nanofibers. The thermal stability of SSQ showed up when differential thermogravimety revealed that all composites had decomposition temperatures higher than chitin nanofibers, and became higher with increasing SSQ content ratios. Thermally stable SSQs serve as a good matrix choice to make carbohydrate polymer chitin more resistant to high temperatures.

Similarly to the study assessing the changes in optical transparency of cellulose nanofibers caused by differences between the refractive indexes of nanofibers and matrix resin, chitin nanofibers were used to reinforce an array of different acrylic and methacrylic resins having varied refractive indexes.[32] The regular transmittances at 600 nm of the resulting composites were all over 70%, with the refractive index of the matrix ranging from 1.468 to 1.54. Even though the nanofiber content was around 40 wt%, the composites transparency peaked at a matrix refractive index of 1.50 to 1.51, in some cases showing a transmittance drop from the neat resin by just 1.3 percentage points. In addition, CTEs were reduced from high values between 98.3 and $185.4 \times 10^{-6}\,K^{-1}$ of resins, to lower values in the range of 29.5 to $17.5 \times 10^{-6}\,K^{-1}$ for the corresponding composites.

An important characteristic of chitin nanofibers, due to the presence of hydroxyl and acetamido groups, is the possibility to be chemically modified to expand their potential applications. Chitin nanofibers sheets were acetylated by a simple process.[33] Oven-dried sheets were dipped in a mixture of 5 mL acetic anhydride and 0.1 mL of 60% perchloric acid solution under stirring and kept for the intended treatment time. Afterwards, sheets were Soxhlet extracted with methanol to remove the non-reacted compounds. The degree of substitution increased from 0.99 to 2.96 for a reaction time of 50 min, an almost complete replacement of hydroxyl groups by acetyl groups in the chitin nanofibers, as confirmed by infra-red spectrometry. The X-ray diffraction measurements confirmed that acetylation proceeds from the surface to the core of chitin nanofibers, preserving the original morphology. Observations by SEM showed that the nanofibers are kept intact after acetylation, with only a gradual increase in average diameter from 21.6 nm of the untreated nanofibers to 32.1 nm for those at DS 2.96, due to the introduction of bulkier acetyl groups inside the nanofibers.

Transparent composites produced by impregnation of acrylic resin into acetylated chitin nanofiber sheets resulted in light transmittances at 700 nm wavelength, of 73% for the completely acetylated DS 2.96 sample. As the untreated sample exhibited transmittance of 77%, the 4 percentage point reduction was considered small compared to 20 points in the case of composites with acetylated BC nanofibers.[8] Acetylation decreases the refractive index of chitin and the larger mismatch with the refractive index of the resin matrix causes the drop in transparency. However this phenomenon is less pronounced in the case of chitin nanofibers since their diameters are an order of magnitude smaller than the wavelength of visible light.

Similarly to cellulose, one of the major problems of chitin is its inherent hygroscopicity, and acetylation is an effective chemical modification to replace hydrophilic hydroxyl groups by hydrophobic acetyl groups. When kept at an environment with 75.1% relative humidity and room temperature, untreated chitin nanofiber composites showed a moisture content of 4% while the neat matrix resin showed only 0.33%. The acetylation at a DS of 1.3 brought the moisture content of composites to 2.2% as the hydrophobic character of the surface of the acetylated chitin nanofibers enhanced the compatibility with the acrylic matrix, decreasing the adsorption of water molecules to the interfaces.

At the atomic scale, Young's modulus and thermal expansion are both dependent on the depth of the atomic bond energy function, and it is especially accurate when considering crystalline materials. As chitin nanofibers consist of an antiparallel crystalline structure, and the Young's modulus is 41 GPa and constant over a temperature range from -190 to 150°C,[34] the CTE is expected to be very small. The CTE of composites with 25% nanofiber content was restricted to just $2.3 \times 10^{-5}\,K^{-1}$, almost a third of the neat acrylic resin at $6.4 \times 10^{-5}\,K^{-1}$. Even though acetylation could increase the thermal expansion of chitin nanofibers, a DS of 1.3 delivered nanocomposites with CTE slightly higher than that of untreated nanofibers. In fact, the CTE increases linearly with DS values as the progressive acetylation of the bulk of the nanofibers reduces its crystallinity, gradually increasing thermal expansion.

An extraction process that does not require acidic medium to fibrillate was also developed.[35] Since crab shells are mostly composed of coarse bundles of chitin nanofibers, a finer structure was identified in prawn shells. After the usual purification treatment to remove proteins and minerals, chitin was maintained in a wet state and fibrillated by a single pass through a grinder. The authors attribute the easy fibrillation without the aid of charged nanofiber surfaces, to the differences in the structures of prawn shells and carapaces of crab. The exoskeleton of crab is mostly comprised of a coarse endocuticle whereas the prawn shell is made primarily of a finer exocuticle with thinner fibril bundles. Trials with other prawn species showed that the same 10 to 20 nm wide chitin nanofibers can be easily obtained by the process, suggesting that it is extensible to any species of prawns. This finding makes again chitin nanofibers even easier to extract than the cellulose counterparts.

A simpler approach was then explored when Shams *et al.*[36] transformed crab carapaces into transparent ones, without changing the natural anatomy (Fig. 12.6). This was accomplished by carefully chemically removing minerals, proteins, and pigments, but keeping the shape integrity of a carapace made exclusively of a nanoporous chitin nanofiber structure. The nanocavities left by the removed matrix substances were thoroughly filled with acrylic resin, and an optically transparent carapace was produced. As an experimentation to show that chitin in nature is made of a nanofibrous framework, it also encouraged the authors to prepare transparent nanocomposites from macroscopic purified chitin particles 0.35 mm in size.

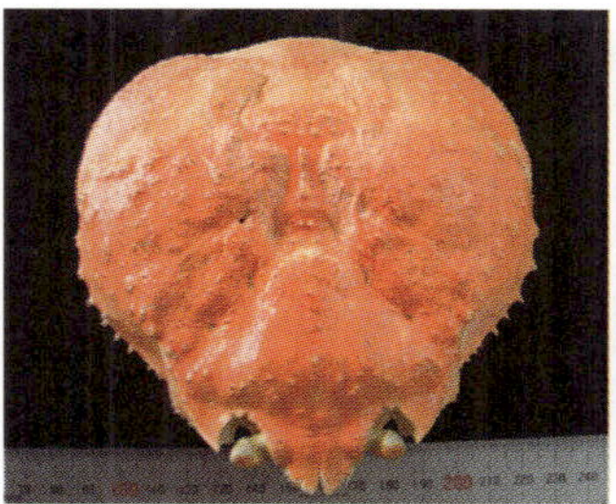
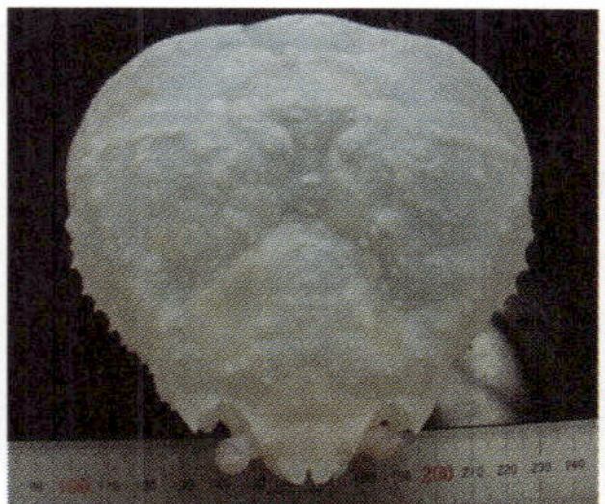

Fig. 12.6. Crab shell turning transparent. The original carapace (left), after removal of matrix substances (center), and after impregnation with acrylic resin (right).

Fig. 12.7. Optically transparent paper before (left) and after (right) acrylic resin impregnation.[37]

These materials showed optical transparency and CTEs very much similar to those prepared with nanofibrillated chitin. The regular transmittance dropped by about 10 percentage points relative to that of resin, whereas the CTE was reduced from 213 to $25 \times 10^{-6}\,K^{-1}$. The experiment revealed that a careful procedure to take advantage of the nanostructured chitin nanofibers as they are found in nature brings the possibility to produce nanocomposites by skipping the costly and time consuming fibrillation step. A similar approach is proposed by Yano.[37] Instead of disassembling pulp fibers to extract cellulose nanofibers, a pulp fiber sheet was acetylated, dried, and impregnated with acrylic resin to make transparent composites. In other words, an ordinary piece of paper, composed of micron-sized highly purified cellulose fibers, had the constituent nanofibers acetylated and composed with a transparent resin. As a transparent paper was obtained (Fig. 12.7), it implies that the nanofibrillation step is not necessary and the dewatering time to obtain the sheets is significantly shortened, reducing both cost and time to produce the composites. These are alternative ways to make transparent nanocomposites skipping the costly fibrillation process.

12.4. Conclusions and prospects

Cellulose and chitin are both abundant natural carbohydrates with many similarities as well as differences. Cellulose nanofibers are significantly stiffer than chitin, with a Young's modulus rivaling that of man-made aramid fibers. However, when composed with optically transparent resins, both nanofibers give similar properties in terms of light transmittance and low thermal expansion. The transparency may be attributed to the uniform nanoscalar size of both nanofibers, but chitin seems to show better affinities with hydrophobic matrixes owing to the less hydrophilic character compared to cellulose. And this more favorable affinity with the resin may be responsible for the thermal expansion restriction as effective as cellulose, despite the significantly lower stiffness of chitin. Add these to the faster dewatering times required to obtain nanofiber sheets and chitin places itself in a slight advantage over cellulose.

Another point to consider is the extraction of nanofibers. The currently available processes are costly and time consuming, even if the raw materials are inexpensive. The simpler planar multiply structures of chitin in contrast to helicoidal structures of cellulose in plant fibers, added to the interfibrillar electrostatic repulsion brought about by cationization of amino groups might give chitin another head start over cellulose. Such factors considered, a different approach is needed, and the transparent crab shell may give a hint to further developments. Avoiding the fibrillation step could be a smarter strategy to reduce costs and streamline the production process of transparent composites. At this point, it is difficult to tell which material is better, an answer left to the requirements of each specific application. In other words, the needs will dictate the choices.

References

1. G.P. Crawford (Ed.) *Flexible flat panel displays*, New York: Wiley, Society for Information Display (2005).
2. J.E. Gordon, *The new science of strong materials,* New Jersey: Princeton University Press (1976).
3. T. Nishino, K. Takano and K. Nakamae, Elastic modulus of the crystalline regions of cellulose polymorphs, *J. Polym. Sci. Pol. Phys.* **33**(11) (1995) 1647–1651.
4. H. Yano, J. Sugiyama, A.N. Nakagaito, M. Nogi, T. Matsuura, M. Hikita and K. Handa, Optically transparent composites reinforced with networks of bacterial nanofibers, *Adv. Mater.* **17**(2) (2005) 153–155.
5. M. Nogi, K. Handa, A.N. Nakagaito and H. Yano, Optically transparent bionanofiber composites with low sensitivity to refractive index of the polymer matrix, *Appl. Phys. Lett.* **87**(24) (2005) 243110.
6. M. Nogi, S. Ifuku, K. Abe, K. Handa, A.N. Nakagaito and H. Yano, Fiber content dependency of the optical transparency and thermal expansion of bacterial nanofiber reinforced composites, *Appl. Phys. Lett.* **88**(13) (2006) 133124.
7. M. Nogi, K. Abe, K. Handa, F. Nakatsubo, S. Ifuku and H. Yano, Property enhancement of optically transparent bionanofiber composites by acetylation, *Appl. Phys. Lett.* **89**(23) (2006) 233123.

8. S. Ifuku, M. Nogi, K. Abe, K. Handa, F. Nakatsubo and H. Yano, Surface modification of bacterial cellulose nanofibers for property enhancement of optically transparent composites: dependence on acetyl-group DS, *Biomacromolecules* **8**(6) (2007) 1973–1978.
9. M. Nogi and H. Yano, Transparent nanocomposites based on cellulose produced by bacteria offer potential innovation in the electronics device industry, *Adv. Mater.* **20**(10) (2008) 1849–1852.
10. S. Iwamoto, A.N. Nakagaito, H. Yano and M. Nogi, Optically transparent composites reinforced with plant fiber-based nanofibers, *Appl. Phys. A-Mater.* **81**(6) (2005) 1109–1112.
11. Y. Shimazaki, Y. Miyazaki, Y. Takezawa, M. Nogi, K. Abe, S. Ifuku and H. Yano, Excellent thermal conductivity of transparent cellulose nanofiber/epoxy resin, *Biomacromolecules* **8**(9) (2007) 2976–2978.
12. K. Abe, S. Iwamoto and H. Yano, Obtaining cellulose nanofibers with a uniform width of 15 nm from wood, *Biomacromolecules* **8**(10) (2007) 3276–3278.
13. S. Iwamoto, A.N. Nakagaito and H. Yano, Nano-fibrillation of pulp fibers for the processing of transparent nanocomposites, *Appl. Phys. A-Mater.* **89**(2) (2007) 461–466.
14. S. Iwamoto, K. Abe and H. Yano, The effect of hemicelluloses on wood pulp nanofibrillation and nanofiber network characteristics, *Biomacromolecules* **9**(3) (2008) 1022–1026.
15. Y. Okahisa, A. Yoshida, S. Miyaguchi and H. Yano, Optically transparent wood-cellulose nanocomposite as a base substrate for flexible organic light-emitting displays, *Compos. Sci. Technol.* **69**(11–12) (2009) 1958–1961.
16. M. Nogi, S. Iwamoto, A.N. Nakagaito and H. Yano, Optically transparent nanofiber paper, *Adv. Mater.* **21**(16) (2009) 1595–1598.
17. J. Sugiyama, R. Vuong and H. Chanzy, Electron diffraction study on the two crystalline phases occurring in native cellulose from an algal cell wall, *Macromolecules* **24**(14) (1991) 4168–4175.
18. M. Nogi and H. Yano, Optically transparent nanofiber sheets by deposition of transparent materials: a concept for a roll-to-roll processing, *Appl. Phys. Lett.* **94**(23) (2009) 233117.
19. E.S. Stevens, *Green plastics: an introduction to the new science of biodegradable plastics*, Princeton, NJ: Princeton University Press (2002) p. 86.
20. D. Raabe, P. Romano, C. Sachs, H. Fabritius, A. Al-Sawalmih, S. Yi, G. Servos and H.G. Hartwig, Microstructure and crystallographic texture of the chitin-protein network in the biological composite material of the exoskeleton of the lobster Homarus americanus, *Mat. Sci. Eng, A-Struct.* **421**(1–2) (2006) 143–153.
21. P.Y. Chen, A.Y.M. Lin, J. McKittrick and M.A. Meyers, Structure and mechanical properties of crab exoskeletons, *Acta Biomater.* **4**(3) (2008) 587–596.
22. M.M. Giraudguille, Fine structure of the chitin-protein system in the crab cuticle, *Tissue Cell* **16**(1) (1984) 75–92.
23. D. Raabe, C. Sachs and P. Romano, The crustacean exoskeleton as an example of a structurally and mechanically graded biological nanocomposite material, *Acta Mater.* **53**(15) (2005) 4281–4292.
24. S. Ifuku, M. Nogi, K. Abe, M. Yoshioka, M. Morimoto, H. Saimoto and H. Yano, Preparation of chitin nanofibers with a uniform width as alpha-chitin from crab shells, *Biomacromolecules* **10**(6) (2009) 1584–1588.
25. K.G. Nair and A. Dufresne, Crab shell chitin whisker reinforced natural rubber nanocomposites. 1. Processing and swelling behavior, *Biomacromolecules* **4**(3) (2003) 657–665.

26. J.N. BeMiller and R.L. Whistler, Alkaline degradation of amino sugars, *J. Org. Chem.* **27**(4) (1962) 1161–1164.
27. Y.M. Fan, T. Saito and A. Isogai, Preparation of chitin nanofibers from squid pen beta-chitin by simple mechanical treatment under acidic conditions, *Biomacromolecules* **9**(7) (2008) 1919–1923.
28. S. Ifuku, M. Nogi, M. Yoshioka, M. Morimoto, H. Yano and H. Saimoto, Fibrillation of dried chitin into 10–20 nm nanofibers by a simple grinding method under acidic conditions. *Carbohydr. Polym.* **81**(1) (2010) 134–139.
29. M.I. Shams, S. Ifuku, M. Nogi, T. Oku and H. Yano, Fabrication of optically transparent chitin nanocomposites. *Appl. Phys. A-Mater.* **102**(2) (2011) 325–331.
30. M. Wada and Y. Saito, Lateral thermal expansion of chitin crystals. *J. Polym. Sci. Polym. Phys.* **39**(1) (2001) 168–174.
31. S. Ifuku, A. Ikuta, T. Hosomi, S. Kanaya, Z. Shervani, M. Morimoto and H. Saimoto, Preparation of polysilsequioxane-urethaneacrylate copolymer film reinforced with chitin nanofibers, *Carbohydr. Polym.* **89**(3) (2012) 865–869.
32. S. Ifuku, S. Morooka, A.N. Nakagaito, M. Morimoto and H. Saimoto, Preparation and characterization of optically transparent chitin nanofiber/(meth)acrylic resin composites, *Green Chem.* **13**(7) (2011) 1708–1711.
33. S. Ifuku, S. Morooka, M. Morimoto and H. Saimoto, Acetylation of chitin nanofibers and their transparent nanocomposite films, *Biomacromolecules* **11**(5) (2010) 1326–1330.
34. T. Nishino, R. Matsui and K. Nakamae, Elastic modulus of the crystalline regions of chitin and chitosan, *J. Polym. Sci. Pol. Phys.* **37**(11) (1999) 1191–1196.
35. S. Ifuku, M. Nogi, K. Abe, M. Yoshioka, M. Morimoto, H. Saimoto and H. Yano, Simple preparation method of chitin nanofibers with a uniform width of 10–20 nm from prawn shell under neutral conditions, *Carbohydr. Polym.* **84**(2) (2011) 762–764.
36. M.I. Shams, M. Nogi, L. A. Berglund and H. Yano, The transparent crab: preparation and nanostructural implications for bioinspired optically transparent nanocomposites, *Soft Matter* **8**(5) (2012) 1369–1373.
37. H. Yano and F. Nakatsubo, A paradigm in nanocellulose materials — from nanofibers to nanostructured fibers, in *Proceedings 11th Pacific Rim Bio-based Composites Symposium,* Shizuoka, Japan (2012) pp. 121–124.

Chapter 13

Responsive Nanocellulose Composites

Norma E. Marcovich[1], María L. Auad[2] and Mirta I. Aranguren[1]
[1]*INTEMA, Facultad de Ingeniería,*
Universidad Nacional de Mar del Plata, Argentina
National Scientific and Technical Research Council (CONICET), Argentina
[2]*Polymer and Fiber Engineering Department,*
Auburn University, Alabama, USA

The outstanding mechanical properties of nanocellulose, added to its wide availability, biodegradability, and wide number of alternatives for chemical modification, have been the driving force for its utilization as reinforcement in responsive polymers. In particular, segmented polyurethanes (SPUs) with shape memory behavior have been utilized in the production of nanocellulose-reinforced responsive composites. Additionally, the hygroscopic nature of cellulose has led to the investigation of nanocomposites with water-responsive behavior, displaying a dramatic reduction of the material rigidity in the presence of the liquid or switchable swelling behavior. More recent developments in the area of nanocellulose responsive materials were addressed to growing active polymers or functional nanoparticles on the surface of cellulosic crystals or fibrillar structures. These developments are opening new roads for activating different material responses by indirect or non-contacting methods.

13.1. Introduction

Responsive materials are those that react with a variation of one or more of their properties due to a change in the surrounding environment. These responses can be triggered by variations in temperature (the most used activation factor), stress, light, humidity, pH, etc. Furthermore, these materials can respond to these variations with remarkably large changes in properties, such as light transmittance, electrical conductance, and mechanical properties, which can also lead to notable changes of the material shape.

Actually, the so-called "shape memory behavior" is an outstanding demonstration of the responsive actuation of materials. It consists on "fixing" a temporary shape on the sample and the later autonomous recovery of a "remembered" shape when it senses an external change in the surrounding conditions. This function is activated via external stimuli, often a temperature change, usually involving physical contact between the material and the agent that induces the alteration.[1–9] There are also other activation methods that have been investigated:

light irradiation, application of electrical or magnetic fields, and chemical activation (contact with a given solvent, changes in the pH or the ionic strength in the media). The wide number of alternatives opens the field for different potential applications in biomedicine (mechanically adaptive devices,[6,10–14] bioseparation), aerospace, adhesives and optical devices, sensors,[15] actuators and thus have attracted tremendous attention, especially in the last two decades.

The existence of this behavior is related to the existence of a discrete reversible phase transition of the polymer, usually the glass–rubber transition or the melting of a crystalline phase of the material.[16] The most frequently studied behavior is the thermo-responsive type, based on the heating/cooling of the material through a transition temperature. Interesting approaches of indirect heating have been proposed that include inductive heating using oscillating magnetic fields,[17,18] radio frequency irradiation,[19] and electric fields.[20] There is also an interest in developing materials that respond to stimuli other than heating to activate a response, pH,[21] and light activation.[20,22–23]

The outstanding properties of nanocellulose, as microfibrillated cellulose or as cellulose nanocrystals (CNCs) (low density, high modulus, easy chemical modification) have pushed the interest in its use as nanoreinforcement in different polymeric matrices. Thus, it has also been considered in the case of the reinforcement of shape memory polymers (SMPs), which are usually not highly rigid materials.[24] It is considered that while low concentration of microparticles could damage the shape recovery properties of SMP, an equal amount of nanoparticles may improve the shape recovery stress of these polymers.[25] Most of the research published in this area has been focused on smart polyurethanes (PUs), which are very compatible with polar cellulose.[2,4,26]

The hygroscopic nature of nanocellulose has also been the origin for novel materials that take advantage of this property to generate responsive behaviors activated by water sorption.[16]

More recent attempts have been reported for activating cellulose nanocomposites by other means, including the photoresponse of coated nanocellulose aerogels.[27] Additionally, more preliminary works have been aimed to generate an electroresponse in shape memory cellulose nanocomposites.[28,29]

In the following sections, the responsive behavior of nanocellulose composites that react to variations in temperature or humidity in their surrounding environment are considered, since they are presently the two most investigated examples of these types of materials. A brief presentation of a light-activated responsive material is also included.

13.2. Nanocellulose as reinforcing phase in responsive polymers

Although different types of SMPs have been investigated, segmented polyurethanes (SPUs) have been frequently considered because of the superior properties that arise from the phase-separated structures of soft and hard segments (HSs), or the thermally reversible phase and the frozen phase, respectively.[6,30–33] The frozen phase is generated by the HSs that form physical cross-links arising from polar interactions, hydrogen bonding, and crystallization. The reversible phase is formed by soft segments, which are long molecular chains with high mobility in the rubbery state.[2,4,6,26,34–36] The thermally reversible phase exhibits a lower phase transition temperature (T_{trans}), which is the melting temperature of the soft segments in this case. This phase serves as a "molecular switch" and enables the fixation of a temporary shape.

On the other hand, the frozen phase is responsible for memorizing the permanent shape of the specimen and shows a high thermal transition temperature (T_{perm}), which in the case of the SPU is the melting temperature of the HSs. When a SMP is heated to a temperature between these mentioned transitions ($T_{\text{trans}} < T < T_{\text{perm}}$), large deformations can be easily developed in the low-modulus polymer. Such deformations can be subsequently fixed if the SMP is cooled at a temperature below T_{trans} because of the dramatic increase in the modulus occurring as the soft segments crystallize.

Then, the original shape can be recovered by reheating up to a temperature above T_{trans}, because of the entropy-governed response of the elastic polymer chains in the soft segments that remain anchored to the permanent physical cross-links given by the HSs. A thermo-mechanical cycle that illustrates the shape memory effect can be summarized as follows:

(1) *Deformation*: deforming the SMP in the temperature range $T_{\text{trans}} < T < T_{\text{perm}}$.
(2) *Fixing*: fixing the temporary shape of the SMP by cooling to below T_{trans} and releasing the constraint.
(3) *Recovery*: heating the SMP above T_{trans} to recover the original shape.

One negative aspect of the SMPs (including SPUs) is their low recovery force. If the applied stress is higher than a relatively low value (i.e., 4 MPa) SPUs losses its shape-recovery property.[37–39] Therefore, SMPs are usually utilized in cases where only free recovery or very low recovery force and mechanical strength are required. Another significant drawback of PU-based shape memory materials is their low stiffness compared to metals and ceramics.[2,4,26,39,40] Thus, for practical applications, where increasing stiffness is critical for enhancing recoverable stress levels, various fillers (nano or micro size) have been considered in the past as potential reinforcement for SMPs.[2,4–6,39] However, it has been observed that there is a trade-off between enhancement of modulus and reduction of recoverable strain ratio. Generally, fillers exert negative impact on recoverable strain due to their size and substantially higher stiffness compared to the matrix polymer. In some cases, they even disturb the polymer network responsible for shape memory functions, especially at high loading levels. Research efforts to strike a balance between the recovery stress and the recovery strain through the use of fillers are still evolving, but presently it is clear that the shape memory behavior of composites strongly depends on the shape, size, and concentration of the filler used.[2,4,41,42]

13.2.1 *Nanocellulose–SPU shape memory thermo-responsive composites*

During recent years, attention has been devoted to the use of cellulose crystals, and important studies have been published.[2,4,24,26,27] CNCs (nanowhiskers) are nanofillers with high aspect ratios that have polar groups that can interact with polar polymers, leading to good interfacial adhesion, which is essential to obtain a material with enhanced properties.

SPUs are not the exception and several studies on the effect of the incorporation of CNCs to responsive PU have been reported in the literature.[2,4,26,29]

A first report from Auad *et al.*[8] illustrated the good dispersion achieved in SPU–CNC systems, which could be inferred from scanning electron microscopy (SEM) images of cryogenic fracture surfaces as well as from the overall thermo-mechanical performance of the nanocomposites. Figure 13.1 shows that the featureless fracture surface of the SPU changes to a rough surface spanning the whole sample with the addition of only 1 wt% of cellulose crystals.

The dispersed cellulose strongly interacts with the polymer, which is reflected in the shift toward a higher temperature of the soft segments' melting point (1–2°C), and the more remarkable increase of the melting heat (about 80% increase with the addition of only 1 wt% cellulose). It was concluded that cellulose was acting as a nucleating agent for the crystallization process, in agreement with observations made by other authors.[43] Again, this effect can be traced to the very good dispersion obtained in this system, which favors matrix–fibril interactions and the phase separation between soft and hard segments. Wide angle X-ray scattering (WAXS) patterns of the SPU and the composites confirm the strong interactions as can be seen in Fig. 13.2.

The SPU shows the typical amorphous pattern added to a crystalline peak that appears at $2q = 24.1°$. The addition of just 1 wt% cellulose crystals result in the appearance of a new peak at 26.3°. On the other hand, the peak of the SPU crystalline phase is shifted toward larger angles, which is consistent with a smaller

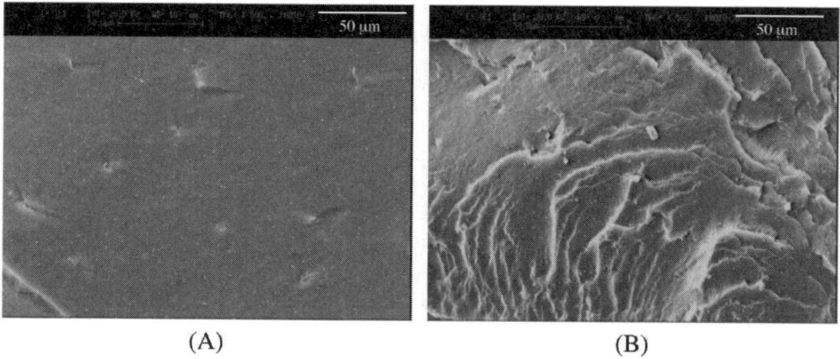

Fig. 13.1. SEM micrographs of cryo-fractured samples: (A) unreinforced PU; (B) 1 wt% nanocellulose-reinforced PU. (Reprinted from Ref. 2, with permission.)

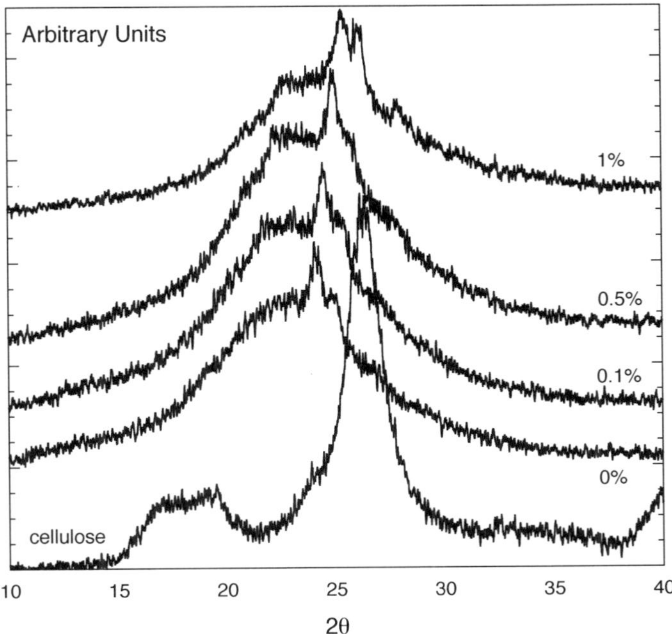

Fig. 13.2. WAXS curves of nanocellulose, unreinforced PU, and nanocellulose-reinforced PU (percentages expressed by weight). (Reprinted from Ref. 2, with permission.)

d-spacing, probably due to the formation of more perfect crystals resulting from a better phase separation.

The effect of CNCs on the SPU glass transition temperature was also analyzed from dynamical mechanical results, which showed a slight increase in the position of the tan δ peak as the nanocellulose content increased. The height of the peak decreased in a proportion larger than expected from a simple dilution effect, which was explained by the large interfacial interaction between the nanocellulose and the SPU, resulting in some extent of chain mobility reduction.

In regards to the mechanical properties, the addition of small amounts of cellulose led to interesting effects. The unreinforced PU shows the behavior of a largely extensible thermoplastic elastomer. As more cellulose was incorporated, the curves separate from each other at strains of 20–30% and a clear yield point appears for the 1 wt% sample. While large stretching was possible for the composites, the modulus is noticeably increased (41.1 MPa, 53.3 and 62.7 MPa for the 0, 0.5, and 1 wt% samples). This improvement also results in reduced creep, as evidenced by how the minimal addition of 0.1 wt% cellulose produces a measurable reduction of the deformation (15% reduction in 1 h-creep at room temperature).

On the other hand, the increased rigidity did not have any substantial effect on the shape fixity or recovery of the material: the CNC composites essentially displayed the same shape memory behavior as the neat PU. The shape recovery was quite large and repeatable after the second cycle for all the materials (neat PU and nanocomposites). This indicates that the SPU controls this behavior under the conditions of temperature and deformation imposed in the tensile cycling. Moreover, the addition of nanocrystals does not inhibit the shape memory response, while improving the mechanical properties of the nanocomposites.

In order to further investigate the effect of the incorporation of the CNCs on the development of the semicrystalline structure of SPUs, a SPU was synthesized to better control the chemistry of the system.[4] The polymer was obtained from 4,4-methylene-bisphenyldiisocyanate (MDI) and poly(tetramethyleneglycol) (PTMG) (molar mass = 2,000 Da); 1,4-butanediol (BD) was used as chain extender. CNCs (up to 1 wt%) were introduced during the SPU polymerization, which resulted in nanocelluloses covalently linked to the polymer structure. The interactions between the nanocellulose and the matrix led to changes in the microstructure of the PUs, with the HS phase being more affected by these interactions. This result is consequence of the chemical reaction between cellulose and MDI, which had been previously reported and investigated by the same authors using Fourier transform infrared spectroscopy (FTIR) techniques.[44] As cellulose can chemically react with the isocyanate, it strongly interferes in the formation of hard domains, because it reduces the regularity of the molecular structures, and the material is unable to show shape memory behavior. On the other hand, if the cellulose is incorporated after the synthesis (as it was done with the commercial SPU), the shape memory behavior is still possible.

In general, the addition of nanocellulose favored the phase separation between the soft and hard domains, generating an upward shift in the melting temperature of the soft segment crystalline phase and an increase in the storage and Young's modulus of the material, as it was observed for the commercial SPU. This effect is illustrated in Fig. 13.3, where the shift of the melting point of the soft segments can be observed, as well as the reduction of the melting endotherm corresponding to the hard crystalline phase. Thus, the incorporation of cellulose deleteriously affects the formation of hard domains, which will translate in difficulties in "remembering" the original shape (recovery).

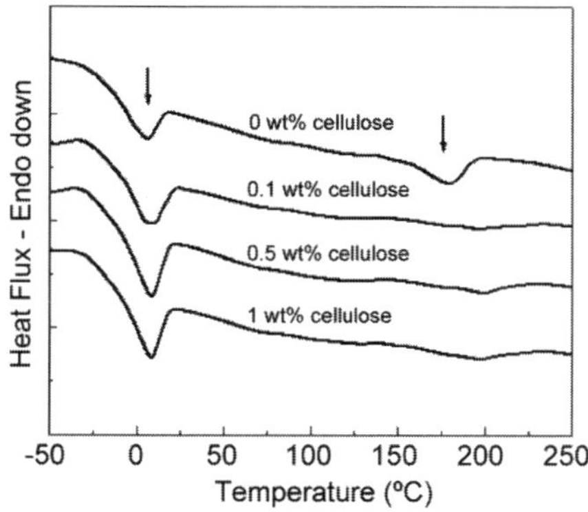

Fig. 13.3. Differential scanning calorimetry (DSC) curves of the PU made from PTGM 2000 with 40% HSs containing different concentrations of cellulose. (Reprinted from Ref. 4, with permission.)

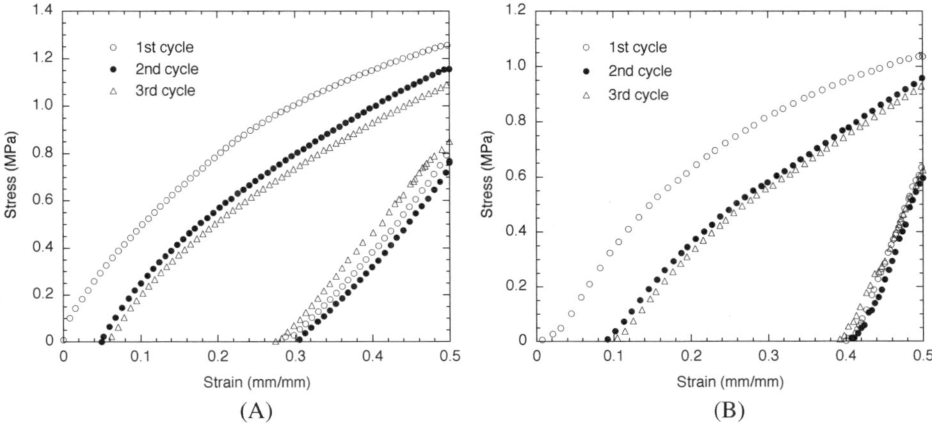

Fig. 13.4. Tensile thermal cycles for (A) neat PU made from PTGM 2000 with 23% HSs and (B) composite prepared with 1 wt% nanocellulose added after reaction. (Reprinted from Ref. 4, with permission.)

At the high temperature part of the shape memory cycles (45°C, in this case), the low-ordered HS domains of the composites are easily perturbed. As it can be observed in Fig. 13.4, the recovery is reduced, while exhibiting higher fixity than the base polymer.[4]

A more recent work of Auad and coworkers[26] dealt with the study of the dependence of the shape memory behavior of a SPU based on a polyether glycol (PTMG) (TERATHANE®) with molar mass = 2900 g/mol, tolylene 2,4-diisocyanate (TDI), and BD, containing 32% HSs and variable amounts of nanocellulose (up to 2 wt%) with testing conditions. In this opportunity, nanocomposites were prepared by mixing a suspension of CNCs in *N*,*N*-dimethylformamide (DMF) with the thermoplastic PU dissolved in the same organic solvent.

Figure 13.5 further illustrates the effect of the addition of CNCs to this presynthesized SPU matrix.[26] The preferential interaction of the cellulose with the HSs results in reduced recovery, an effect that increases (although very slightly) as more crystals are incorporated. Other authors noticed that relatively high loading of nanoparticles, i.e., above 3 wt% also decreases the shape recovery ratio.[5,45] On the other hand, the fixity of the samples was clearly improved by addition of CNCs, although the values did not change appreciably with cellulose concentration.

The fixity of the composites was almost not affected by cycling and was considerably high in all the studied cases (>96%). The authors not only related this increase to the higher modulus of the crystals, but also to the interaction of the nanocellulose with hard and soft segments,[4] which contributed to the disordering of the hard phase and the crystallization of the soft segments. By disordering the HS, the soft segments had a higher mobility to align in the deformation direction, and thus presented a higher fixity after the cooling step.

As it was observed on other properties, the largest incremental change occurred by addition of a minimum cellulose concentration and further changes at increasing

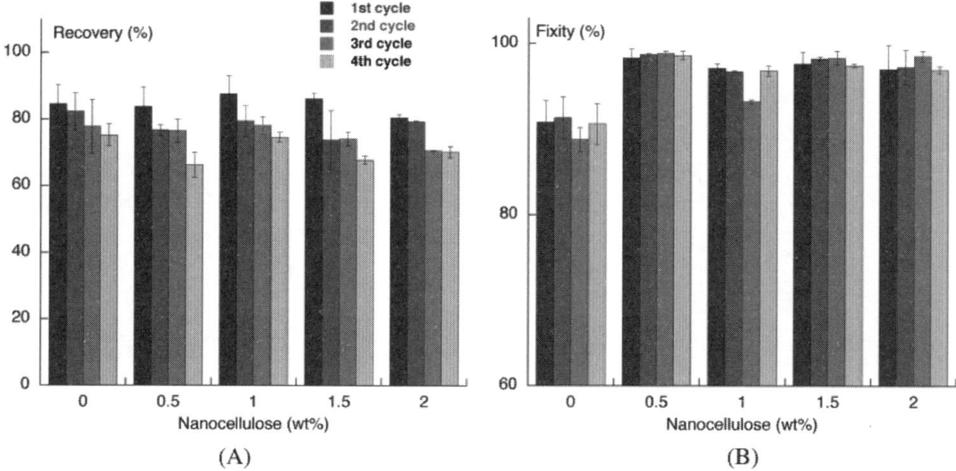

Fig. 13.5. Shape memory behavior of shape memory polyurethane (SMPU) composites. Testing conditions: $\varepsilon_{\max} = 100\%$, elongation rate = 20 mm/min, Ts = 60°C, time recovery = 5 min. (Reprinted from Ref. 26, with permission.)

concentration were less important. This suggests that the interfacial surface of the nanocrystals became less accessible as more fibers are introduced, in direct correlation with the increasing difficulty of obtaining perfect dispersions. The increased soft segment crystallinity with cellulose addition led to stiffer materials that easily maintained the stretched length once the tensile load was removed. However, as pointed out by Cao and Jana,[46] the deformed shape was not fully retained since the amorphous soft segment phase remained in rubbery state after cooling and thus underwent immediate shrinkage after release of the tensile load, while the crystalline soft segment phase added to cellulose rigid reinforcement and helped retain the stretched shape.

Koerner *et al.*[20] also found increases in the fixity and stress recovery for the nanocomposites prepared with carbon nanotubes (CNTs), compared with both the neat PU matrix and composites made with comparable (and higher) carbon black content. They remarked that the addition of four times more carbon black than CNT is necessary to impart comparable fixity. These differences emphasize the importance of filler aspect ratio; general conclusions imply a reduction in shape memory characteristics, which probably arise from varying degrees of dispersion (micrometer-scale agglomerates rather than uniform distribution of primary particles)[29] and associated impact on crystallization behavior (reduction rather than increase of HS crystallite content and nucleation sites for soft segment crystallization).[5]

Current progress in this area includes extruding and blending these smart polymers by spinning fibers or yarns, for knitting, and weaving shape memory materials.[47–49] For example, Auad *et al.*[50] produced SMPU fibers reinforced with cellulose nanofibers (CNFs) by an extrusion process. The interesting result from the spun CNF-reinforced fibers was the significant increase in the breaking tenacity and Young modulus over the unreinforced PU fibers. Fibers prepared with the addition of 0.5 wt% CNF showed an increase in breaking tenacity of 110% over that of the control matrix, while the addition of 1 wt% CNF fibers showed an increase as large as 260%. As in the case of the films, this behavior was not only attributable to the nanoreinforcements, but also to their impact on the behavior of the soft segment crystallization, which allowed for larger deformations.[51] The values of elongation at break for these fibers exceeded the range of values reported for Spandex (ranging from 500–600%) and showed a much lower linear density.[52]

13.3. Nanocellulose as responsive phase in smart materials

13.3.1 *Nanocellulose as water-responsive phase in composite materials*

The development of a new class of mechanically adaptive nanocomposites has been inspired by biological creatures, such as sea cucumbers, which have the ability to reversibly change the stiffness of their dermis. Several recent studies have related this dynamic mechanical behavior to the distinctive nanocomposite architecture of their collagenous tissue, in which interactions among rigid collagen fibrils, embedded in a viscoelastic matrix of fibrillin microfibrils, are regulated by neurosecretory proteins.[53]

This new family of artificial polymer nanocomposites that mimic the architecture and the mechanic adaptability of the sea cucumber dermis are based on low-modulus matrix polymers that are reinforced with a percolating CNF network. Owing to the abundance of surface hydroxyl groups, the CNFs display strong interactions between themselves, causing the evenly dispersed percolated nanocomposites to display high stiffness. The nanofiber–nanofiber interactions can be largely switched off by the introduction of a chemical regulator that allows for competitive hydrogen bonding, resulting in a significant decrease in the stiffness of the material.[53]

One of the advantages of a synthetic system is the ability to introduce other effects into the mimetic biological design. This was achieved quite recently by Shanmuganathan *et al.*[24] who developed stimuli-responsive nanocomposites by introducing percolating networks of CNCs isolated from cotton (CNC) into a poly(vinyl acetate) (PVAc) matrix. They indicated that below the glass-transition temperature ($T_g \sim 63°C$), the tensile storage moduli (E_0) of the dry nanocomposites increased twofold, from 2 GPa for the neat polymer to 4 GPa for a nanocomposite with 16.5% v/v crystals. The relative reinforcement was more significant above T_g, where E_0 was increased nearly 40-fold, from ~1.2 MPa to ~45 MPa. Upon exposure to emulated physiological conditions (immersion in artificial cerebrospinal fluid at 37°C), all nanocomposites showed a pronounced decrease in E_0, for example to 5 MPa for the 16.5% v/v crystal nanocomposites with only about 28% w/w swelling. The authors indicated that this is a significant reduction in the amount of swelling required to decrease the E_0, compared to earlier material versions based on cellulose crystals with higher surface charge density; for example, this 28% swelling is only ~1/3 of that of a corresponding tunicate crystal nanocomposite.[54] Apart from being derived from a different source and having different dimensions, the major difference between the tunicate and cotton cellulose crystals used in their studies is the density of charged sulfate groups (*vide supra*). Thus, authors indicated, in view of the well-established correlation between water uptake and concentration of ionic groups in ionic polymer membranes,[55] that it is plausible that the very desirable, low degree of swelling displayed by the cotton CNC nanocomposites was related to the lower sulfate charge density of the crystals employed. As an example, this decreased swelling may be a considerable advantage for the intended use of these materials as adaptive substrates for intracortical electrodes and other biomedical applications.[24] In summary, nanocomposites of PVAc and cotton CNCs demonstrate a mechanically adaptive behavior in response to thermal and chemical stimuli. Moreover, these nanocomposites can be produced by using CNCs isolated from cotton, which is a more readily accessible cellulose source than tunicates.[24]

Of course, there is still a long way to go to achieve some of the properties of the natural model.[53] For example, the switching rate in the sea cucumber dermis is less than a second, while the current generations of synthetic mimics are diffusion controlled and so switching is much slower. This critical issue was further explored by Hess *et al.*[56] very recently. In this case, they tried to found whether the chemo-responsive switchable stiffness behavior demonstrated previously in bulk samples, based on a tunicate cellulose nanofiber (TW) network encased in a PVAc matrix with dimensions much larger than typical intracortical probes, scales to microelectromechanical systems (MEMS) dimensions. To answer the question, they used a direct-write CO_2 laser to cut the bulk sample, achieving a spot size of $25\,\mu$m. The micromachined tensile testing structures fabricated from the nanocomposite displayed a reversible and switchable stiffness comparable to bulk samples, with a Young's modulus of 3420 MPa when dry, reducing to ~20 MPa when wet. In contrast to the bulk material, which required 15 min to complete the switch at 37°C,[57] PVAc–TW microstructures switch from stiff to flexible in approximately 5 min at room temperature, presumably due to the much larger surface area-to-volume ratio. This mechanically dynamic behavior is particularly attractive for the development of adaptive intracortical probes that are sufficiently stiff to insert into the brain without buckling, but become highly compliant upon insertion. Moreover, the authors found that the microstructures display anisotropic swelling behavior, with 8–12 times more swelling through the film thickness than across the film. Along these lines, a micromachined neural probe incorporating parylene insulating/moisture barrier layers and Ti/Au electrodes was prepared from the nanocomposite using a fabrication process designed specifically for this chemical- and temperature-sensitive material. It was found that the parylene layers only slightly increased the stiffness of the probe in the wet state in spite of its much higher Young's modulus. Furthermore, the Ti/Au electrodes exhibited impedance comparable to Au electrodes on conventional substrates. The anisotropic swelling behavior of the nanocomposite led to excellent adhesion to the parylene layers and to negligible deformation through-plane of the probes when deployed in deionized water.

Mendez *et al.*[16] further investigated the water-activated shape memory behavior of a material whose active phase corresponds to a percolating nanocellulose network included in an elastomeric PU matrix. CNCs obtained by acidolysis of cotton cellulose were used to produce films with up to 20% v/v concentration by solution blending followed by compression molding. As observed in other publications, the incorporation of nanocrystals led to an increase of the storage modulus of the dry nanocomposites that followed a percolation model at concentrations above the threshold (~7% v/v), and the Halpin Kardos model below that concentration.

When water was absorbed in the material, the value of the modulus dropped from 1 GPa to 144 MPa (20% v/v nanocomposite) and the shape memory response of the nanocomposite was observed. In this case, the repetition of shape memory cycles was investigated. The cycles consisted in deforming the samples in the wet state, followed by drying to freeze the temporary shape (a small recovery occurred, however, in this stage). After that step, the sample was re-wetted and the material recovered the original shape (incomplete recovery was observed). Figure 13.6 shows the cycling of the neat PU and two nanocomposites (10 and 20% v/v). After an initial first arrangement (after an initial deformation and recovery), the cyclic response observed is remarkably repetitive. In particular, the fixity ratio is 13.3% for the neat PU, while it increases to 60.7 and 74.3% for the 10 and 20% v/v nanocomposites, respectively. This large, qualitative difference illustrates that the smart response of the material is due to the development of a nanocrystal network that percolates the material. Absorbed water is the chemical regulator that competes for the H-bonding with the surface OH groups of the cellulose and reduces the strong interaction between the crystals. The material swells (up to 12% swelling in the 20% v/v sample, instead of the 4% swelling of the neat PU) and the cellulose network is destroyed; consequently, the modulus of the material drops.

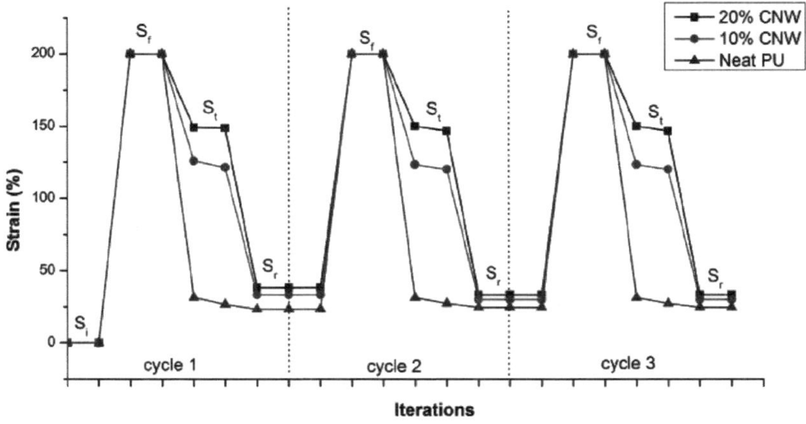

Fig. 13.6. Strain as a function of iterations of shape memory experiments conducted with PU/CNC composite films comprising 0, 10, 20% v/v cotton nanocrystals, where S_i is the initial shape with length l_i, S_f is the intermediate shape under stress with length l_f, S_t is the temporary shape (stress-free) with length l_t, and S_r is the recovered shape with length l_r. (Reprinted from Ref. 16, with permission.)

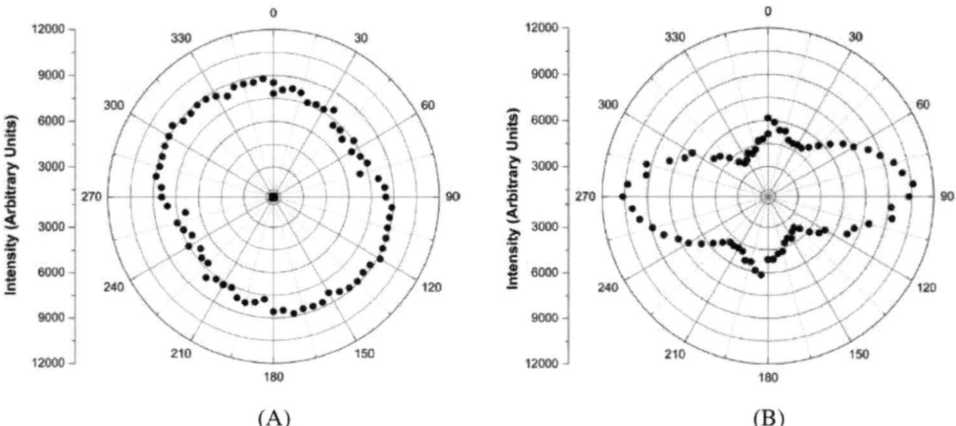

Fig. 13.7. Polar plots of the intensity of the Raman band located at located at $1095\,\text{cm}^{-1}$ as a function of rotation angle of PU/CNC composite film comprising 20% v/v cotton nanocrystals (A) undeformed; (B) wetted, stretched to a strain of 70% for 5 min and recovered (wet) after 5 min to strains of 65%. (Reprinted from Ref. 16, with permission.)

Additionally, the rearrangement of the cellulose orientation during the cycle was studied by polarized Raman spectroscopy. Figure 13.7 shows the polar plot of the intensity of the Raman band located at $1095\,\text{cm}^{-1}$, corresponding to the C–O ring stretch mode of the cellulose, for the undeformed PU–CNCs (20% v/v). In this case, the intensity of the band as a function of the rotating angle shows that there is no preferential orientation of the nanocrystals. Once the material is wetted, the reduction of the crystal–crystal interactions allows reordering of the cellulose network while deforming (stretching) the material. Then, the specimen is dried in this deformed state and the orientation is partially retained. This can be observed in Fig. 13.7, which shows the polar plot of the Raman $1095\,\text{cm}^{-1}$ intensity of the material in this condition. The angular dependence of the intensity of the band is a clear consequence of the preferential (uniaxial) orientation of the crystals in the stretching direction. This effect can be related to the H-bonds that are established between the crystals during drying, which help to stabilize the temporary shape.

13.3.2 *Nanocellulose-containing responsive hydrogels*

As well as in Sec. 13.3.1, the responsive characteristic of the nanocellulose-containing hydrogels is based on the hydrophilicity of the cellulosic phase, but the emphasis in the behavior of these materials is focused on the water absorption swelling behavior. In addition to their high water uptake, stimuli-responsive swelling capabilities and biocompatibility are some of the features that render them suitable for biological and biomedical applications.[58–60]

Historically, the fabrication by conventional chemical cross-linking of mechanically strong hydrogels with good water uptake and stimuli responsiveness properties has been a challenge.[60] To reach this goal, Karaaslan and coworkers[60] designed novel hydrogels with improved mechanical properties utilizing hemicellulose and cellulose crystals derived from wood. Hemicellulose was functionalized using a "grafting from" approach to attach methacrylic functional groups for polymerization and network formation (PHEMA). Utilizing its inherent affinity, hemicellulose was then deposited on the surface of individual cellulose crystals to create multifunctional cross-linking sites. Network structure, mechanical, viscoelastic, and swelling properties of the synthesized PHEMA hydrogels were dependent on the degree of modification of the hemicellulose and the CNC content. Toughness, extensibility, as well as recovery properties of nanoreinforced PHEMA hydrogels were superior to those prepared using conventional cross-linking agents. Overall, the characteristics of the produced PHEMA hydrogels, such as water holding capacity, mechanical properties, and viscoelasticity, appeared to be similar to load-bearing natural tissue having hydrogel-like characteristics. Therefore, the authors proposed that these hydrogels could be considered as potential replacement materials for articular cartilage.

In the same line, Spagnol and coworkers[61–63] used cellulose crystals derived from cotton fibers and different polymers to produce superabsorbent hydrogel composites. These superabsorbent hydrogels are defined as water-insoluble polymers capable of absorbing large amounts of aqueous fluids and retaining them even under pressure. Therefore, they have great advantages over traditional water-absorbing materials and for this reason, they are widely applied in several fields, such as hygienic products, horticulture, gel actuators, drug delivery systems, water blocking tapes, and coal dewatering.[64–69] Regarding the superabsorbent hydrogel composites based on chitosan–graft–poly(acrylic acid) copolymer, the authors found that the 40% and 30% of the swelling capacity of samples is due to the cross-linker (N,N'-methylenebisacrylamide) and filler (cellulose nanofibrils), respectively. The addition of nanocrystals provided faster equilibrium conditions as well as improved the swelling capacity in *ca.* 100 units, from 381 to 486. Their addition also led to the increase of the pores' average dimension. Both hydrogels, chitosan–graft–poly(acrylic acid) and chitosan–graft–poly(acrylic acid)/cellulose nanofibrils, showed responsive behavior in relation to pH and to the salt solution; thus, they presented good potential for application as devices in the controlled release of solutes, as shown in Fig. 13.8.[61]

The same research group[62] synthesized another kind of superabsorbent hydrogel composites, this time using poly(acrylamide-*co*-acrylate) as the matrix and up to 20 wt% CNCs as filler. A superabsorbent hydrogel with $W_{eq} > 1195\,g\,H_2O/g\,gel$ was obtained with a percentage of 10 wt% of CNCs and 0.05 mol% of cross-linking agent. The hydrogel seemed to be responsive to the pH variation (2–12) and also to the presence of salts (NaCl, KCl, $MgCl_2$, and $CaCl_2$). The hydrogel composites synthesized in their work proved to be more efficient in water absorption than the commercial ones or poly(acrylamide-*co*-acrylate) hydrogels without CNCs. Moreover, in their latest publication,[63] they dealt with hydrogel nanocomposites based on starch-*g*-poly(sodium acrylate) and found that the incorporation of CNCs up to 10 wt% provided an improvement in the swelling capacity of the hydrogel due to

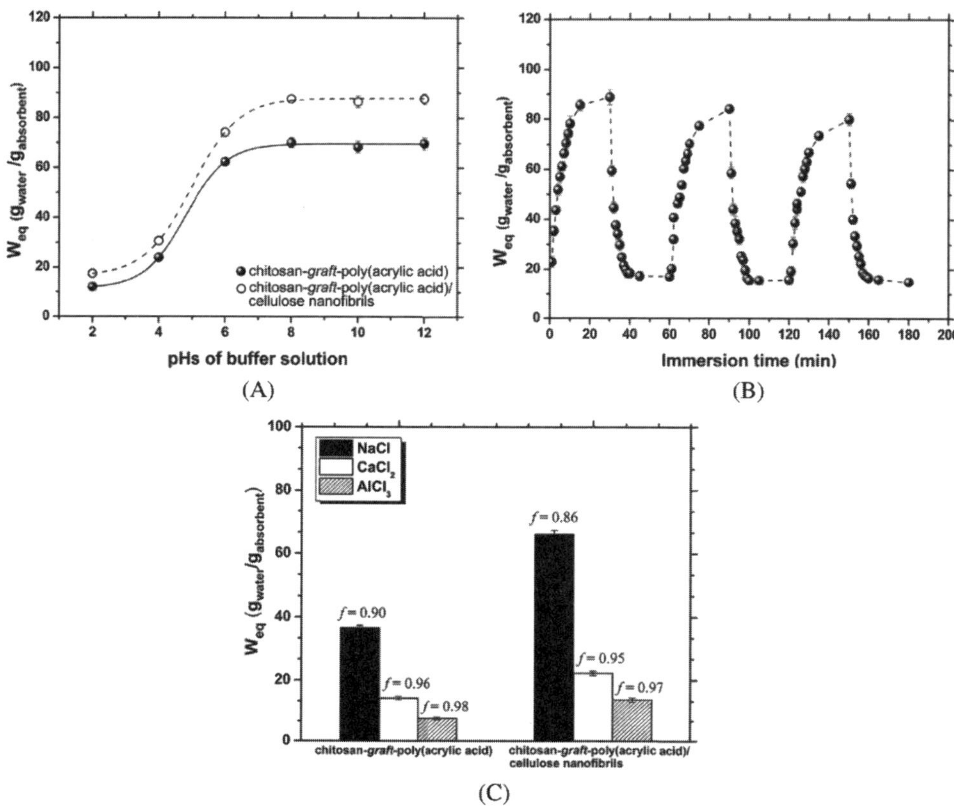

Fig. 13.8. (A) Effect of pH in the swelling properties of chitosan–graft–poly(acrylic acid) and chitosan–graft–poly(acrylic acid)/CNC hydrogel composite with 10 wt% of cellulose nanocrystals. (B) On–off switching behavior as reversible pulsatile swelling (pH 8.0) and deswelling (pH 2.0) of the chitosan–graft–poly(acrylic acid)/CNCs superabsorbent with 10 wt% of CNCs. The time interval between the pH changes was 30 min. (C) W_{eq} of chitosan–graft–poly(acrylic acid) and chitosan–graft–poly(acrylic acid)/CNCs hydrogel composites at aqueous solutions from different salts (conc. equal to 0.15 mg L^{-1}). (Reprinted from Ref. 61, with permission.)

the hydrophilic groups from cellobiose units. Additionally, the CNCs improved the mechanical properties and led to an increase in the average pores size of composites.

13.4. Final considerations

The wide availability, outstanding mechanical properties, low density, biodegradability, and easy chemical modification of nanocellulose make it a prime source for the development of novel materials. This chapter has shown that these properties can be further exploited in the fabrication of responsive/smart composites that incorporate cellulose as nanoreinforcement or in its utilization as the continuous phase in functional nanocomposites.

The work of many researchers around the world has resulted in the incorporation of nanocellulose in different polymers (not limited to water soluble or dispersible ones) by using polar solvents or chemically modifying the cellulose surface. Most of the published work on responsive nanocomposites has dealt with the use of SMPs or with the intelligent manipulation of the hygroscopic nature of a nanocellulose gel. The deposition/decoration of functional nanoparticles or grafting of responsive polymers onto nanocellulose to impart responsive behavior to the derived nanocomposites is also being considered.

Several authors have investigated the grafting of different polymer chains to the surface of the cellulose crystals using various techniques. The results are very exciting since different functionalities can be added to the nanocelluloses, varying hydrophilicity, inducing self-organization, and incorporating optical responses. Coating or bulk incorporation of functional nanoparticles (or polymers) to induce responsive behavior through changes that originate in these additives is also an alternative (e.g., synthesis of superparamagnetic ferrite onto cellulose nanofibers to be used in protein separation[70]).

Given the exponential growth that the research on nanocellulose has experienced in the last years, it is foreseen a large increase in the development of new functional applications.

Novel responsive materials can be expected to appear that will benefit not only from the excellent mechanical properties and easy chemical modification of nanocellulose, but also from its self-organizing properties that can be tailored by chemical modification or carefully selected conditions of the initial suspension media. The challenge is to tie these exciting properties into industrially processable and commercially cost-effective applications.

References

1. P.T. Mather, X. Luo and I.A. Rousseau, Shape memory polymer research, *Annu. Rev. Mater. Res.* **39** (2009) 445–471.
2. M.L. Auad, V.S. Contos, S. Nutt, M.I. Aranguren and N.E. Marcovich, Characterization of nanocellulose-reinforced shape memory polyurethanes, *Polym. Inter.* **57**(4) (2008) 651–659.
3. F. Li, X. Zhang, J. Hou, M. Xu, X. Luo, D. Ma and B.K. Kim, Studies on thermally stimulated shape memory effect of segmented polyurethanes, *J. Appl. Polym. Sci.* **64** (1997) 1511–1516.
4. M.L. Auad, M.A. Mosiewicki, T. Richardson, M.I. Aranguren and N.E. Marcovich, Nanocomposites made from cellulose nanocrystals and tailored segmented polyurethanes, *J. Appl. Polym. Sci.* **115**(2010) 1215–1225.
5. I.S. Gunes and S.C. Jana, Shape memory polymers and their nanocomposites: a review of science and technology of new multifunctional materials, *J. Nanosci. Nanotechnol.* **8**(4) (2008) 1616–1637.
6. A. Lendlein and S. Kelch, Shape-memory polymers, *Angew Chem. Int. Ed. Engl.* **41** (2002) 2034–2057.
7. S.H. Ajili, N.G. Ebrahimi and M. Soleimani, Polyurethane/polycaprolactane blend with shape memory effect as a proposed material for cardiovascular implants, *Acta Biomaterial.* **5** (2009) 1519–1530.
8. S. Oprea, Effect of composition and hard-segment content on thermo-mechanical properties of crosslinked polyurethane copolymers, *High Perform. Polym.* **2** (2009) 353–370.

9. P.T. Knight, K.M. Lee, H. Qin and P.T. Mather, Biodegradable thermoplastic polyurethanes incorporating polyhedral oligosilsesquioxane, *Biomacromolecules* **9** (2008) 2458–2467.
10. C.M. Yakacki, R. Shandas, C. Lanning and K. Gall, Free recovery effects of shape-memory polymers for cardiovascular stents, *Materials Research Society Symposium Proceedings*, 898E (Mechanical Behavior of Biological and Biomimetic Materials) (2006) Paper #: 0898-L15-01.
11. M. Brojan, D. Bombac, F. Kosel and T. Videnic, Shape memory alloys in medicine, *RMZ – Material. Geoenviron.* **55** (2008) 173–189.
12. A. Lendlein and R. Langer, Biodegradable, elastic shape-memory polymers for potential biomedical applications, *Science* **296** (2002) 1673–1676.
13. A. Lendlein, Solving a knotty problem - surgical sutures from shape memory polymers, *Mater. World* **10**(7) (2002) 29–30.
14. W. Sokolowski, A. Metcalfe, S. Hayashi, L.H. Yahia and J. Raymond, Medical applications of shape memory polymers, *Biomed. Mater.* **2** (2007) S23–S27.
15. H.S. Tzou, H.J. Lee and S.M. Arnold, Smart materials, precision sensors/actuators, smart structures, and structronic systems, *Mech. Ad. Mater. Struct.* **11** (2004) 367–393.
16. J. Mendez, P.K. Annamalai, S.J. Eichhorn, R. Rusli, S.J. Rowan, E.J. Foster and C. Weder, Bioinspired mechanically adaptive polymer nanocomposites with water-activated shape-memory effect, *Macromolecules* **44** (2011) 6827–6835.
17. T.J. Lu and A.G. Evans, Design of a high authority flexural actuator using an electrostrictive polymer, *Sens. Actuators, A: Physical* A **99**(3) (2002) 290–296.
18. R. Buckley Patrick, H. McKinley Gareth, S. Wilson Thomas, W. Small, J. Benett William, P. Bearinger Jane, W. McElfresh Michael and J. Maitland Duncan, Inductively heated shape memory polymer for the magnetic actuation of medical devices, *IEEE Trans. Biomed. Eng.* **53** (2006) 2075–2083.
19. C.S. Hazelton, S.C. Arzberger, M.S. Lake and N.A. Munshi, RF actuation of a thermoset shape memory polymer with embedded magnetoelectroelastic particles, *Materials Research Society Symposium Proceedings* **947E** (Responsive Soft Matter) (2007) Paper #: 0947-A01-08.
20. H. Koerner, G. Price, N.A. Pearce, M. Alexander and R.A. Vaia, Remotely actuated polymer nanocomposites-stress-recovery of carbon-nanotube-filled thermoplastic elastomers, *Nat. Mater.* **3**(2) (2004) 115–120.
21. K.D. Harris, C.W.M. Bastiaansen and D.J. Broer, Physical properties of anisotropically swelling hydrogen-bonded liquid crystal polymer actuators, *J. Microelectromech. Syst.* **16**(2) (2007) 480–488.
22. A. Lendlein, H. Jiang, O. Juenger and R. Langer, Light-induced shape-memory polymers, *Nature* **434**(7035) (2005) 879–882.
23. H. Jiang, S. Kelch and A. Lendlein, Polymers move in response to light, *Adv. Mater.* **18**(11) (2006) 1471–1475.
24. K. Shanmuganathan, J.R. Capadona, S.J. Rowan and C. Weder, Bio-inspired mechanically-adaptive nanocomposites derived from cotton cellulose whiskers, *J. Mater. Chem.* **20** (2010) 180–186.
25. Q. Meng and J. Hu, A review of shape memory polymer composites and blends, *Compos. Part A* **40**(11) (2009) 1661–1672.
26. M.L. Auad, T. Richardson, M. Hicks, M.A. Mosiewicki, M.I. Aranguren and N.E. Marcovich, Shape memory-segmented polyurethanes: behavior dependence on nanocellulose addition and testing conditions, *Polym. Int.* **61**(2) (2012) 321–327.
27. H. Jin, M. Kettunen, A. Laiho, H. Pynnönen, J. Paltakari, A. Marmur, O. Ikkala and R.H.A. Ras, Superhydrophobic and superoleophobic nanocellulose aerogel membranes as bioinspired cargo carriers on water and oil, *Langmuir* **27**(5) (2011) 1930–1934.

28. M.L. Auad, T. Richardson, W.J. Orts, E.S. Medeiros, L.H.C. Mattoso, M.A. Mosiewicki, N.E. Marcovich and M.I. Aranguren, Polyaniline modified cellulose nanofibrils as reinforcement of a smart polyurethane, *Polym. Int.* **60**(4) (2011) 743–750.
29. U.M. Casado, R.U. Quintanilla, M.I. Aranguren and N.E. Marcovich, Composite films based on shape memory polyurethanes and nanostructured polyaniline or cellulose-polyaniline particles, *Synth. Metals* **162** (2012) 1654–1664.
30. J.L. Hu, F.L. Ji and Y.W. Wong, Dependency of the shape memory properties of a polyurethane upon thermomechanical cyclic conditions, *Polym. Int.* **54** (2005) 600–605.
31. B.K. Kim and S.Y. Lee, Polyurethanes having shape memory effects, *Polymer* **37** (1996) 5781–5793.
32. B.K. Kim and S.Y. Lee, Polyurethane ionomers having shape memory effects, *Polymer* **39** (1998) 2803–2808.
33. H.M. Jeong, S.Y. Lee and B.K. Kim, Shape memory polyurethane containing amorphous reversible phase, *J. Mater. Sci.* **35** (2000) 1579–1583.
34. B.S. Lee, B.C. Chun, Y.C. Chung, K.I. Sul and J.W. Cho, Structure and thermomechanical properties of polyurethane block copolymers with shape memory effect, *Macromolecules* **34** (2001) 6431–6437.
35. J.H. Yang, B.C. Chun, Y.C. Chung and J.H. Cho, Comparison of thermal/mechanical properties and shape memory effect of polyurethane block-copolymers with planar or bent shape of hard segment, *Polymer* **44**(11) (2003) 3251–3258.
36. J.W. Cho, J.W. Kim, Y.C. Jung and N.S. Goo, Electroactive shape-memory polyurethane composites incorporating carbon nanotubes, *Macromol. Rapid Commun.* **26** (2005) 412–416.
37. Z.G. Wei and R. Sandström, Review. Shape-memory materials and hybrid composites for smart systems. Part I. Shape-memory materials, *J. Mater. Sci.* **33** (1998) 3743–3762.
38. E. Wornyo, K. Gall, F. Yang and W. King, Nanoindentation of shape memory polymer networks, *Polymer* **48** (2007) 3213–3225.
39. I.S. Gunes, F. Cao and S.C. Jana, Evaluation of nanoparticulate fillers for development of shape memory polyurethane nanocomposites, *Polymer* **49** (2008) 2223–2234.
40. J. Xu, W. Shi and W. Pang, Synthesis and shape memory effects of Si–O–Si cross-linked hybrid polyurethanes, *Polymer* **47** (2006) 457–465.
41. K. Gall, M.L. Dunn, Y. Liu, D. Fich, M. Lake, N.A. Munshi, Shape memory polymer nanocomposites, *Acta Mater.* **50** (2002) 5115–5126.
42. Z.S. Petrovic, Y.J. Cho, I. Javni, S. Magonov, N. Yerina, D.W. Schaefer, J. Ilavsky and A. Waddon, Effect of silica nanoparticles on morphology of segmented polyurethanes, *Polymer* **45**(12) (2004) 4285–4295.
43. R.C.R. Nunes, R.A. Pereira, J.L.C. Fonseca and M.R. Pereira, X-ray studies on compositions of polyurethane and silica, *Polym. Test.* **20** (2001) 707–712.
44. N.E. Marcovich, N.E. Bellesi, M.L. Auad, S.R. Nutt and M.I. Aranguren, Cellulose micro/nanocrystals reinforced polyurethane, *J. Mater. Res.* **21**(4) (2006) 870–881.
45. M.S. Kim, J.K. Jun and H.M. Jeong, Shape memory and physical properties of poly(ethyl methacrylate)/Na-MMT nanocomposites prepared by macroazoinitiator intercalated in Na-MMT, *Compos. Sci. Technol.* **68**(7–8) (2008) 1919–1926.
46. F. Cao and S.C. Jana, Nanoclay-tethered shape memory polyurethane nanocomposites, *Polymer* **48**(13) (2007) 3790–3800.
47. Q. Meng, J. Hu and Y. Zhu, Shape-memory polyurethane/multiwalled carbon nanotube fibers, *J. Appl. Polym. Sci.* **106**(2) (2007) 837–848.
48. S.P.L. Tang and G.K. Stylios, An overview of smart technologies for clothing design and engineering, *Int. J. Cloth. Sci. Technol.* **18**(2) (2006) 108–128.

49. F. Ji, Y. Zhu, J. Hu, Y. Liu, L. Yeung and G. Ye, Smart polymer fibers with shape memory effect, *Smart Mater. Struct.* **15** (2006) 1547–1554.
50. T. Richardson, M.A. Mosiewicki, N. Marcovich, M.I. Aranguren, F. Kilinc-Balci, R.M. Broughton, Jr. and M. L. Auad, Study of nano-reinforced shape memory polymers processed by casting and extrusion, *Polym. Compos.* **32**(3) (2011) 455–463.
51. W. Chen, M.L. Auad, R.J.J. Williams and S.R. Nutt, *Eur. Polym. J.* **42** (2006) 1082.
52. Q. Meng and J. Hu, Investigation of sorption and diffusion of supercritical carbon dioxide in polycarbonate, *J. Appl. Polym. Sci.* **109** (2008) 1661–1666.
53. K. Shanmuganathan, J.R. Capadona, S.J. Rowan and C. Weder, Biomimetic mechanically adaptive nanocomposites, *Prog. Polym. Sci.* **35** (2010) 212–222.
54. K. Shanmuganathan, J.R. Capadona, S.J. Rowan and C. Weder, Stimuli-responsive mechanically adaptive polymer nanocomposites, *ACS Appl. Mater. Interfaces* **2**(1) (2010) 165–174.
55. T.S. Zhao, K.D. Kreuer and T.V. Nguyen (Eds.) *Advances in fuel cells.* Elsevier (2007).
56. A.E. Hess, J.R. Capadona, K. Shanmuganathan, L. Hsu, S.J. Rowan, C. Weder, D.J. Tyler and C.A. Zorman, Development of a stimuli-responsive polymer nanocomposite toward biologically optimized, MEMS-based neural probes, *J. Micromech. Microeng.* **21** (2011) 054009 (9 pp).
57. J.R. Capadona, S.K Shanmuganathan, D.J. Tyler, S.J. Rowan and C. Weder, Stimuli-responsive polymer nanocomposites inspired by the sea cucumber dermis, *Science* **319** (2008) 1370–1374.
58. N.A. Peppas, J.Z. Hilt, A. Khademhosseini and R. Langer, Hydrogels in biology and medicine: from fundamentals to bionanotechnology, *Adv. Mater.* **18**(11) (2006) 1345–1360.
59. B.V. Slaughter, S.S. Khurshid, O.Z. Fisher, A. Khademhosseini and N.A .Peppas, Hydrogels in regenerative medicine, *Adv. Mater.* **21**(32–33) (2009) 3307–3329.
60. M.A. Karaaslan, M.A. Tshabalala, D.J. Yelle and G. Buschle-Diller, Nanoreinforced biocompatible hydrogels from wood hemicelluloses and cellulose whiskers, *Carbohydr. Polym.* **86** (2011) 192–201.
61. C. Spagnol, F.H.A. Rodrigues, A.G.B. Pereira, A.R. Fajardo, A.F. Rubira and E.C. Muniz, Superabsorbent hydrogel composite made of cellulose nanofibrils and chitosan-graft-poly (acrylic acid), *Carbohydr. Polym.* **87** (2012) 2038–2045.
62. C. Spagnol, F.H.A. Rodrigues, A.G.V.C. Neto, A.G.B. Pereira, A.R. Fajardo, E. Radovanovic, A.F. Rubira and E.C. Muniz, Nanocomposites based on poly(acrylamide-co-acrylate) and cellulose nanowhiskers, *Eur. Polym. J.* **48** (2012) 454–463.
63. C. Spagnol, F.H.A. Rodrigues, A.G.B. Pereira, A.R. Fajardo and A.F. Rubira, Superabsorbent hydrogel nanocomposites based on starch-*gg*-poly (sodium acrylate) matrix filled with cellulose nanowhiskers, *Cell* **19**(4) (2012) 1225–1237.
64. A. Pourjavadi, M. Sadeghi, M.M. Hashemi and H. Hosseinzadeh, Synthesis and absorbency of gelatin-graft-poly (sodium acrylate-co-acrylamide) superabsorbent hydrogel with salt and pH-responsiveness properties, *E-Polymers* **57** (2006) 1–15.
65. P. Rokhade, S.A. Patil and T.M. Aminabhavi, Synthesis and characterization of semi-interpenetrating polymer network microspheres of acrylamide grafted dextran and chitosan for controlled release of acyclovir, *Carbohydr. Polym.* **67** (2007) 605–613.
66. Y.Q. Xia, T.Y. Guo, B.H. Zhang and B.L. Zhang, Hemoglobin recognition by imprinting in semi-interpenetrating polymer network hydrogel based on polyacrylamide and chitosan, *Biomacromolecules* **6** (2005) 2601–2606.
67. H. Omidian, J.G. Rocca and K. Park, Advances in superporous hydrogels, *J. Control. Release* **102** (2005) 3–12.
68. A. Li, R. Liu and A. Wang, Preparation of starch-graft-poly(acrylamide)/attapulgite superabsorbent composite, *J. Appl. Polym. Sci.* **98** (2005) 1351–1357.

69. R.J.J. Hill, Electric-field-enhanced transport in polyacrylamide hydrogel nanocomposites, *Colloid. Interface Sci.* **316** (2007) 635–644.
70. R.H. Marchessault, G. Bremner and G. Chauve, Fishing for proteins with magnetic cellulosic nanocrystals. In *Polysaccharides for drug delivery and pharmaceutical applications*, Washington, DC: American Chemical Society (2006), ACS Symposium Series Chapter 1. pp. 3–17.

Chapter 14

All-cellulose Composites

Takashi Nishino[1] and Ton Peijs[2]
[1] *Graduate School of Engineering, Kobe University, Japan*
[2] *School of Engineering and Materials Science, Queen Mary, University of London, UK*

The area of self-reinforced polymer composites is one of the fastest growing areas in engineering polymers, but until now, these materials have been mainly developed on the basis of thermoplastic fibers of moderate performance. In this work, we review a new type of self-reinforced composites based on cellulosic fibers to produce *all*-cellulose composites or *all*-cellulose nanocomposites, in which both the fibers and matrix are cellulose. Natural cellulose boasts a high elastic modulus and high tensile strength, implying that cellulose possesses the potential to replace glass fibers. The concept of *all*-cellulose composites allows for the production of composites with higher fiber contents than traditional fiber-reinforced plastics. Moreover, since the matrix and reinforcement phases of these biocomposites are completely compatible with each other, *all*-cellulose composites allow for efficient stress transfer and adhesion at their interface. Under optimized processing conditions, the mechanical and thermal properties of the cellulose fibers in these *all*-cellulose composites can be retained, while the excellent interface can bring optical transparency to these composites. Fabrication, structure, and properties of such *all*-cellulose composites are reviewed.

14.1. Introduction

Composite materials, typically based on glass or carbon fibers embedded into thermosetting resins such as epoxy or unsaturated polyester, show excellent mechanical and thermal properties; thus, they are widely used in various applications ranging from aerospace to vehicles to sports utensils. However, these materials are difficult to recycle and cause environmental problems when disposing them by incineration. Consequently, there are growing demands for environmentally-friendly composites. Paradigm shift from energy consuming materials to sustainable materials has brought increasing importance of biomass utilization. Of these, biofibers are among the most keenly researched materials of the 21st century.

Alternative routes to environmentally-friendly polymer composites have recently focused on approaches following mono-material-based eco-design concepts; so-called "*all*-polymer composites" or "self-reinforced polymer composites". For example, fully recyclable *all*-polypropylene (*all*-PP) or self-reinforced polypropylene

(SR-PP) composites have been proposed to replace traditional glass fiber-reinforced plastics for a number of applications, notably the automotive industry.[1–5] Following the success of these *all*-PP composites, *all*-cellulose composites have recently been introduced.[6–8]

In this chapter, processing methodologies, structure, and properties of *all*-cellulose composites and *all*-cellulose nanocomposites will be reviewed.

14.2. *All*-cellulose composites

In a similar manner to *all*-polymer composites, two different types of approaches can be followed for the creation of self-reinforced cellulose composites: (i) conventional impregnation methods of cellulose matrix into cellulose fibers[6] and (ii) novel selective dissolution methods where the cellulose fiber skins are selectively dissolved to form a matrix phase that bonds fibers together.[7]

Figure 14.1 shows these two different types for the creation of *all*-cellulose composites.

The impregnation method has been used to create *all*-cellulose composites based on ligno-cellulose fibers such as ramie,[6,8] rice husk,[9] bacterial cellulose (BC),[10] and cellulose nanowhiskers.[11] Nishino and coworkers[6] created *all*-cellulose composites in which both the fibers and matrix are cellulose, by distinguishing the solubility of the matrix cellulose into the solvent from that of the fibers through a pretreatment of the fibers.

Natural cellulose can be dissolved into N,N-dimethyl acetamide (DMAc) containing LiCl through the coordination of Li ions to the hydroxyl groups of cellulose. These studies showed great promise. For example, the stress–strain curve of an *all*-cellulose composite based on ramie fibers is shown in Fig. 14.2, together with those of ramie single fiber and cellophane (model for cellulose matrix).

In the case of plant cellulose fibers (ramie), exceptionally high mechanical properties were obtained using LiCl/DMAc as a solvent. Unidirectional composites with

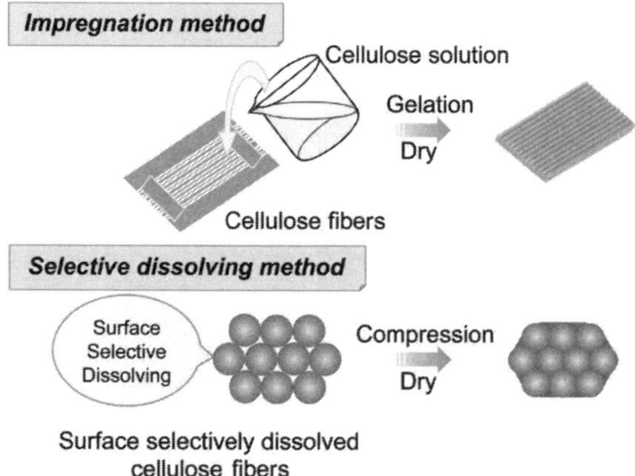

Fig. 14.1. Schematic preparation methodologies for the creation of *all*-cellulose (nano)composite with impregnation method (top) and selective dissolution method (bottom).

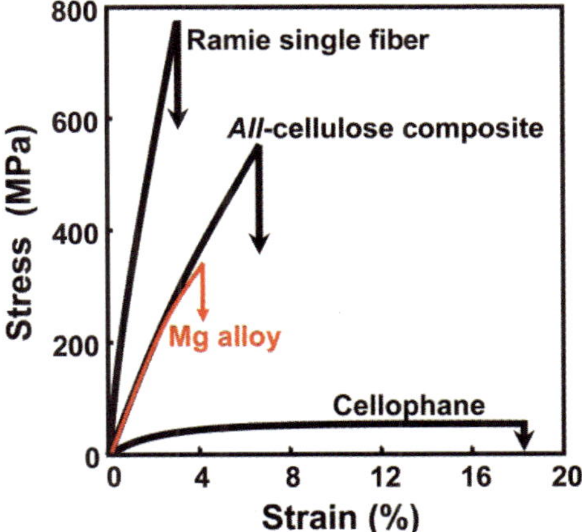

Fig. 14.2. Stress–strain curves of *all*-cellulose composite based on ramie fibers, together with those of ramie single fiber and cellophane (model for matrix).

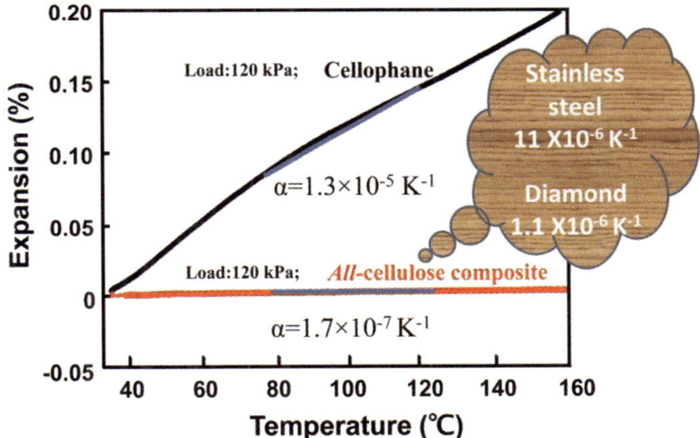

Fig. 14.3. Thermal expansion behavior of *all*-cellulose composite, together with that of cellophane.

a Young's modulus of around 30 GPa and tensile strengths of 550 MPa[6,8] were created (compared to a ramie fiber strength of 700 MPa). As such, these composites sometimes out-performed those of traditional natural fiber composites by a factor of two and are among the highest ever reported for a natural fiber-reinforced composite.[13–20] Young's modulus of *all*-cellulose composites resembles that of Mg alloy, frequently used as light-weight metal alloy for electric devices, with the tensile strength being even higher. Another excellent property of *all*-cellulose composites is their thermal behavior, where the thermal expansion coefficient is shown in Fig. 14.3 together with that of cellophane.

The thermal expansion coefficient (α) of cellophane is in the order of 10^{-5} K^{-1}, which is common for conventional polymers. On the contrary, the α value of *all*-cellulose composite is in the order of 10^{-7} K^{-1}. This is almost one-tenth of that of diamond. However, the processing method employed for these materials meets the difficulty of impregnation especially due to the very high viscosity of cellulose solutions even at low concentrations.

Alternatively, various studies have reported the manufacture of *all*-cellulose composites using selective dissolution of the skins of the cellulose fibers. During composite preparation, rather than selectively melting of fiber surfaces as in the case of thermoplastic *all*-PP composites,[1,2,21–24] here the surface skin layer of the cellulose fibers is selectively dissolved to form the matrix phase of the *all*-cellulose composites. Meanwhile, the remaining cellulose fiber cores maintain their original structure and impart a reinforcing effect to the composite. This method constitutes not only a simplification of the composite's preparation but also provides a significantly improved fiber/matrix interface. The surface selective dissolution method results not only in very high fiber volume fractions but also in a gradual change in properties of the fiber, forming an interphase or interfacial region which minimizes voids and stress concentrations as in sharp, well-defined fiber/matrix interfaces.

This surface dissolution concept for creating *all*-cellulose composites has been explored for a wide range of cellulose materials, including wood pulp fibers,[25–27] filter and Kraft paper,[7,28–30] microcrystalline cellulose (MCC),[31–35] sisal,[36,37] ramie,[38] canola straw,[39] regenerated cellulose (Lyocell), and cellulose fibers spun from an anisotropic phosphoric acid solution (Bocell),[40] electrospun fibers,[41] and BC.[42] In fact, as the process relies on selective dissolution of the outer fiber skins, the process works extremely well with natural cellulose, where the outer layers of the fibers, which are dissolved, mainly consist of disordered cellulose, while the core, which remains, consists of highly oriented cellulose.[43,44] Similar concepts have been applied for creating *all*-cellulose nanocomposites as will be discussed in Sec. 14.3.

14.3. *All*-cellulose nanocomposites

There are several top-down approaches to isolate cellulose nanofibers from the plant cell wall. They include acid hydrolysis,[45] homogenizing,[46,47] grinding,[48] enzymatic hydrolysis,[49] 2,2,6,6-tetramethylpiperidine-1-oxyl radical (TEMPO)-mediated oxidation.[50] Among them, the mechanical grinding method is described here.

First of all, the processing procedure for making cellulose nanofiber is shown in Fig. 14.4 with scanning electron micrographs (SEMs) of fibers (a) before and (b) after grinding.

Refined cellulose microfibers (average diameter of 30 μm) from bast of kenaf were dispersed in water, which was fed into a central hole of a grinding machine from the hopper. The grinding machine is composed of a pair of patterned millstones, which rotate at high speed (*ca.* 1500 rpm). By passing the suspension once, micrometer fibers were downsized into cellulose nanofibers with diameters of *ca.* 30 nm after grinding.

Figure 14.5 shows the X-ray diffraction profiles of cellulose microfiber and nanofiber from kenaf bast fiber. Both fibers show the profiles assigned with

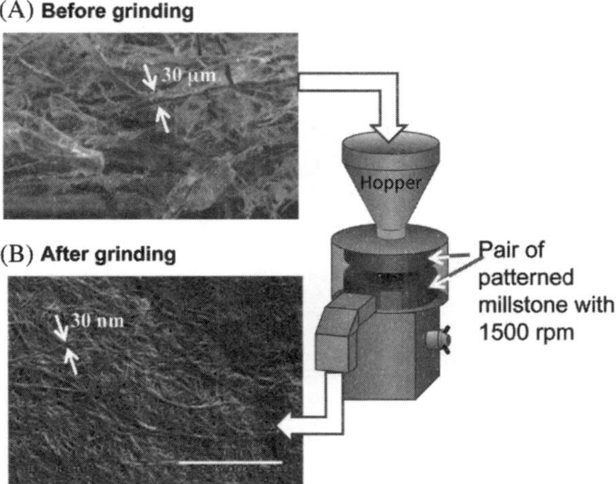

Fig. 14.4. Schematic illustration of an apparatus for grinding, and SEM images of kenaf bark fiber (A) before and (B) after grinding.

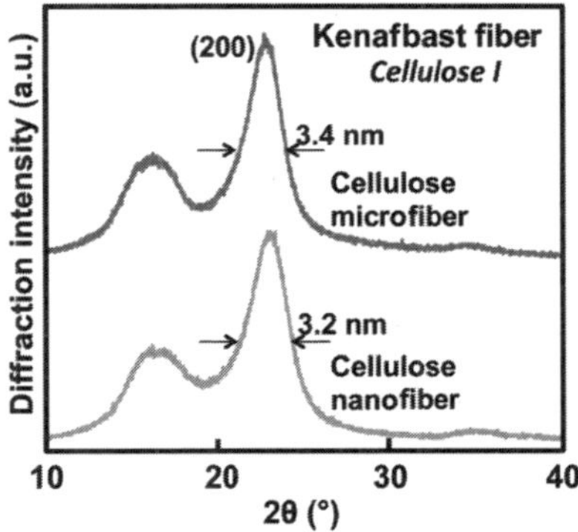

Fig. 14.5. X-ray diffraction profiles of kenaf bast fiber before and after grinding, showing retention of cellulose crystallites in nanofibers after grinding.

cellulose I crystal modification as usual for plant fiber. The crystallite size for the 200 reflection did not change so much around 3 nm even after grinding. This indicates that the crystallites were not damaged while the fiber diameter decreased three magnitudes.

Figure 14.6 shows the stress–strain curves of the paper-like sheet from cellulose microfibers (micropaper) and that from nanofibers (nanopaper).

The curve of micropaper resembles that of normal paper like notebook, photocopy paper, where the paper breaks gradually with fibrillation. In contrast, the curve for the nanopaper showed brittle fracture. In addition, Young's modulus, tensile strength, and elongation at break increased drastically for the nanopaper. Especially, Young's modulus increased by a factor of two, while work-to-fracture increased 13 times for the nanopaper compared with those of micropaper. This

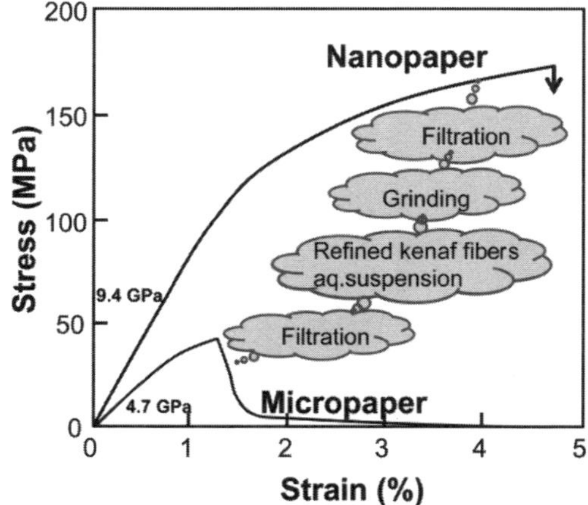

Fig. 14.6. Stress–strain curves of the paper-like sheet from cellulose microfibers (micropaper) and that from nanofibers (nanopaper).

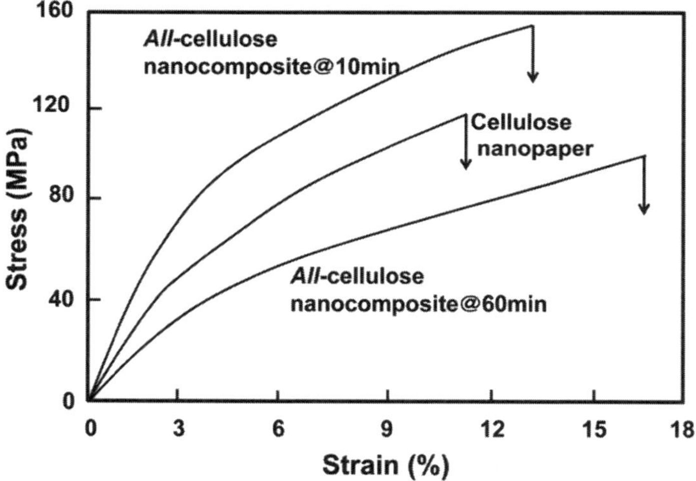

Fig. 14.7. Stress–strain curves of nanopaper, and *all*-cellulose nanocomposites prepared by immersion of canola nanofibers into DMAc/LiCl using the selective dissolution method.

increase is mainly due to the increased network and nanofiber entanglement density, which contribute to the macroscopic properties.

Figure 14.7 shows the stress–strain curves of nanopaper, and *all*-cellulose nanocomposites prepared by immersing these nanofibers into DMAc/LiCl using the selective dissolving method.[39]

In this case, the starting material is canola fibers, which are ground into nanofibers using the same process as described above. The tensile strength is known to strongly depend on the molecular weight of cellulose.[51] The differences in the stress–strain curves of the nanopapers shown in Figs. 14.6 and 14.7 may depend on variability in the molecular weight of the starting fibers. Tensile strength, Young's

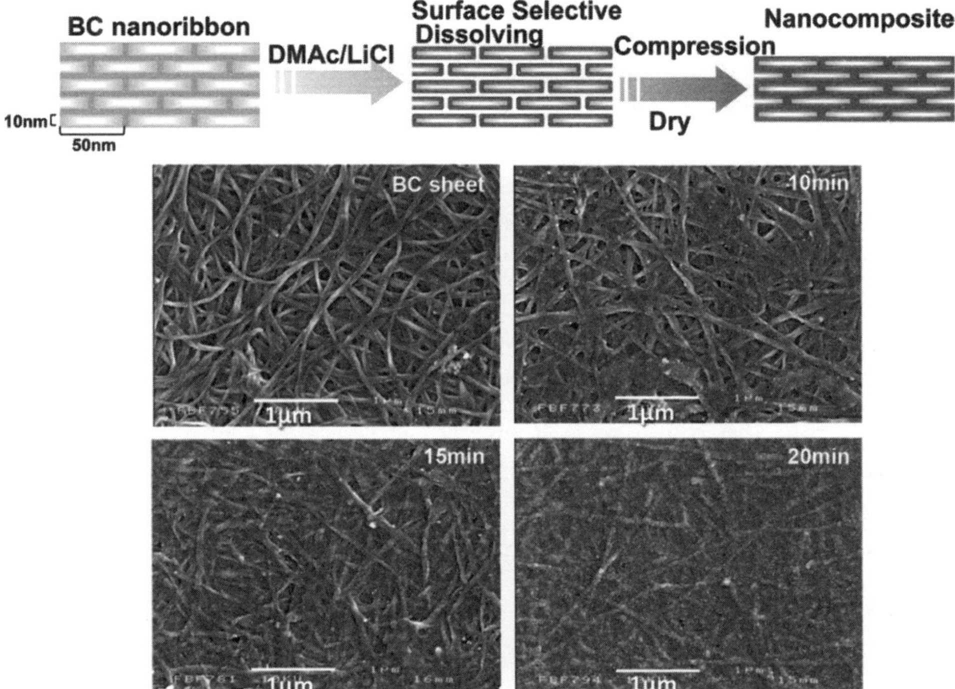

Fig. 14.8. Schematic illustrations of BC nanoribbons and *all*-cellulose nanocomposite cross section, together with SEM images of the *all*-cellulose nanocomposites at various immersion times in DMAc/LiCl.

modulus, and elongation-at-break increased for *all*-cellulose nanocomposite after immersing nanofibers into the solvent for 10 min, followed by compression and drying. However, all of these mechanical properties decreased after longer immersions for 60 min. This reveals that there is an optimum processing window for preparing high-performance *all*-cellulose nanocomposites.

All-cellulose nanocomposites based on MCC[31–35] and BC[42] have also been prepared by these methods. Figure 14.8 shows the microstructure of a BC sheet and a BC composite prepared by Soykeabkaew *et al.*[42] Besides being the cell-wall component of plants, cellulose is also secreted extracellularly as synthesized cellulose fibers by some bacterial species, such as *Acetobacter xylinum*[52] which is called as BC or more popularly as Nata-de-Coco, a familiar dessert. BC presents a unique network structure of a random assembly of ribbon-shaped nanofibers. The fibrils extruded from the bacterial cell are bundled together, forming nanofibers with rectangular cross sections, typically with dimensions of roughly 10 nm (thickness) × 50 nm (width). Using the selective dissolution method, the surface of these nanoribbons can be dissolved, which fills the gap between the nanofibers to form the matrix of the nanocomposites.

Figure 14.9 shows the stress–strain curves of these BC nanocomposites, where with increasing immersion time, after an initial small increase in strength, an obvious

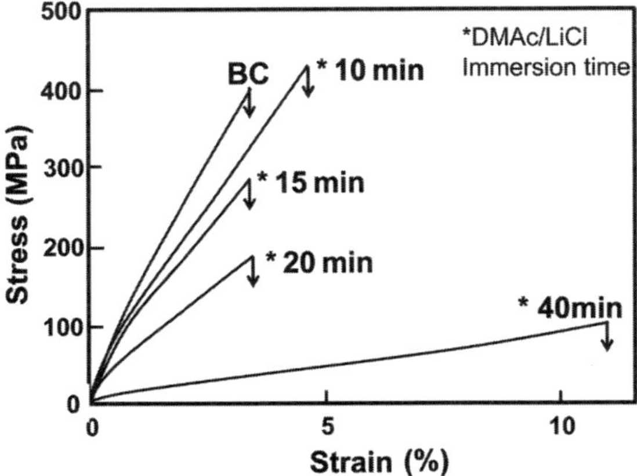

Fig. 14.9. Stress–strain curves of BC sheet and *all*-cellulose nanocomposites prepared with BC at various immersion times.

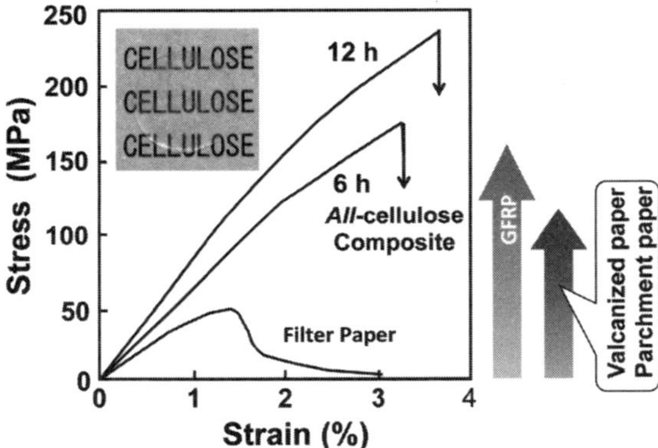

Fig. 14.10. Stress–strain curves of *all*-cellulose composites prepared with 6 and 12 h immersion time, together with that of starting filter paper based on microsize cellulose fibers.

reduction in the composites' tensile properties is apparent. In the case of BC, optimum processing conditions using LiCl/DMAc as a solvent allowed for the preparation of nanocomposites with tensile strengths of 410 MPa and Young's moduli of 18 GPa. Depending on the processing time, the *all*-cellulose nanocomposite showed also remarkable high toughness characteristic possessing a work-of-fracture as high as 16 MJm^{-3}. Interestingly, compared to the *all*-cellulose nanocomposites prepared by partial dissolution of MCC powder of Gindl and Keckes[31] and the nanopaper prepared from cellulose nanofibrils by Henriksson *et al.*,[51] BC nanocomposites exhibit nearly twice the tensile strength, while similar high values of work-to-fracture can be obtained.

In light of these results, it is interesting to compare these data for *all*-cellulose composite from nanosize cellulose fibers of canola and BC, with those based on micron-sized cellulose fibers. Nishino and Arimoto[7] developed an isotropic *all*-cellulose composite using filter paper as a cellulose source as shown in Fig. 14.10.

Unlike BC and canola nanocomposites, these *all*-cellulose microcomposites showed a strong improvement in tensile strength with immersion time (almost five-fold increase in strength from 50 to 240 MPa). In comparison to BC and canola nanopaper, filter paper consists of a loosely formed, much weaker, micro-size cellulose fiber network with the appearance of larger voids and less hydrogen and van der Waals bonded fiber–fiber interactions. After the surface-selective dissolution process, these voids are filled with cellulose matrix leading to a stronger interface and better stress transfer capability and as a result a marked increase in strength of these *all*-cellulose composites. Again, the mechanical properties of these *all*-cellulose composites are far superior compared to traditional isotropic natural fiber mat composites based on flax/PP with typical tensile strengths of 50 MPa,[53,54] short glass fiber-reinforced composites, or vulcanized and parchment paper. The improvement in the composite's interface with increasing immersion time was also evident through the observed improvement in optical transparency with immersion time as shown in Fig. 14.10.

Nanocellulose like BC, on the other hand, has already a very strong network structure[55] based on high-modulus nanosize cellulose ribbons, which allows them to form more extensive van der Waals and hydrogen bonding as schematically shown in Fig. 14.11.

In the case of BC nanocomposites, only a slight enhancement in the network structure caused by the improved bonding from the newly created matrix is obtained. BC sheets have already a high initial level of inter-fiber bonding through extensive hydrogen bonding of the continuous nanoribbons and no significant further improvements are observed with further dissolution times. These results are a further indication of the very strong initial van der Waals or hydrogen bonded network that can be created by nanosize cellulose fibers such as in BC, which do not require further strengthening through a cellulose matrix.

An alternative procedure for fabricating *all*-cellulose nanocomposites is the use of an ionic liquid. Some kinds of ionic liquid are well known to dissolve cellulose under mild condition.[57] Yousefi *et al.*[58] fabricated *all*-cellulose

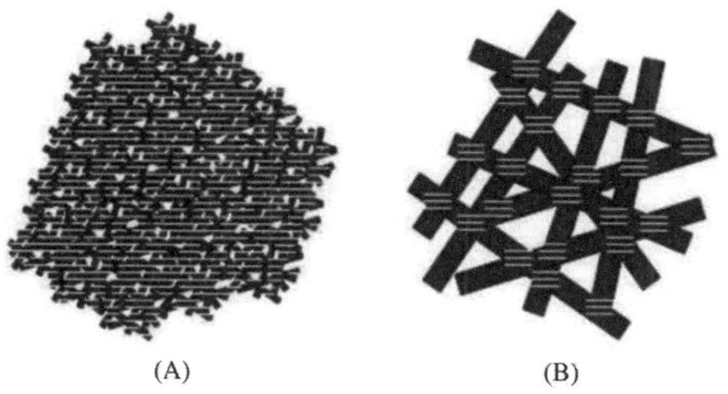

Fig. 14.11. Schematic illustration of extensive hydrogen bonding in (A) BC (nanosize network) compared to (B) cellulose micropaper (microsize network).

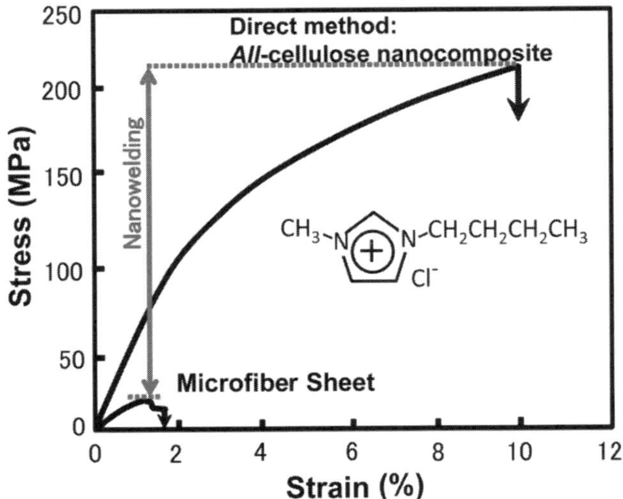

Fig. 14.12. Stress–strain curves of micropaper and *all*-cellulose nanocomposite at a dissolution time of 8 h into ionic liquid BMIMCl.

nanocomposite directly from cellulose microfiber using the ionic liquid, 1-butyl-3-methylimidazolium chloride (BMIMCl), for selective dissolution. Here, cellulosic microfibers were just immersed into BMIMCl, followed by solvent exchange using methanol, compression, and drying.

Figure 14.12 shows the stress–strain curves of cellulose micropaper and *all*-cellulose nanocomposite prepared using this ionic liquid (its chemical formula is given in the figure).

Clearly, the nanocomposite possessed far superior mechanical properties than the micropaper. Tensile strength, Young's modulus, and strain-at-break for the nanocomposite increased to 208 MPa, 20 GPa, and 9.8%, respectively. In other words, the values for the *all*-cellulose nanocomposite after a dissolution time of 8 h increased 9, 4, and 7 times compared with those of micropaper, respectively

Figure 14.13 shows the SEM images of (a) starting canola fiber, (b) fiber after 5 min partial dissolution in BMIMCl, and (c) the tensile fracture surface of the *all*-cellulose nanocomposite prepared by partial dissolution in BMIMCl.

Starting canola fiber possesses a diameter of 26 μm (a), some nanostructures (nanofibers) appeared on the surface of the microfiber after an initial dissolution time of 5 min (b). It is seen that the ionic liquid partly penetrated the gaps among the nanofibers and began to separate them into individual cellulose nanofibers. Here, nanofibers less than 100 nm in diameter are observed. After solvent exchange, compression, and drying, nanofibers were completely dismantled, and a unique structure had been created as shown in the tensile fracture surface (c). Here, partly dissolved nanofibers were surrounded by non-crystalline cellulose.

Figure 14.14 shows a schematic cross-sectional model of *in situ* fabrication of *all*-cellulose nanocomposites from cellulose microfibers by partial dissolution and nanowelding.

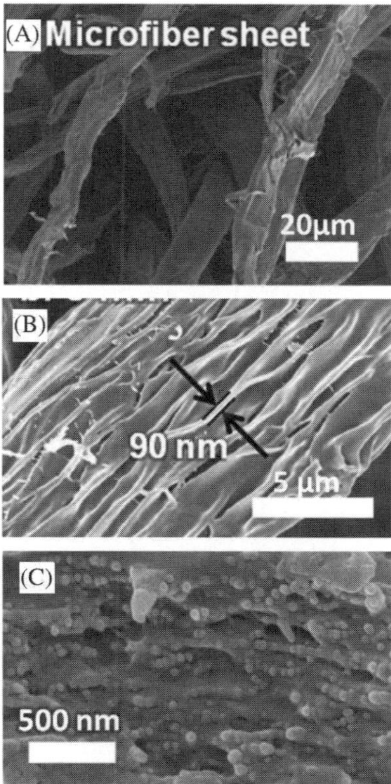

Fig. 14.13. SEM images of (A) starting canola fiber, (B) fiber after 5 min partial dissolution in BMIMCl, and (C) the tensile fracture surface of the *all*-cellulose nanocomposite prepared by partial dissolution in ionic liquid BMIMCl

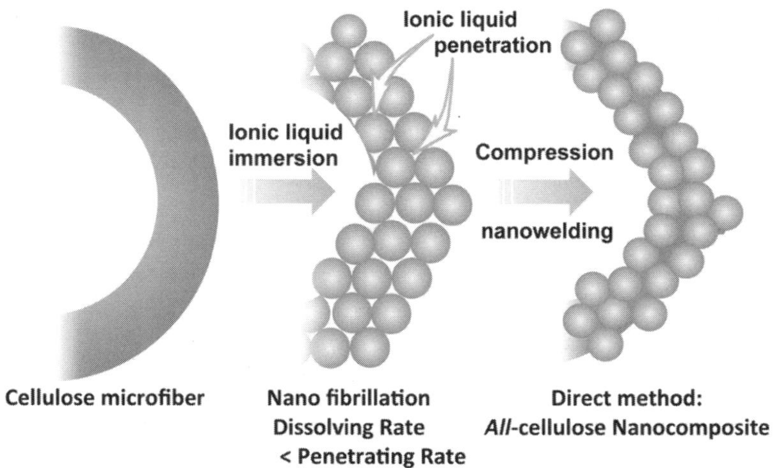

Fig. 14.14. Cross-sectional schematic model of *in situ* fabrication of *all*-cellulose nanocomposites from cellulose microfibers by partial dissolution and nanowelding using ionic liquid BMIMCl.

It is well-known that the gaps among cellulose nanostructures make a permeable path for fluids such as water and various solvents. Therefore, BMIMCl penetrated the gaps and dissolved the skin part of the nanostructures. The amount of non-crystalline phase increased as a result of the partial dissolution process. When this occurs, the cellulose chains are free to move in the solvent and entangle with other similarly dissolved chains from the other components. During the solvent exchange, compression, and drying processes, the chains lose their mobility: as they are brought into intimate contact, entanglements result in a weld[59] This leaves a resolidified mass of entangled cellulose chains, which constitutes a nanowelded interface/interphase contact. The key of using ionic liquids is that the penetrating rate of the ionic liquid into the cellulose microfiber is much higher than the dissolution rate, which enables the direct nanofibrillation of the microfibers. The welding of nanostructures by non-crystalline phases has been previously applied to carbon nanotubes[60] The weld layer plays here the role of a matrix, encapsulating the residual undissolved nanofibers, filling the voids, and joining adjacent nanofibers together.

Figure 14.15 shows the optical transmittance between 200 and 1000 nm and a distinct difference in the transparency of the micropaper and that of the *all*-cellulose nanocomposite.

The optical transmittance at 800 nm was 0.3% and 76% for the micropaper and the nanocomposite, respectively. In other words, transparency increased 250-fold with the partial dissolution and nanowelding process of microfibers to create the nanocomposite. As discussed, the high transparency can be attributed to the finer reinforcement dimension,[61] the interface free structure and reduction in volume of voids, as well as the improved surface roughness of the nanocomposite film compared to micropaper.

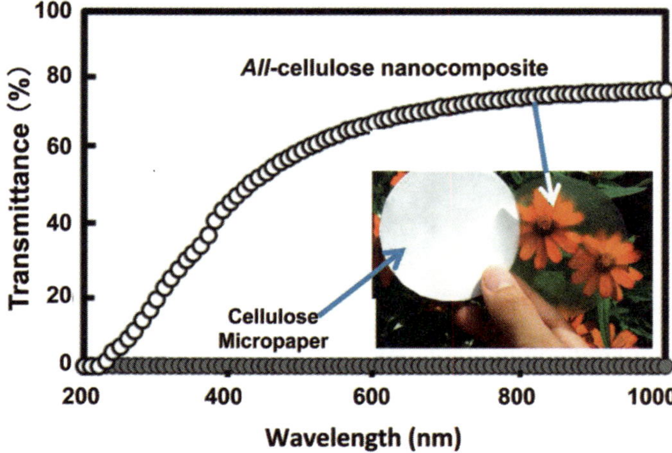

Fig. 14.15. Optical transmittance between 200 and 1000 nm and the distinct difference in transparency of micropaper and *all*-cellulose nanocomposite.

14.4. Conclusions

Traditional arguments for using natural fibers as replacements for glass fibers in polymer composites are often based on improved mechanical properties and reduced cost. However, the benefits of natural fibers are quite different to those of glass, and they should be used to create novel multifunctional eco-materials. As one possible solution, the fabrication using both impregnation and selective dissolution methods, structure and properties (mechanical, thermal, optical) of *all*-cellulose composites and nanocomposites based on different cellulosic resources (ramie, kenaf, canola straw, regenerated cellulose, filter paper, and BC) are reviewed in this chapter. These composites are totally composed of sustainable cellulosic resources, and hence they are fully biodegradable after service, giving them advantages with regard to end-of-life disposal through composting or incineration. These composites not only overcome environmental issues but their good interfacial properties bring additional optical transparency. Moreover, these composites possess excellent mechanical and thermal properties, which are superior to traditional natural fiber-reinforced plastics. Their promising properties make *all*-cellulose composites and nanocomposites high-performance multifunctional materials that have the potential to be used for lightweight structures, biomedical engineering, aerospace, sports equipment, and highly flexible electronic and magnetic devices.

References

1. I.M. Ward and P.J. Hine, Novel composites by hot compaction of fibers, *Polym. Eng. Sci.* **37**(11) (1997) 1809–1814.
2. I.M. Ward and P.J. Hine, The science and technology of hot compaction, *Polymer* **45** (2004) 1413–1427.
3. N.N. Cabrera, B. Alcock, J. Loos and T. Peijs, Processing of all-polypropylene composites for ultimate recyclability, *Proc. Inst. Mech. Eng Part L* **218** (2004) 145–155.
4. B. Alcock, N.O. Cabrera, N.-M. Barkoula, J. Loos and T. Peijs, The mechanical properties of unidirectional all-polypropylene composites, *Compos. Part A* **37**(5) (2006) 716–726.
5. B. Alcock, N.O. Cabrera, N.-M. Barkoula, A.B. Spoelstra, J. Loos and T. Peijs, The mechanical properties of woven tape all-polypropylene composites, *Compos. Part A* **38** (2007) 147–161.
6. T. Nishino, I. Matsuda and K. Hirao, All-cellulose composite, *Macromol.* **37** (2004) 7683–7687.
7. T. Nishino and N. Arimoto, All-cellulose composite prepared by selective dissolving of fiber surface, *Biomacromol.* **8**(9) (2007) 2712–2716.
8. C. Qin, N. Soykeabkaew, N. Xiuyuan and T. Peijs, The effect of fibre volume fraction and mercerization on the properties of all-cellulose composites, *Carbohydr. Polym.* **71**(3) (2008) 458–467.
9. Q. Zhao, R.C.M. Yam, B.Q. Zhang, Y.K. Yang, X.J. Cheng and R.K.Y. Li, Novel all-cellulose ecocomposites prepared in ionic liquids, *Cellulose* **16**(2) (2009) 217–226.
10. W. Gindl and J. Keckes, Tensile properties of cellulose acetate butyrate composites reinforced with bacterial cellulose, *Compos. Sci. Technol.* **64**(15) (2004) 2407–2413.
11. T. Pullawan, A.N. Wilkinson and S.J. Eichhorn, Influence of magnetic field alignment of cellulose whiskers on the mechanics of all-cellulose nanocomposites, *Biomacromol.* **13**(8) (2012) 2528–2536.

12. R. Heijenrath and T. Peijs, Natural-fibre-mat-reinforced thermoplastic composites based on flax fibres and polypropylene, *Adv. Compos. Lett.* **5** (1996) 81–85.
13. S. Luo and A.N. Netravali, Mechanical and thermal properties of environment-friendly "Green" composites made from pineapple leaf fibers and poIy (hydroxybutyrate-co-valerate) resin, *Polym. Compos.* **20** (1999) 367–378.
14. M.J.A. Van Den Oever, H.L. Bos and M.J.J.M. Van Kemenade, Influence of the physical structure of flax fibres on the mechanical properties of flax fibre reinforced polypropylene composites, *Appl. Compos. Mater.* **7** (2000) 387–402.
15. K. Oksman, L. Wallstrom, L.A. Berglund and R.D.T. Filho, Morphology and mechanical properties of unidirectional sisal–epoxy composites, *J. Appl. Polym. Sci.* **84** (2002) 2358–2365.
16. B. Madsen and H. Lilholt, Physical and mechanical properties of unidirectional plant fibre composites — an evaluation of the influence of porosity, *Compos. Sci. Technol.* **63** (2003) 1265–1272.
17. W. Liu, M. Misra, P. Askeland, L.T. Drzal and A.K. Mohanty, 'Green' composites from soy based plastic and pineapple leaf fiber: fabrication and properties evaluation, *Polymer* **46** (2005) 2710–2721.
18. I. Van de Weyenberg, T.C. Truong, B. Vangrimde and I. Verpoest, Improving the properties of UD flax fibre reinforced composites by applying an alkaline fibre treatment, *Compos. Part A* **37** (2006) 1368–1376.
19. S.B. Brahim and R.B. Cheikh, Influence of fibre orientation and volume fraction on the tensile properties of unidirectional Alfa-polyester composite, *Compos. Sci. Technol.* **67** (2007) 140–147.
20. C.C. Gao, L. Yu, H.S. Liu and L. Chen, Development of self-reinforced polymer composites, *Prog. Polym. Sci.* **37** (2012) 767–780.
21. T. Peijs, Composites for recyclability, *Mater. Today* **6** (2003) 30–35.
22. N. Cabrera, B. Alcock, J. Loos and T. Peijs, Processing of all-polypropylene composites for ultimate recyclability, *Proc. Inst. Mech. Eng. Part L* **218** (2004) 145–155.
23. B. Alcock, N.O. Cabrera, N.-M. Barkoula, J. Loos and T. Peijs, The mechanical properties of unidirectional all-polypropylene composites, *Compos. Part A* **37** (2006) 716–726.
24. B. Alcock, N.O. Cabrera, N.-M. Barkoula, A.B. Spoelstra, J. Loos and T. Peijs, The mechanical properties of woven tape all-polypropylene composites, *Compos. Part A* **38** (2007) 147–161.
25. H. Matsumura, J. Sugiyama and W.G. Glasser, Cellulosic nanocomposites. I. Thermally deformable cellulose hexanoates from heterogeneous reaction, *J. Appl. Polym. Sci.* **78** (2000) 2242–2253.
26. H. Matsumura and W.G. Glasser, Cellulosic nanocomposites. II. Studies by atomic force microscopy, *J. Appl. Polym. Sci.* **78** (2000) 2254–2261.
27. H. Nilsson, S. Galland, P.T. Larsson, E.K. Gamstedt, T. Nishino, L.A. Berglund and T. Iversen, A non-solvent approach for high-stiffness all-cellulose biocomposites based on pure wood cellulose, *Compos. Sci. Technol.* **70** (2010) 1704–1712.
28. A. Gandini, A.A.S. Curvelo, D. Pasquini and A.J. de Menezes, Direct transformation of cellulose fibres into self-reinforced composites by partial oxypropylation, *Polymer* **46** (2005) 10611–10613.
29. A.J. de Menezes, D. Pasquini, A.A.S. Curvelo and A. Gandini, Self-reinforced composites obtained by the partial oxypropylation of cellulose fibers. 1. Characterization of the materials obtained with different types of fibers, *Carbohydr. Polym.* **76** (2009) 437–442.
30. A.J. de Menezes, D. Pasquini, A.A.S. Curvelo and A. Gandini, Self-reinforced composites obtained by the partial oxypropylation of cellulose fibers. 2. Effect of catalyst on the mechanical and dynamic mechanical properties, *Cellulose* **16** (2009) 239–246.
31. W. Gindl and J. Keckes, All-cellulose nanocomposite, *Polymer* **46** (2005) 10221–10225.

32. W. Gindl, K.J. Martinschitz, P. Boesecke and J. Keckes, Structural changes during tensile testing of an all-cellulose composite by in situ synchrotron X-ray diffraction, *Compos. Sci. Technol.* **66** (2006) 2639–2647.
33. W. Gindl and J. Keckes, Drawing of self-reinforced cellulose films, *J. Appl. Polym. Sci.* **103** (2007) 2703–2708.
34. B.J.C. Duchemin, R.H. Newman and M.P. Staiger, Phase transformations in microcrystalline cellulose due to partial dissolution, *Cellulose* **14** (2007) 311–320.
35. B.J.C. Duchemin, R.H. Newman and M.P. Staiger, Structure–property relationship of all-cellulose composites, *Compos. Sci. Technol.* **69** (2009) 1225–1230.
36. L. Xun, Q.Z. Ming, Z.R. Min and C.Y. Gui, Enzyme degradability of benzylated sisal and its self-reinforced composites, *Polym. Adv. Technol.* **14** (2003) 676–685.
37. X. Lu, M.Q. Zhang, M.Z. Rong, D.L. Yue and G.C. Yang, The preparation of self-reinforced sisal fiber composites, *Polym. Compos.* **12** (2004) 297–307.
38. N. Soykeabkaew, N. Arimoto, T. Nishino and T. Peijs, All-cellulose composites by surface selective dissolution of aligned ligno-cellulosic fibres, *Compos. Sci. Technol.* **68** (2008) 2201–2207.
39. H. Yousefi, M. Faezipour, T. Nishino, A. Shakeri and G. Ebrahimi, All-cellulose composite and nanocomposite made from partially dissolved micro- and nanofibers of canola straw, *Polym. J.* **43**(6) (2011) 559–564.
40. N. Soykeabkaew, T. Nishino and T. Peijs, All-cellulose composites of regenerated cellulose fibres by surface selective dissolution, *Compos. Part A* **40** (2009) 321–328.
41. W.L.E. Magalhães, X.D. Cao, M.A. Ramires and L.A. Lucia, Novel method for Inducing the alignment of cellulose nanocrystals-reinforced cellulose nanofibers, *Tappi J.* **10** (2011) 19–25.
42. N. Soykeabkaew, C. Sian, S. Gea, T. Nishino and T. Peijs, All-cellulose nanocomposites by surface selective dissolution of bacterial cellulose, *Cellulose* **16**(3) (2009) 435–444.
43. D. Klemm, H.-P. Schmauder and T. Heinze, Cellulose, *Biopolymers* **6** (2003) 275–287.
44. X. Lu, M.Q. Zhang, M.Z. Rong, G. Shi and G.C. Yang, Self-reinforced melt processable composites of sisal, *Compos. Sci. Technol.* **63** (2003) 177–186.
45. B.G. Ranby, Aqueous colloidal solutions of cellulose micelles, *Acta. Chem. Scand.* **3** (1949) 649–650.
46. A.F. Turbak, F.W. Snyder and K.R. Sandberg, Microfibrillated cellulose, a new cellulose product: properties, uses, and commercial potential, *J. Appl. Polym. Sci. Appl. Polym. Symp.* **37** (1983) 815–827.
47. M. Henriksson, G. Henriksson, L.A. Berglund and T. Lindström, An environmentally friendly method for enzyme-assisted preparation of microfibrillated cellulose (MFC) nanofibers, *Eur. Polym. J.* **43** (2007) 3434–3441.
48. T. Taniguchi and K. Okamura, New films produced from microfibrillated natural fibres, *Polym. Int.* **47** (1998) 291–294.
49. S. Janardhnan and M.M. Sain, Isolation of cellulose microfibrils — an enzymatic approach, *Bioresources* **1** (2006) 176–188.
50. T. Saito, S. Kimura, Y. Nishiyama and A. Isogai, Cellulose nanofibers prepared by TEMPO-mediated oxidation of native cellulose, *Biomacromol.* **8** (2007) 2485–2491.
51. M. Henriksson, L.A. Berglund, P. Isaksson, T. Lindström and T. Nishino, Cellulose nanopaper structures of high toughness, *Biomacromol.* **9** (2008) 1579–1585.
52. Y. Nishi, M. Uryu, S. Yamanaka, K. Watanabe, N. Kitamura, M. Iguchi and S. Mitsuhashi, The structure and mechanical properties of sheets prepared from bacterial cellulose Part 2 Improvement of the mechanical properties of sheets and their applicability to diaphragms of electroacoustic transducers, *J. Mater. Sci.* **25**(6) (1990) 2997–3001.
53. R. Heijenrath and T. Peijs, Natural-fibre-mat-reinforced thermoplastic composites based on flax fibres and polypropylene, *Adv. Compos. Lett.* **5**(3) (1996) 81–85.

54. S. Luo and A.N. Netravali, Mechanical and thermal properties of environment-friendly "Green" composites made from pineapple leaf fibers and poly (hydroxybutyrate-co-valerate) resin, *Polym. Compos.* **20** (1999) 367–378.
55. S. Gea, F.G. Torres, O.P. Troncoso, C.T. Reynolds, F. Vilasecca, M. Iguchi and T. Peijs, Biocomposites based on bacterial cellulose and apple and radish pulp, *Int. Polym. Process* **22** (2007) 497–501.
56. I. Van de Weyenberg, J. Ivens, A. De Coster, B. Kino, E. Baetens and I. Verpoest, *1st International eco comp conference*, Elsevier Sci Ltd., London, United Kingdom (2001) p. 1241.
57. R.P. Swatloski, S.K. Spear, J.D. Holbrey and R.D. Rogers, Dissolution of cellulose with ionic liquids, *J. Am. Chem. Soc.* **124** (2002) 4974–4975.
58. H. Yousefi, T. Nishino, M. Faezipour, G. Ebrahimi and A. Shakeri, Direct fabrication of all-cellulose nanocomposite from cellulose microfibers using ionic liquid-based nanowelding, *Biomacromol.* **12** (2011) 4080–4085.
59. L.M. Haverhals, W.M. Reichert, H.C.D. Long and P.C. Trulove, Natural fiber welding, *Macromol Mater Eng.* **295** (2010) 425–430.
60. S. Possidonio, L.C. Fidale and O.A. El Seoud, Microwave-assisted derivatization of cellulose in an ionic liquid: an efficient, expedient synthesis of simple and mixed carboxylic esters, *J Polym Sci. Part A Polym Chem.* **48** (2010) 134–143.
61. M. Nogi, S. Iwamoto, A.N. Nakagaito and H. Yano, Optically transparent nanofiber paper, *Adv Mater* **21** (2009) 1595–1598.

Chapter 15

Bacterial Nanocellulose for Medical Applications: Potential and Examples

Dieter Klemm[1], Friederike Kramer[2], Hannes Ahrem[1,2], Victoria Kopsch[1,2], Thomas Richter[1], Wolfgang Fried[1], Ulrike Udhardt[1], Anja Sterner-Kock[3], Maximilian Scherner[4], Stephanie Reutter[4] and Jens Wippermann[4]

[1]*Polymet Jena Association, Jena, Germany*

[2]*Jenpolymer Materials. Ltd. & Co. KG, Jena, Germany*

[3]*Center for Experimental Medicine, University Hospital of Cologne, Cologne, Germany*

[4]*Department of Cardiothoracic Surgery, University Hospital of Cologne, Cologne, Germany*

Bacterial nanocellulose (BNC) represents an important green nanomaterial and, in particular, a promising natural biomaterial. It is produced biotechnologically from low-molecular-weight sugars like D-glucose by using non-pathogenic bacteria such as *Gluconacetobacter* species. In this process, BNC is generated as a dimensionally stable hydrogel. The unique valuable properties of BNC are mainly based both on the collagen-like hierarchical three-dimensional nanofiber network structure and its exciting controllability and design directly during biosynthesis. The great potential for medical applications ranges from wound dressings up to innovative implants for visceral and cardiovascular surgery. This chapter assembles initially the current knowledge in biocompatibility, bioactivity, and mechanical properties of BNC. The following are examples for the *in situ* design of channeled BNC membranes and of BNC tubes, the latter as bioactive implants for small-diameter ($\leq$5 mm) blood vessels. Using a matrix technology, the walls of these BNC tubes consist of several layers, firmly connected and arranged rotationally symmetric around the longitudinal axis of the tubes — similar to natural blood vessels. As substitutes of the carotid artery of sheep, such tubes with a length of 100 mm show good surgical feasibility and functional performance during the evaluation period of 3 months. They are specifically colonized with endothelial cells, smooth muscle cells, and fibroblasts.

15.1. Introduction

Over the past decade, and in particular in the last five years, a growing number of research groups worldwide have reported on the creation and utilization of celluloses with fibrils or crystals widths in the nanometer range. Nanofibrillated cellulose (NFC) from wood, nanocrystalline cellulose (NCC) from highly crystalline celluloses by removing of amorphous parts, as well as bacterial nanocellulose (BNC) biotechnologically produced from simple sugars such as D-glucose represent a new family of natural nanomaterials. This nanocellulose family is the subject of a current review article.[1] The state of knowledge on this topic was further discussed in 2012.[2,3] The nanocelluloses combine, in an exciting manner, significant structural elements and properties of cellulose[4–9] with the unique characteristics of nanoscale materials. In particular, the special features of the biotech BNC expand the application profile of cellulose materials principally and significantly. These features, summarized below, include the manner of synthesis and its control, the material structure, properties and their modifiability as well as the good handling of the BNC products. They are an excellent basis for the development and use of BNC as an emerging biomaterial with great potential as wound and burn dressings, scaffolds for tissue regeneration and, as a particular challenge, as novel types of natural and bioactive implants.

- Both, the highly controllable bottom-up production of BNC from low-molecular sugars and the direct formation of three-dimensional cellulosic objects by this process are unique in the manufacturing of cellulose products. In contrast, all other cellulose materials are isolated from plants or other cellulosic resources and can only be shaped into bodies by postprocessing/regeneration steps.[10]
- The *in situ* control of the BNC biofabrication opens up amazingly diverse possibilities for designing shape, surfaces, and nanofiber network structures (fiber diameter 20–100 nm) of the BNC products. These parameters dominate the properties which are decisive for the application.
- Flat BNC materials such as films, patches, and membranes as well as hollow bodies equipped with pores or channels and different types of BNC tubes can be designed. This diversity opens up a tremendous potential for applications of BNC in numerous areas of medicine.
- The natural hierarchical three-dimensional nanofiber network of BNC is highly hydrated and structurally comparable with the extracellular matrix of the living body.
- The BNC bodies are formed as dimensionally and mechanically stable hydrogels of pure cellulose with high molecular weight (4000–10,000 anhydroglucose units)[11] and high crystallinity (80–90%).[12] They prove to be transparent for liquids, ions, and small molecules.
- The BNC materials and its properties can be effectively modified by postprocessing steps such as controlled dewatering, variously perforating, and generating of different types of composites.
- The BNC products can be effectually purified (removal of bacteria and their degradation products) by heating with aqueous sodium hydroxide solutions up to a concentration of 2.0 M. They can be sterilized with all known methods, are storable over years, and non-degradable in the mammalian body.

Under the usual conditions of static cultivation, for example in an Erlenmeyer flask, BNC will be formed as a flat hydrogel body (fleece) at the interface of the liquid culture medium to the ambient air. This allows the production of the abovementioned films, patches, and membranes of varying thickness and in the shape and the dimensions of the culture vessel. For producing hollow bodies, matrices can be placed in the culture medium. These matrices and optionally their individual components (templates) keep open appropriate cavities in the forming BNC bodies according to their shape and dimensions. Today, there are various types of matrix technologies which differ primarily in the type of matrices/templates and their arrangement in the reactor as well as in the supply of the nutrient medium and the air (see examples in Secs. 15.3 and 15.4).

The main fields of targeted medical applications of BNC are wound care and novel types of bioactive implants. In the case of wound dressings, first products are on the market.[13] The development of medical implants ranges from the design of materials for bone and cartilage repair to the development of tubular prototypes as grafts for vascular surgery. In all cases, BNC is active as a three-dimensional support for *in vitro* and *in vivo* tissue growth.[14–20]

This chapter describes initially the current knowledge regarding biocompatibility, bioactivity, and mechanical behavior of BNC as essential benefits for applications as a biomaterial in medicine. The following will show examples of the *in situ* production of perforated BNC membranes and, in particular, the development of biomimetic tubular BNC implants for small-diameter blood vessels (inner diameter less than or equal to 5 mm) and their evaluation in sheep studies.

15.2. Benefits of BNC as a biomaterial

In this section, the most important references concerning the advantages of BNC regarding its medical application are compiled and discussed.

15.2.1 *Biocompatibility*

All characteristics, already been tested for BNC, giving the material a high potential for medical applications. But up to now, the clinical use of BNC is restricted to wound dressings as well as cosmetic practice. In the context of the development of these external applications, many *in vivo* and *in vitro* cell and animal studies were performed to consider the usability of BNC for possible future implantation into the human body. A good biocompatibility of the material was observed during these studies indicating no harmful interactions with surrounding cells and tissue and causing no foreign body reactions. The implantation of tubular or flat-shaped BNC grafts in rats for up to 12 weeks indicates no inflammatory responses.[20,21] *In vitro* studies with commercially available and isolated primary cell lines support these investigations at the cellular level.[22–24] To avoid inflammatory responses, the purity of the BNC materials is important. Given the biotechnological production of BNC with bacteria, the material must undergo special purification steps to eliminate pyrogens, especially endotoxins (for example lipopolysaccharides), as residues from the bacterial membrane. A treatment of BNC with aqueous sodium hydroxide solution (0.1–2.0 M at 60–120°C) for adapted time periods (in relation to the size and dimension of the BNC body) reduces the amount of endotoxins. Using 1.0 M sodium hydroxide solution, the purity levels are below the limit value of 0.5 EU/mL given by the US Food and Drug Administration (FDA). The purified material can be tested by the limulus amebocyte lysate assay (LAL-assay).[25]

15.2.2 *Bioactivity*

The mentioned similarity of the BNC nanofiber network to the extracellular matrix of the living body results in a mimicking effect of BNC causing the attachment of cells onto the surface of the material. It is important to note that BNC inhibits dedifferentiation of seeded cells[25] which may offer a significant characteristic in regenerative medicine. Furthermore, the architecture of the network enables the transport of nutrients and oxygen throughout the material and supports the survival of cells. Because of the surface chemistry of BNC fibers and their structural composition, it is possible to modify the properties of the material regarding the bioactivity of BNC. A broad range of modifications affecting the adhesion, differentiation, or proliferation of cells is described in the literature. For example, a phosphorylation or sulfation of the BNC surface applied to mimic glucosaminoglycans of native cartilage, leads to an increased adhesion of bovine chondrocytes.[26] Other authors indicated that a binding of arginine–glycine–aspartic acid (RGD) peptides onto the fiber surface affects the adhesion of endothelial cells.[27] By mimicking extracellular matrix binding sites, the introduction of RGD sequences leads to a significant higher attachment of endothelial cells in comparison to unmodified BNC. A further modification described in the literature is a precipitation of hydroxyl apatite nanocrystals on the BNC fibers to mimic mineral components in native bone.[28,29]

Due to the fact that humans and other mammalian species do not have cellulases in the target tissue, BNC is not biodegradable after implantation into the body. This was confirmed after macroscopical evaluation of implanted BNC specimens into mice or rats for a time period of up to 90 days.[20,30] No evidences for *in vivo* degradation were observed, but the implantation time may be too short for detectable structural changes.

Own long-term studies in rats showed a stability of BNC implants over a test period of up to one year. These implants concern cuffs for nerve surgery and tubes as vascular interponates. Upon dissection of nervus ischiadicus of rat and subsequent reconnection by microsurgical suture, a protective BNC cuff prevents connective tissue from growing into the nerve gap and facilitates an early regeneration. After the observation time from 1 to 6 months postoperatively, the BNC cuff was stable and covered with connective tissue and small vessels within. Neither an inflammatory reaction nor an encapsulation of the implant was observed.[11] The application of BNC tubes as a microvessel prosthesis was investigated in the case of a typical end-to-end anastomosis using the carotid artery of the rat. Upon reconnection of the dissected carotid with the BNC tube, the internal surface of the BNC implant becomes completely covered by an endothelial layer. After 12 months in the body, the implant is still stable and branches of fibroblasts have penetrated the cellulose network and produce collagen.[11]

Nevertheless, until now, it is not known in detail what happens with BNC on the molecular and supramolecular level during the interaction with body tissue and body fluids. BNC could possibly be degraded by the interaction with chemically reactive substances in the body fluids under cleavage of the β-1,4-glycosidic bonds of the polymer chains. However, BNC has been proven as stable over periods which are essential for the reconstruction of damaged tissue. Therefore, the material can bridge problems associated with biodegradable implants, for example the synchronization of reconstruction of body tissue and degradation of the used biomaterial.

15.2.3 *Mechanical properties*

While the single fibers of BNC have mechanical properties which are comparable to steel or Kevlar, the mechanics is based mainly on the water-holding capacity of the hydrogel as a whole and its loss of water during compression. This capacity in turn is influenced largely by the BNC network structure and its anisotropic architecture. This means that the densities of the BNC fiber distribution are different in different regions causing direction-dependent mechanical properties. The wide-meshed lower part of BNC fleeces, biosynthesized during static culture, is not as stable as the bulk material because of decreased BNC fiber density resulting in a higher water binding. In case of implants, a sufficient mechanical stability is pivotal for successful surgical handling and fixation. It also needs to be able to withstand the forces present at the site of implantation. Most current studies deal with the biomechanics of BNC and its optimization in comparison to native tissue. Such an optimization of the mechanical properties will be achieved by changing the biosynthesis parameters for BNC production as well as by postprocessing of the resulting BNC materials. In comparison to collagen matrices used in tissue engineering, the Young's modulus of a native BNC hydrogel is higher. But when compared to those of the cartilage of menisci, it will not fulfill the requirements of mechanical strength.[31] The improvement in compression and tensile strengths of BNC hydrogels during post-processing is mainly achieved through the formation of double-network gels up to now. These composites can be formed by soaking the BNC in solutions of network creators (for example, gelatin, sodium alginate, or i-carrageenan) followed by their cross-linking.

15.3. Perforated BNC membranes from growing cultures of *Gluconacetobacter* species

As mentioned before, under static cultivation conditions and without the use of an additional matrix, the biosynthesis of BNC takes place in form of a fleece at the air/culture medium interface.[32,33] The new layer builds up horizontally on the top of the already formed BNC. Because the place of the biosynthesis remains fixed, the BNC fleece moves vertically downward into the medium during cultivation.[34] This method leads to initially wet BNC products with a natural "homemade" pore system. The pore diameters are in the range of 2–10 μm.[1]

For special applications, for example as scaffold material in the field of tissue engineering, the pore system of the BNC has to be modified. Microporous BNC membranes with an increased adequate pore size, with pore interconnectivity as well as with openings to the outside of the support structure are necessary to enable cell seeding and the effective diffusion of liquids. Two principal possibilities are described in the literature to introduce microporosity into the BNC materials during the cultivation: Firstly, the construction of globular voids (pores) and, on the other hand, the insertion of channels.

The usual static cultivation method of BNC (see Sec. 15.1) was used to introduce channel structures. Rambo *et al.*[35] have described the biosynthesis of BNC films containing *in situ* produced unidirectional microchannels. Here, the channel formation was attained through the use of pin templates with diameters varying from 60 to 300 μm. Polystyrene pins (300 μm diameter) or optical fibers (60 μm diameter) were vertically immersed in the culture medium and the BNC biosynthesis occurred around the pins leaving tiny channels vertically through the BNC membrane.

In WO2008/079034 A2,[36] Bielecki *et al.* have mentioned the production of unidirectional perforated BNC fleeces with the help of glass rods also vertically placed in the culture medium. The diameters of the large channels are about 0.08 or 0.5 cm.

The use of matrix technologies for BNC cultivation (see Sec. 15.1), based on the subsequently described patented biosyntheses of BNC hollow bodies, allows the creation of pores distributed across the body and of channels running in different directions within the BNC.

Bäckdahl *et al.*[37] presented a method to prepare BNC scaffolds with controlled microporosity following WO 2008/040729 A2.[38] For this purpose, globular particles made from paraffin wax (90–500 μm in diameter) or starch (5–100 μm in diameter) are used as porogens. These particles can be incorporated in different areas of BNC during its biosynthesis on the outer side of an oxygen-permeable silicon tube. After removal of the porogen particles, the scaffolds show a network structure with increased pore sizes and interconnectivity.

Bertholdt patented in WO 2008/098942 A2[39] a process for the production of BNC bodies containing a plurality of internal pores or tubular channels with branches. Some of the branches converge again at a different location. The pores and the channels are separated from one another but can be connected with openings at the outside of the material. Cells are able to enter and nutrient medium or blood can flow and supply the cells. The invention comprises the preparation of the said bodies using an outer mold and a mold core as matrix for shaping the pores and channels. The BNC formation takes place around the mold core in the form of suspended globular wax drops (50–100 μm in diameter) or horizontally positioned wax strands. Then, the demolding of the core is carried out by melting, evaporation, chemical, or mechanical treatment. The general principle of the support-aided BNC preparation is presented in WO 07/093445 A1.[40]

In WO 01/61026 A1,[41] a matrix technology for the production of BNC tubes is described. This technology proved to be very suitable for the *in situ* construction of

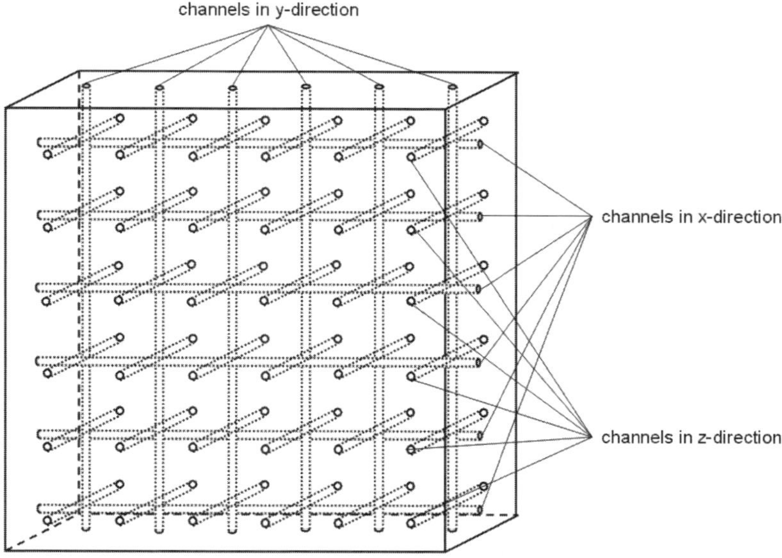

Fig. 15.1. Schematic representation of a BNC membrane with channels extending in several directions.

BNC membranes with interconnected channels in several directions. For this purpose, the matrix consists in two flat and interconnected panels of polymer materials such as polytetrafluoroethylene (ePTFE). In the space between the panels, thread-shaped placeholders are fixed. The so-prepared matrix is immersed vertically in the culture medium so that the inoculated nutrient solution is drawn into the area between the walls of the matrix where the cultivation takes place. During the biosynthesis, the placeholders are incorporated into the newly forming BNC membrane and can be removed mechanically after matrix dismantling and prior to the cleaning process. Figure 15.1 illustrates schematically and as an example the arrangement of interconnected channels in all three spatial direction of a membrane. It should be emphasized that all channels have external openings.

According to this procedure, we were successful in preparing microstructured membranes. Such membranes have, for example, dimensions of about $10 \times 10 \times 2$ mm and channel diameters of $50\,\mu$m.

15.4. Biomimetic BNC implants for small-diameter (≤ 5 mm) blood vessels

The nanofiber network structure of BNC and its design during biosynthesis enable the development of implants, which are biomimetically composed analogous to the functional layer structure of the endogenous target organs such as blood vessels. In this section, the urgent need for novel types of blood vessel implants is highlighted followed by the description of a biomimetic approach for the development of vascular BNC implants (protected name BActerial SYnthesized Cellulose BASYC®).

Further focal points are the design of small-diameter blood vessel prostheses by a matrix technology, the demonstration of the achieved biomimetically layered structure of the BNC tubes, and the evaluation of their suitability, stability as well as compatibility as carotid artery implants in sheep trials. In this context, the *in vivo* colonization with autologous cells is a key question.

15.4.1 *Need for novel blood vessel implants*

Atherosclerotic disease is the major cause of mortality in industrialized societies. Every year, there is a need for millions of coronary artery bypass implants worldwide. Arterial occlusive disease can take place in all arteries of the body, including the brain and limbs. However, the biggest demand for further innovation is for tubes with an inner diameter of 5 mm or less, given that currently there are no implants for such small-sized blood vessels available.

There are ongoing efforts to develop synthetic implants for this purpose based on ePTFE, polyester (Dacron), and polyurethane. Currently, all these synthetic tubes do not represent a convincingly superior alternative. When used in the small-diameter bypass, clinically poor medium-term to long-term patency of the synthetic implants (ePTFE, Dacron) was observed. For example, the five-year patency of ePTFE implants across the knee joint is less than 30%.[42]

So far, autologous vessels (of the patient) have to be used exclusively for the cardiac bypass surgery. This necessitates an additional surgical procedure during the operation to remove required vessels. Moreover, many patients do not have vessels suitable for use due to vascular disease, amputation, or previous harvest. Replacement of the damaged vessel area by a man-made tubular implant in a good functional condition could offer a sustainable solution for these problems.

In view of this situation, bioactive implants with a high potential for helpful interaction with living tissue become more and more important. BNC proved to be a very promising biomaterial for this purpose. Hence, we started an interdisciplinary ongoing research project to develop tubular implants based on BNC. These investigations include the preparation and characterization as well as the evaluation of the BNC tubes in cell ingrowth tests, proliferation assays, and in animal studies (rat, pig, sheep).[1,11,21,43,44]

15.4.2 *Preparation and characterization of the BNC tubes*

In this section, details of the biotechnological synthesis and the characterization methods for the materials and properties are described. These data serve as basis for the presentation and discussion of the results in Secs. 15.4.3 and 15.4.4.

15.4.2.1 *Biofabrication*

Applying the described matrix technology, BNC tubes were prepared using the cellulose-forming bacterium *Gluconacetobacter xylinus*, strain ATCC 53 582 (ATCC: American Type Culture Collection). The cultivation was carried out in a nutrient medium according to Hestrin and Schramm,[45] containing 20.00 g D-glucose, 5.00 g yeast extract, 5.00 g peptone, 3.40 g disodium hydrogen phosphate monohydrate, and 1.15 g citric acid dihydrate per 1000 mL water. The medium was inoculated with a preculture of the bacteria in a volume ratio of 20:1. Cultivation takes place within 5–10 days at 25–28°C using a matrix which contains cylindrical templates in the dimensions of the lengths and inner diameter of the manufactured tubes.

At the end of the production, the completed BNC tubes (after removal of the templates) were treated with 1.0 M aqueous sodium hydroxide solution for 30 min at 100°C to remove microbial contaminants and culture media components. Afterwards, the tubes were neutralized by repeated rinsing with deionized water and stored in deionized water until further use. For medical application, washing with endotoxin- and pyrogen-free water may be necessary as well as a transfer into physiological saline solution followed by steam sterilization.

15.4.2.2 *Suture retention strength*

Loops of surgical suture material (Prolene 5/0) are placed through the wall of the BNC tube similar to a surgical "single-knob"-suture. The loops lead from the outside through the wall toward the lumen, are pulled through, and knotted thereafter. This procedure was done six times to have six loops around the circumference of each tube. In a tensile testing machine (TIRA-test 2850, TIRA GmbH, Schalkau, Germany), the opposite site of the tube was fixed and one loop was linked with a hook connected to the force sensor. The mechanism was moving in the opposite direction to the fixation of the tube. The suture retention strength was determined until a strong decrease of the measured value could be observed (ripping of the loop out of the tube wall). This procedure was repeated six times (using all the six loops) for each of five tubes. The mean value of the test results provides the fixed suture retention strength.

15.4.2.3 *Burst strength*

A BNC tube was horizontally fixed on tube clips using clamps. One of the tube clips was connected with a valve, the other one with a water reservoir which can be pressurized using compressed air. To measure the burst strength, the air pressure into the reservoir was increased to around 200 mbar to remove the air from the system and fill it with water. Then the valve was closed and the pressure was increased further (0.1–0.2 bar/s) until a burst effect on the BNC tube could be observed. The pressure was read of a manometer. The measurement was repeated three times using three samples from the same production charge and the mean value was given as bursting strength.

15.4.2.4 *Scanning electron microscopy*

Scanning electron microscope (SEM) (LEO-1450 VP, Zeiss Oberkochen, Germany) investigations were performed to show the biomimetic structure of the BNC tubes.

The selected samples were freeze dried over 24 h, mounted on an aluminum stub using an adhesive carbon tab, and sputtered with gold.

15.4.2.5 *Animal studies — experimental setup*

Never-dried BNC tubes with a length of 100 mm, an inner diameter of 4–5 mm, and a wall thickness of 2–3 mm were used to replace one of the carotid arteries of 4 sheeps for up to 3 months.

The evaluation of the BNC implant properties was carried out by preoperative measurements of suture retention strength, burst strength, subjective impressions of performing surgeons, and Doppler ultrasonography as well as by histological investigations of explants.

15.4.2.6 *In vivo ultrasound*

In vivo ultrasound measurements of the tubular BNC implants were performed after 1, 2, and 3 months post-surgery in comparison to the contralateral corresponding artery without administration of anesthesia. Implants were evaluated along a longitudinal and transversal axis via transcutaneous ultrasonography using the CX50 ultrasound machine (Phillips, Andover, Mass., USA). Lumens of the BNC implants were evaluated to assess tube patency, stenosis/occlusion, or aneurismal dilatation. Blood flow velocity (V_{max}) was measured using Doppler ultrasonography.

15.4.2.7 *Staining of explants: Extracellular matrix stains and immunohistological staining*·

Cross sections (thickness 5 μm were histologically prepared from explants, taken from the first third part following proximal anastomosis. Three different immunohistological staining methods were used for the specific detection of autologous cells in the explants. Staining for von Willebrand factor (vWF) (factor VIII), blood vessel staining kit, alkaline phosphatase conjugated (Merck, Darmstadt Germany), was used as proof of endothelial cells. To analyze vascular smooth muscle cell (VSMC) colonization, smooth-muscle cell α-actin (DCS, Berlin, Germany) was selected. Elastin (formed collagen) could be detected by van Gieson staining. The investigations are discussed in Sec. 15.4.4.

15.4.3 *The biomimetic approach for blood vessel implants*

Figure 15.2 shows schematically and simplified the layered architecture of a natural arterial blood vessel. The inner layer (intima, contact to the blood) is formed by a monolayer of endothelial cells. The middle layer (media) contains smooth muscle cells as well as elastic and collagen fibers, dependent on the position and function of the vessel in the blood stream. The anchoring of the blood vessel in the body area via connecting tissue (mainly collagen fibers, elastine) is effected by the outer layer (adventitia). That means, the structural elements of a natural blood vessel are

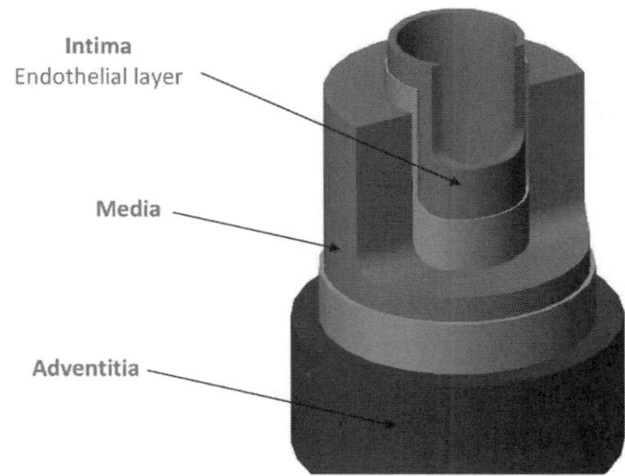

Fig. 15.2. Schematic representation of the architecture of natural arterial blood vessels.

particularly and specifically characterized by the presence and function of specific cell types as well as by thin elastic membranes between intima and media, media and adventitia as well as inside the media.

The previously described layered structure of BNC and the mentioned collagen-like nanofiber network leads to the exciting questions: Is it possible to use these specific features of BNC to design biomimetically structured BNC blood vessel implants? Can these implants be colonized with living cells in animal experiments? This raises the more wide-ranging issue whether such a type of implant would be "vitalized" in the sense of an *in vivo* and multifunctional scaffold.

The starting point for the successful synthesis of BNC tubes as nature-based blood vessel implants was the method described in WO01/61026 A1.[41] This biotechnological fabrication enables the design of BNC tubes with an inner diameter in the range of 0.5–5.0 mm and ensures a very high and reproducible quality of the blood-contacting surface. The formation of the BNC tubes takes place on cylindrically-shaped glass templates. The templates are partially submerged into the bacteria-containing nutrient liquid. They guarantee that a lumen remains open during the formation of the emerging BNC bodies. The operations in the reactor (subject of pending patents) ensure that culture medium and oxygen — essential for the BNC synthesis — are present at the site of cellulose generation. At the end of the cultivation, the BNC-covered templates are taken out of the reactor. After removing the template from the BNC body, the BNC tube is obtained. As described in Sec. 15.4.2, the tubes were subsequently purified with an aqueous sodium hydroxide solution, neutralized with deionized water, sterilized, and stored in deionized water or physiological saline solution.

The matrix technology allows the preparation of BNC tubes in a wide range of lengths, inner diameters, wall thicknesses, surface structures as well as nanofiber network architectures. The properties and functions of the tubular BNC implants can be adjusted and adapted to the respective medical application and requirements. Figure 15.3 shows a photograph of such a typical BNC tube. The free lumen (blood stream is simulated by a red glass rod) and the hydrogel character of the wall can

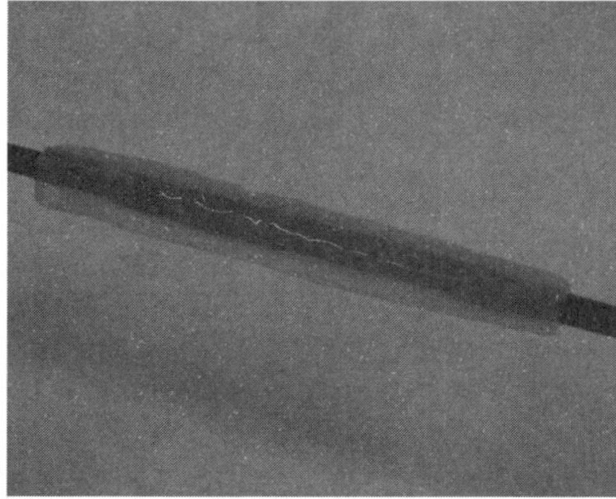

Fig. 15.3. Photograph of a typical small-diameter BNC tube. It is made by the matrix technology and used for the sheep studies. Length 100 mm, inner diameter 5 mm. A glass rod simulates the blood stream.

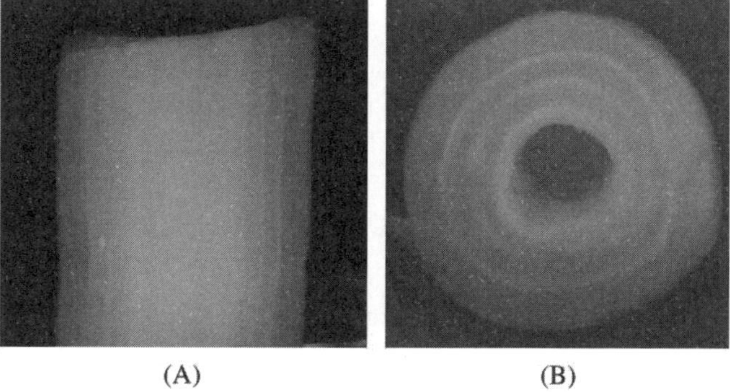

(A) (B)

Fig. 15.4. Microphotographs of a BNC tube, inner diameter 5 mm. (A) side view and (B) top view on cross section.

be seen clearly. These BNC tubes were used for the described sheep studies (see Secs. 15.4.2.5 and 15.4.4).

The measurement of burst strength revealed values in the range of 800 mm Hg. The suture retention strength was determined of about 5 N (mean value of five samples). These mechanical properties showed to be sufficient to expose the BNC tubes to the high pressure blood stream of sheep.

It has to be emphasized that the wall of the described BNC tubes consists of several layers as demonstrated in Fig. 15.4. To show this pattering more clearly, a BNC tube with an increased wall thickness of 6 mm has been used. In Fig. 15.5, a scanning electron micrograph of a cross section confirms the layered structure of the BNC tube. The lumen is to be followed by an inner and a medium layer. A clearly distinguishable outer layer completes the structure. The layers are arranged in a parallel and rotationally symmetric manner around the longitudinal axis of the tube. They are firmly connected and do not show any imperfections or structural

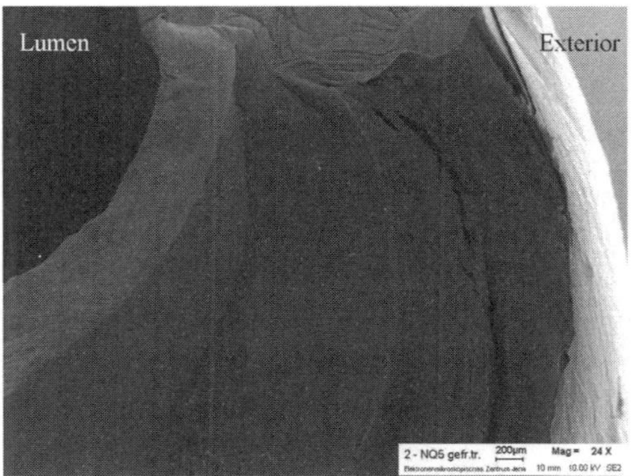

Fig. 15.5. Scanning electron micrograph of a BNC tube, top view on cross section; magnification 24×.

gradients. Moreover, this special structure is homogeneous over the whole length of the tube. This provides implants with a high mechanical stability against deformation as well as with the necessary pressure load connected to sufficient elasticity and tensile strength.

There are several further methods to prepare BNC hollow bodies with potential for surgical applications such as blood vessel or tissue implants. According to the described production and shaping processes, BNC tubes with quite different dimensions, structures, surface types, and resulting properties are obtained. As already stated in Sec. 1, the technologies are mainly different in terms of the tube formation strategy and availability of culture medium and oxygen at the BNC forming site.

As examples, some of these shaping methods should be mentioned. In the first example, the preparation of BNC tubes takes place in the space between two glass tubes (closed at one end) of different diameters and adjusted vertically. The cultivation occurs similarly to the standard horizontal-static cultivation of flat BNC fleeces.[46] In another version, horizontally arranged half shells are used as tube-forming tools.[47] By using oxygen-permeable tubular polymer hoses, the tube synthesis proceeds in a radial-static manner.[48,49]

It can be expected that the BNC tubes manufactured according to the different technologies will show distinct options for application especially as blood vessel implant.

15.4.4 *Evaluation of small-diameter BNC tubes as arterial substitutes in sheep*

This evaluation has been carried out in particular to get answers regarding the surgical feasibility and the functional *in vivo* performance. Furthermore, the assessment of the proinflammatory potential of BNC and insights in its interaction with blood components and living cells were important questions. Finally, the ability of the designed BNC tubes to provide a scaffold for the neoformation of a vascular wall-like structure will be estimated.

Fig. 15.6. Implantation of a BNC tube as interponate in the carotid artery of sheep.

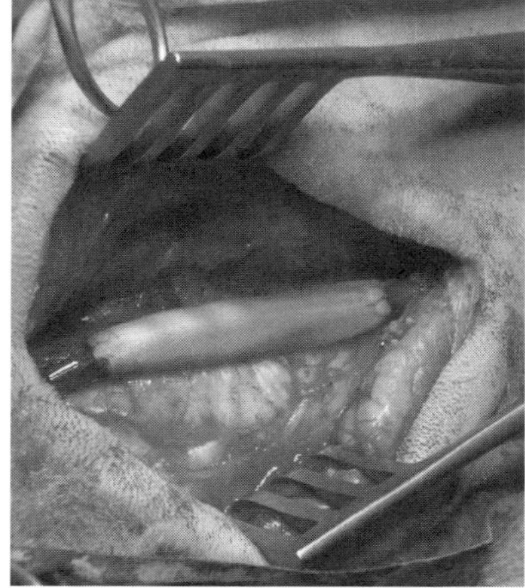

Fig. 15.7. Implanted BNC tube under function. After opening the blood flow, the BNC interponate works.

A part of the sheep carotid artery was replaced by a BNC tube, implanted as an interponate (Fig. 15.6). The BNC tube can be easily handled with forceps and fixed by a single-knob suture technology.

Good surgical handling regarding stability, tear resistance, elasticity, and tensile strength were positive properties which could be determined. Figure 15.7 shows a functioning BNC tube after a successful interposition into the carotid artery. Tube rupture, secondary insufficiency/bleeding, clinical apparent inflammation/infection, or macroscopic signs for inflammatory processes were not found. The Doppler ultrasonographic measurements have shown values for the blood flow velocity in the BNC implants in a range of 30–40 m/s ($V_{\max}$), quite similar to the flow velocity in native carotid artery of sheep. This was found both immediately after surgery and after 1, 2, and 3 months. After 3 months, the BNC implants were removed. The visual evaluation of the explants revealed a very good integration of the BNC tubes in

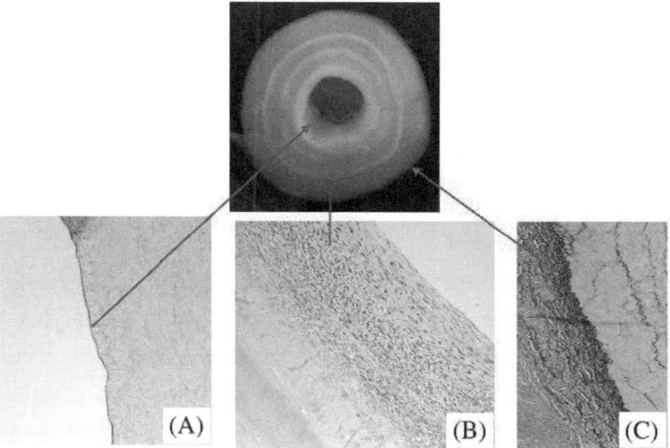

Fig. 15.8. BNC tube cross section illustrating the *in vivo* colonization of BNC implants with cells. Results of histological investigations schematically combined with the layered structure of a BNC tube cross section already presented in Fig. 15.4. Extracellular matrix staining and immunohistological staining confirm specifically (A) endothelial layer; (B) VSMCs; (C) collagen in the newly formed connecting tissue. Magnification 40× (A), (B); 10× (C). For further description, please see text.

the animal body by formation of connecting tissue, including the formation of small blood capillaries.

With regard to the biomimetic approach discussed above, the BNC explants have been investigated by different staining methods. The results of these investigations are exemplified through selected histological findings included in Fig. 15.8. As reliable proof of the deposition of endothelial cells at the surface of the inner layer (blood contact) of the BNC tubes, staining with an antibody against vWF is used. vWF represents a glycoprotein produced and stored in the endothelial cells on the intima of natural blood vessels. The successful detection of this glycoprotein by the antibody test therefore shows the presence of endothelial cells. The search for living cells in the middle layer of the tubular BNC explants was performed using a counter-staining with an antibody against alpha-smooth muscle actin. This actin is a typical component in VSMCs. Therefore, the staining with the antibody indicates specifically the presence of smooth muscle cells besides other cells. In the outer layer of the BNC explants, intact tissue could be identified by a hematoxylin and eosin (HE) staining. Moreover, by use of a specific van Gieson staining, collagen could be detected. As in case of the visual evaluation, connecting tissue with small blood capillaries and channels can also be seen in the outer layer.

Overall, the results of the sheep studies indicate a good mechanical stability and surgical handling of the BNC tubes, no proinflammatory potential, and a specific population of the BNC tubular implants with living cells from the animal body. The cell ingrowth started at the branch/stitch channels and is subject of further investigations.

These results are in good agreement with findings from previous investigations of BNC microtubes in the carotid artery of rats[11] and from studies of BNC patches with a thickness of 0.7 mm for closure of defects (ventricular septum defects) in the pig heart as a model.[50] Continuing work is focused on long-term animal studies and on the adaptation of tubular BNC implants to further applications in the vascular system.

15.5. Conclusions

The current research on biocompatibility, bioactivity, and mechanical behavior of BNC materials demonstrates the unique potential of these natural nanomaterials for medical uses. The exciting possibility to create shape and architecture of the BNC products directly during biotechnological synthesis from simple natural educts opens up a broad field of applications. It ranges from wound dressings to novel types of cardiovascular implants. The described matrix technology has a huge potential to insert pores and channels into the BNC materials and to produce BNC hollow bodies *in situ*, such as tubes for blood vessel implants. The function-determining structures within the bodies and at the surfaces are mainly controlled by the bacterial strain, the matrix material, and the supply of culture medium and air. In connection with the appropriate technical solutions, quite different products can be reached and significant innovations may be expected in near future.

Studies in a sheep model (carotid artery) revealed good surgical handling and stability of the BNC implants, but also a colonization of the tubes with different cells from the body. Accordingly, these implants seem to provide a scaffold for the neoformation of a three-layered vascular wall by ingrowing of smooth muscle cells and deposition of endothelial cells on the inner as well as collagenous fibers and capillaries on the outer surface. Moreover, the good functional performances of the BNC implants compared to native contralateral arteries and the absence of any signs of proinflammatory potential — neither clinically, macroscopically or histologically — are very promising.

The BNC tubes were mainly developed and tested in order to create novel types of implants for the treatment of cardiovascular diseases. The current state indicates that the biomaterial BNC has the potential to achieve this goal.

References

1. D. Klemm, F. Kramer, S. Moritz, T. Lindström, M. Ankerfors, D. Gray and A. Dorris, Nanocelluloses: a new family of nature-based materials, *Angew. Chem. Inter. Ed.* **50** (2011) 5438–5466.
2. M. Gama, P. Gatenholm and D. Klemm, *Bacterial nanocellulose: a sophisticated multifunctional material.* Boca Raton: CRC Press (2012).
3. A. Dufresne, *Nanocellulose: from nature to high performance.* Berlin: Degruyter (2012).
4. D. Klemm, B. Philipp, Th. Heinz, U. Heinze and W. Wagenknecht, *Comprehensive cellulose chemistry volume 1 and 2.* Weinheim: Wiley-VCH (1998).
5. D. Klemm, B. Heublein, H.-P. Fink and A. Bohn, Cellulose: fascinating biopolymer and sustainable raw material, *Angew. Chem. Int. Ed.* **44** (2005) 3358–3393.
6. D.N.-S. Hon and N. Shiraishi, *Wood and cellulosic chemistry*, 2nd edn. New York: Marcel Dekker Inc. (2001).

7. K. Kamide, *Cellulose and cellulose derivatives.* Amsterdam: Elsevier (2005).
8. P. Zugenmaier, *Crystalline cellulose and cellulose derivatives.* Berlin, Heidelberg: Springer (2007).
9. R.M. Brown Jr. and I.M. Saxena, *Cellulose: molecular and structural biology.* Dordrecht: Springer (2007).
10. P. Gatenholm and D. Klemm, Bacterial nanocellulose as a renewable material for biomedical applications, *MRS Bull.* **35** (2010) 208–213.
11. D. Klemm, D. Schumann, F. Kramer, N. Heßler, M. Hornung, H.-P. Schmauder and S. Marsch, Nanocelluloses as innovative polymers in research and application, *Ad. Pol. Sci.* **205** (2006) 49–96.
12. S. Salmon and S.M. Hudson, Crystal morphology, biosynthesis and physical assembly of cellulose, chitin, and chitosan, *J. Macromol. Sci. Rev. Macromol. Chem. Phys.* **C37** (1997) 99–276.
13. Lohmann & Rauscher GmbH & Co.KG, Suprasorb X + PHMB. Available at: http://www.lohmann-rauscher.de/media/archive/4458.pdf.
14. W. Czaja, A. Krystynowicz, S. Bielecki and R.M. Brown, Jr., Microbial cellulose: the natural power to heal wounds, *Biomaterials* **27** (2006) 145–151.
15. W.K. Czaja, D.J. Young, M. Kawecki and R.M. Brown, Jr., The future prospects of microbial cellulose in biomedical applications, *Biomacromolecules* **8** (2007) 1–12.
16. D. Ciechanska, J. Wietecha, D. Kaźmierczak and J. Kaźmierczak, *Fibres Text. East. Eur.* **18**(5) (2010) 98–104.
17. T.J. Webster, Nanotechnology: better materials for all implants, *Mater. Sci. Forum* **539–543** (2007) 511–516.
18. R.M. Brown, Jr., W. Czaja, M. Jeschke and D. Young, Multiribbon nanocellulose as a matrix for wound healing, WOA2007027849 (2007).
19. G. Serafica, C.J. Damien, G.A. Oster, K. Lentz and R. Hoon, Microbial cellulose materials for use in transdermal drug delivery system, WOA2006113796 (2006).
20. G. Helenius, H. Bäckdahl, A. Bodin, U. Nannmark, P. Gatenholm and B. Risberg, In vivo biocompatibility of bacterial cellulose, *J. Biomed. Mater. Res.* A **76** (2006) 431–438.
21. D. Klemm, D. Schumann, U. Udhardt and S. Marsch, Bacterial synthesized cellulose — artificial blood vessels for microsurgery, *Prog. Polym. Sci.* **26** (2001) 1561–1603.
22. Y.M. Chen, T. Xi, Y. Zheng, T. Guo, J. Hou, Y. Wan and C. Gao, In vitro cytotoxicity of bacterial cellulose scaffolds used for tissue-engineered bone, *J. Bioact. Compat. Pol.* **24** (2009) 137–145.
23. Z. Cai and J. Kim, Bacterial cellulose/poly (ethylene glycol) composite: characterization and first evaluation of biocompatibility, *Cellulose* **17** (2010) 83–91.
24. S. Moreira, N.B. Silva, J. Almeida-Lima, H.A.O. Rocha, S.R.B. Medeiros, C. Alves and F.M. Gama, BC nanofibers: in vitro study of genotoxicity and cell proliferation, *Toxicol. Lett.* **189** (2009) 235–241.
25. A. Bodin, S. Bharadwaj, S. Wu, P. Gatenholm, A. Atala and Y. Zhang, Tissue engineered conduit using urine-derived stem cell seeded bacterial cellulose polymer in urinary reconstruction and diversion, *Biomaterials* **31** (2010) 8889–8901.
26. A. Svensson, E. Nicklasson, T. Harrah, B. Panilaitis, D.L. Kaplan, M. Brittberg and P. Gatenholm, Bacterial cellulose as a potential scaffold for tissue engineering of cartilage, *Biomaterials* **26** (2005) 419–431.
27. A. Bodin, L. Ahrenstedt, H. Fink, H. Brumer, B. Risberg, and P. Gatenholm, Modification of nanocellulose with a xyloglucan-RGD conjugate enhances adhesion and proliferation of endothelial cells: implications for tissue engineering, *Biomacromol.* **8** (2007) 3697–3704.
28. H. Jiang, Y. Wang, S. Jia, Y. Huang, F. He and Y. Wan, Preparation and characterization of hydroxyapatite/bacterial cellulose nanocomposite scaffolds for bone tissue engineering, *Key Eng. Mater.* **330–332** (2007) 923–926.

29. Y.Z. Wan, Y. Huang, C.D. Yuan, S. Raman, Y. Zhu, H.J. Jiang, F. He and C. Gao, Biomimetic synthesis of hydroxyapatite/bacterial cellulose nanocomposites for biomedical applications, *Mater. Sci. Eng.* **C27** (2007) 855–864.
30. P.N. Mendes, S.C. Rahal, O.C.M. Pereira-Junior, V.E. Fabris, S.L.R. Lenharo, J. Ferreira de Lima-Neto and L.A. Fernanda da Cruz, In vivo and in vitro evaluation of an Acetobacter xylinum synthesized microbial cellulose membrane intended for guided tissue repair, *Acta Vet. Scand.* **51** (2009) 12.
31. A. Bodin, S. Concaro, M. Brittberg and P. Gatenholm, Bacterial cellulose as a potential meniscus implant, *J. Tissue Eng. Regen. Med.* **1** (2007) 406–408.
32. P. Ross, R. Mayer and M. Benziman, Cellulose biosynthesis and function in bacteria, *Microbiol. Rev.* **55**(1) (1991) 35–58.
33. S. Hestrin, Preparation of cellulose: 2. Bacterial cellulose, *Meth. Carbohydr. Chem.* **3** (1963) 4–9.
34. W. Borzani and S. De Souza, Mechanism of the film thickness increasing during the bacterial production of cellulose on non-agitated liquid media, *Biotechnol. Lett.* **17**(11) (1995) 1271–1272.
35. C.R. Rambo, D.O.S. Recouvreux, C.A. Carminatti, A.K. Pitlovanciv, R.V. Antonio and L.M. Porto, Template assisted synthesis of porous nanofibrous cellulose membranes for tissue engineering, *Mater. Sci. & Eng. C, Biomim. Sypramol. Syst.* **28** (4) (2008) 549–554.
36. S. Bielecki, A. Krystynowicz, M. Kolodziejczyk, J. Bigda, M. Smietanski and J. Jankau, A biomaterial composed of microbiological cellulose for internal use, a method of producing the biomaterial and the use of the biomaterial composed of microbiological cellulose in soft tissue surgery and bone surgery, WO2008/079034 A2 (2008).
37. H. Bäckdahl, M. Esguerra, D. Delbro, B. Risberg and P. Gatenholm, Engineering microporosity in bacterial cellulose scaffolds, *J. Tissue Eng. Regen. Med.* **2** (2008) 320–330.
38. A. Bodin, H. Bäckdahl, P. Gatenholm, L. Gustafsson and B. Risberg, Artificial vessels, WO2008/040729 A2 (2008).
39. G. Bertholdt, Process and support structure for the cultivation of living cells, WO2008/098942 A2 (2008).
40. G. Bertholdt, Process for the production of a long hollow cellulose body, WO2007/093445 A1 (2007).
41. D. Klemm, S. Marsch, S. Schumann and U. Udhardt, Method and device for producing shaped microbial cellulose for use as biomaterial, especially for microsurgery, WO2001/61026 A1 (2001).
42. Y. Banz and R. Rieben, Endothelian cell protection in xenotransplantation: looking after a key player in rejection, *Xenotransplantation* **13** (2006) 19–30.
43. D. Schumann, J. Wippermann, D. Klemm, F. Kramer, D. Koth, H. Kosmehl, T. Wahlers and S. Salehi-Gelani, Artificial vascular implants from bacterial cellulose: preliminary results of small arterial substitutes, *Cellulose* **16** (2009) 877–885.
44. J. Wippermann, D. Schumann, D. Klemm, H. Kosmehl, S. Salehi-Gelani and T. Wahlers, Preliminary results of small arterial substitute performed with a new cylindrical biomaterial composed of bacterial cellulose, *Eur. J. Vasc. Endovas. Surg.* **37** (2009) 592–596.
45. S. Hestrin and M. Schramm, Synthesis of cellulose by Acetobacter xylinum: II. Preparation of freeze-dried cells capable of polymerizing glucose to cellulose, *Biochem. J.* **58** (1954) 345–352.
46. M. Kusakabe, E. Ono, Y. Suzuki, K. Wanatabe and S. Yamanaka, Hollow microbial cellulose, process for preparation thereof and artifical blood vessels formed of said cellulose, EP396344 A3 (1991).

47. G. Berthold, Process for production of a long hollow cellulose body, WO002007093445 A1 (2007).
48. A. Bodin, H. Bäckdahl, P. Gatenholm, L. Gustafsson and B. Risberg, Process for the preparation of hollow cellulose vessels by culturing cellulose producing microorganisms on the surface of a hollow carrier and providing a gas having an oxygen level of at least 35%, WO2008/040729 A3 (2008).
49. A. Yoshihiko, K. Takahisha and N. Koichi, Method for producing cellulose products using cellulose producing bacterium, JP002005320657 A (2005).
50. N. Lang, R. Kozlik-Feldmann, M. Sigler, E. Merkel, F. Fuchs, D. Schumann, F. Kramer, A. Meyer, F. Freudenthal, S. Meyer, Ch. Schroeder, H. Metz and D. Klemm, Evaluation of bacterial cellulose as a new patch material for closure of ventricular septal defects. In *239th ACS National Meeting*, (CELL 193) San Francisco, CA, USA (2010).

Chapter 16

Cellulose in Printed Electronics

Sarute Ummartyotin[1], Mohini Sain[1,2,3] and Pia Qvintus[4]
[1] *Center for Biocomposites and Biomaterial Processing, Faculty of Forestry, University of Toronto, Canada*
[2] *King Abdulaziz University, Jedda, SA*
[3] *Luleå University of Technology, Luleå, Sweden*
[4] *VTT Technical Research Center of Finland, Finland*

During the last decade, printing technique was essentially used for several pages of book, magazines, and newspaper, so-called graphic printer. Nowadays, due to its achievement on rapid, accurate, and reliable concept, it was widely encouraged for electronic community. The use of printer for electronic purpose has been increasing exponentially. The emergence of this concept was strongly evident in many applications such as diode, transistor as well as radiofrequency identification. In this chapter, the overview of printed electronics and its significance and finally, the role of this technique in many applications area are discussed.

16.1. Introduction

Green printed electronics was the term used for a relatively new technology that defines the printing of electronic circuits and components on common substrate media such as plastic, paper, and textile. Common printing roll-to-roll (R2R) techniques or other low-cost equipment suitable for defining patterns on materials were used in the past. Few examples are screen printing, flexography, gravure, offset lithography, and inkjet. The main difference when compared to graphic printing was that instead of using conventional inks, newly developed conductive inks were used to print active devices, typically thin film transistors and thin printed batteries. Although the concept of printed electronics has been known for more than 15 years, recent advances in conductive ink chemistry and flexible substrate development show promise of commercialization for several new applications.

Innovative companies and research institutes are currently using the technology to transform basic circuit elements, such as thin-film transistors, resistors, inductors, diodes, and capacitors into printed batteries, displays, sensors, radiofrequency identification (RFID) tags, interactive packaging, solar panels, and even loud speakers.

The primary advantages of printed electronics included low cost, attractive and flexible form factor, ease of production, and ease of integration. In many industries, printed electronics technology was expected to facilitate widespread development of functional electronic devices which can be useful for applications not typically associated with conventional (i.e., silicon-based) electronics, such as flexible displays, smart labels, animated posters, novelty items, and active clothing.

Moreover, in more conventional industry applications — such as battery-assisted active and passive RFID, and RF-enabled sensor and data logging systems — printed electronics can provide a lower cost alternative to traditional technology. The commercial potential for printed electronics is significant. The trial production site for R2R produced printed electronics (PRINTOCENT) was established in Oulu, Finland during year 2009. IDTechEx, a leading analyst in the field of printed electronics and nanotechnology, estimates that the worldwide market for printed and thin-film electronics will grow to US$48 billion by 2017 and to US$250 billion by 2025.

Many synthetic plastics have been used as the substrate for printed electronics development. Because of increasing concern about environmental issues, urge for sustainability and uncertainty about the cost of oil, as well electronics industry as producers of consumer packaged goods, and packaging industry were looking for new biobased, or recyclable material solutions. Even if paper or cardboard materials can be considered a solution toward bringing sustainable green materials to electronics, the performance of existing cellulose-based products such as cellulose acetate, cellulose acetate butyrate seemed not to be good enough to fulfil the needs of printed electronics in all technical aspects.

In longer term, it was needed that the new biobased substrates could reach the same performance level, smoothness, water repellency, high oxygen barrier, etc., which is offered by today's synthetic polymer substrates in various applications of printed electronics.

16.2. Cellulose-based nanomaterials in printed electronics

Studies related to use of nanocellulose in printed electronics have mainly focused on development of nanocellulose composite films for printing substrates. In general, nanocomposites consisted of two phase materials and one of the materials had at least one dimension in the nanometer range (1–100 nm). The advantages of nanocomposite materials when compared to conventional composites are their superior thermal and mechanical properties as well as their transparency and flexibility.

In printed electronics, plastic substrates have been considered prospective materials due to their inherent flexibility and optical qualities. However, one of the major drawbacks of plastics was the relatively high thermal expansion coefficient. The thermal expansion coefficient of the substrate needs to be close to those of the layers deposited on it. Dissimilarities in thermal expansions of these layers would lead to excessive strain and eventual crack formation during the thermal cycling involved in the printed electronics manufacturing process. One of the proposed solutions to reduce the thermal expansion of plastics without appreciable loss in transparency was to reinforce them with cellulose nanofibers (CNFs). It is important to note that transparent cellulose nanofiber-reinforced urethane resin with a thickness of 100 μm, a light transmittance of 87.8%, and a coefficient of linear thermal expansion of 8.30 ppm/K with a temperature range of 20–150°C can be successfully prepared.[1] The coefficient of linear thermal expansion is about 1/3 that of a PET sheet and closed to glass. Urethane resin was chosen because its refraction index is 1.52–1.60, which is close to the refraction index of cellulose (1.57). In general, the refraction index of resin changed depending on temperature. However, the light transmission of this kind of material did not change much even at a temperature of 90° C because of the fact that the nanocellulose reinforcement reduced the thermal expansion of the resin. The future use application for this kind of materials would be in flexible displays, printed organic diodes (organic light-emitting diodes (OLEDs)), and optical photovoltaics like solar cells.

When used as reinforcement in biopolymer composites, nanocellulose fibers exhibit improved mechanical strength, decreased brittleness, and improved thermal stability. Besides the use of nanocellulose in polymer composites, some research has also been performed in the development of fiber–inorganic filler composites. Novel combination of very high filler content of nanocellulose fibers to a paper-like substrate has given promising results as a material for printed electronics. These structures may offer a cost-effective option for printed electronics substrates with a good surface smoothness, high dimensional stability against heat, and a very good temperature tolerance that only very special plastic films can withstand nowadays. The only disadvantage of the materials is that they were not transparent. Furthermore, the properties of these substrates can be varied within a relatively large range by the selection of raw materials and their relative proportions.

Some applications of cellulose for printed electronics were also found in cellular phone readable application such as a tag printed on the CNF–filler substrate. The substrates performed well in printing tests resulting in the conductivity values of patterns similar to the ones printed on PET film (Mylar A). Interestingly, as high as 220–270°C curing temperatures can be used without causing any damage to the sheets or printed line. These temperatures were much higher than the current substrates for printed electronics could tolerate. Therefore, this high temperature tolerance will make these substrates an attractive alternative once their production is upscaled.

The third option for nanocellulose as a substrate for printed electronics was the development of nanocellulose films. Production of plastic-like translucent nanocellulose films in a R2R process was lately demonstrated by VTT. Produced films can be afterwards surface modified according to the needs of different applications. These films were potential substrates for "green", biobased, low cost applications of printed electronics.

The use of printing technique for manufacturing electronic components constituted to be an emerging class of materials with potential applications including transistors, diodes, energy storage devices, electronic memory devices as well as sensor and actuator. This printing technique offered several advantages over laboratory scale and subtractive batch process, such as photolithography and vacuum processes. Printers for electronic applications generally require more continuous and homogeneous patterns with restrictions on the layer thickness, roughness as well as resolution.

From the structural point of view, electronic components which generally fabricated devices as multilayer thin film, the integration of functional ink deposited on substrate has to be well considered because of the interlayer diffusivity and solubility effects. There are issues that remain unresolved in selection of materials for layer to layer deposition. In order to solve this issue, Ummartyotin *et al.* studied the effect of viscosity on ink solution.[1] The addition of glycerol in ink solution not only increased conductivity but also reduced ink solution viscosity. The rate of flow of ink solution was consequently low. On the other hand, Maattanen *et al.* purposed the use UV irradiation and surface plasma treatment on surface of substrate.[2] The improvement on surface energy and adhesion force among ink layers was successfully achieved.

We also evaluated the print electronic quality. It was difficult to determine the efficiency of device during the process of fabrication. Prior to determination of printed quality, the compatibility between substrate and ink layer was examined as a function of surface roughness, thickness, line widening, line-edge roughness, surface resistivity, volume resistivity as well as all electronic performance.[3,4]

Finally, it was very challenging to select the appropriate printer for good quality printed electronics components. Electronic printing industry typically uses common printing equipment or other low cost equipment suitable for defining patterns on substrate. The example of printer can be generally categorized into different subprocess such as gravure, flexography, offset, screen, and inkjet printing. All subtechniques have their advantages and disadvantages, and thus modifications of the conventional techniques were extensively required in order to print good quality patterns for electronic applications. The selection of printing technique for use needs to be considered based on the printing requirements such as lateral resolution, printed thickness, printing speed, as well as ink properties.

16.3. Printed electronics manufacturing

16.3.1 *Gravure printing*

Gravure, one of the printing processes, has been extensively used for electronic devices such as antennas,[5–8] capacitors,[9] OLEDs,[10] as well as solar cells.[11,12] Generally, gravure equipment comprises of a two-roller system where the coating roller has an engraved pattern. The coating roller was partially inserted into a bath of the ink that was continuously refilled as mentioned in Fig. 16.1. The print patterns were engraved into a metallic roller by a laser or chemical etching and high quality of print pattern can be consequently achieved. This technique is generally appropriated for large print volume. The advantage of gravure was that low viscosity inks worked very well and very high roller speed was possible, 1–10 $\mathrm{m\,s^{-1}}$, whereas, the disadvantage of this technique was attributed to the surface roughness of substrate, which might interrupt the ink transfer to substrate. In addition, the use of this

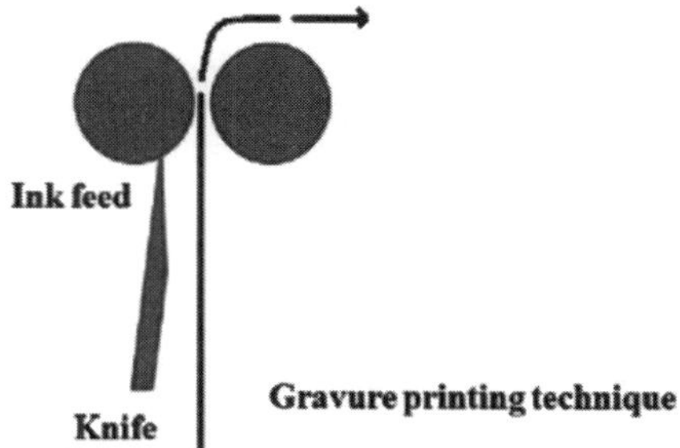

Fig. 16.1. A schematic of gravure printing technique.

technique is preferred over others for low print pressure. High print pressure limits the use of this technique to applications where there is no printed soft layer as a base or the substrate has high roughness of surface.[13]

From the industrial point of view, the gravure quality can be extensively controlled by adjusting the printing speed, pressure, roughness of surface as well as properties of ink.[14] Depending on the device quality and quantity, gravure technique was successfully enrolled to continuous printed pattern for electronic applications.

16.3.2 *Flexographic printing*

Flexographic printing is generally similar to gravure printing. The difference between flexographic printing and gravure is that the printed image stands up on the printing roller, the surface of which is typically made from rubber.[15] The typical flexographic printer consists of fountain rollers that continuously transfer ink into the ceramic anilox roller, which has engraved cells/micro cavities embedded into the exterior. This allows the collection of ink, which transfers to the relief on the printing roller that performs the final transfer to the web. The ink picks out from the anilox corrsesponding to the negative pattern of the motif as shown in Fig. 16.2. For green printed electronics, this printing method is yet to be explored.

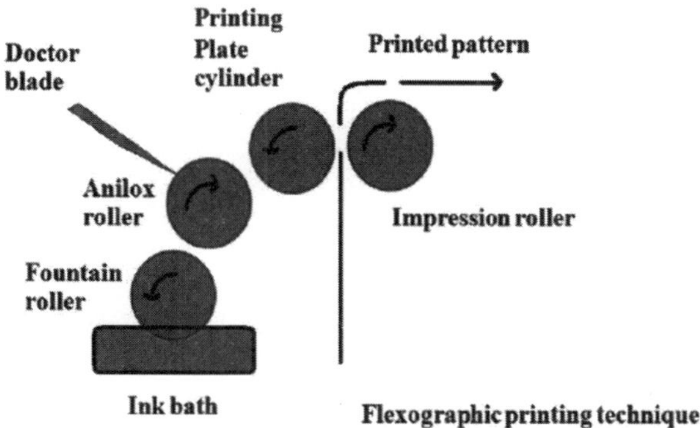

Fig. 16.2. A schematic drawing of flexographic printing technique.

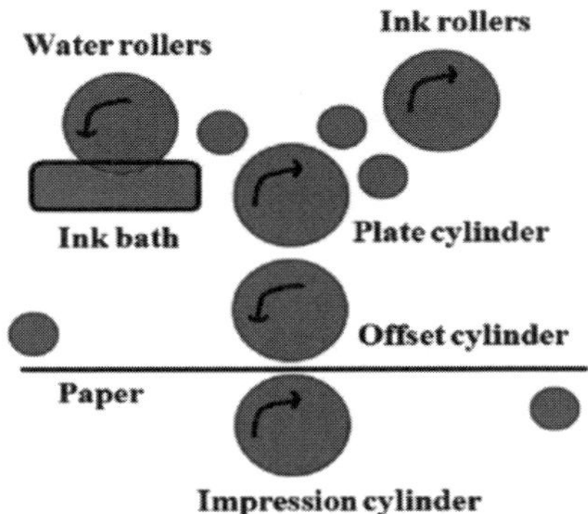

Fig. 16.3. Schematic drawing of offset printing technique.

16.3.3 *Offset printing*

In offset printing, one of the most common printing techniques, the ink solution is transferred (or offset) from a plate to a rubbery blanket, then to the printing surface at a rather high pressure. In particular, the ink conditions for this technique generally require high viscosity (5–200 PaS) with a suitable surface tension (30–60 $\mathrm{mN\,m^{-1}}$). The printed pattern on the offset roll was based on a difference in surface energy, which is generally created by photolithography. Typically, the non-printing areas on the printing roll were wetted with water prior to the ink printed on the surface as exhibited in Fig. 16.3. This technique offers numerous advantages, including consistency of high image, quality for large area of printing, and low cost of processing of printing and low cost of processing.[16,17].

At present, there is hardly any attempt to use green cellulosic substrate for printable electronics using this process. This is mainly attributed to the fact that the transfer mechanism of new polymers into the substrate is not well understood.

16.3.4 *Screen printing*

Screen printing is a very flexible printing technique that allows for full two-dimensional patterning of the printed layer. It was commonly used as a flat printing technique in batch processes, but it can also be adaptable to a R2R process, rotary printing, which enables a rather high throughput. Generally, this technique uses a woven mesh to support an ink-blocking stencil. The attached stencil forms open areas of mesh that transfers ink, which can be pressed through the mesh as a sharp-edged image onto a green nanocellulose substrate as exhibited in Fig. 16.4. A roller is moved across the screen stencil, pushing ink past the threads of the woven mesh in the open areas. Ink solution requires high viscosity of 1–50 PaS, the print resolution is around 100 μm and a wet layer thickness of up to 100 μm can be printed.[18–20] The advantage of this technique is essentially no loss of coating solution during printing.

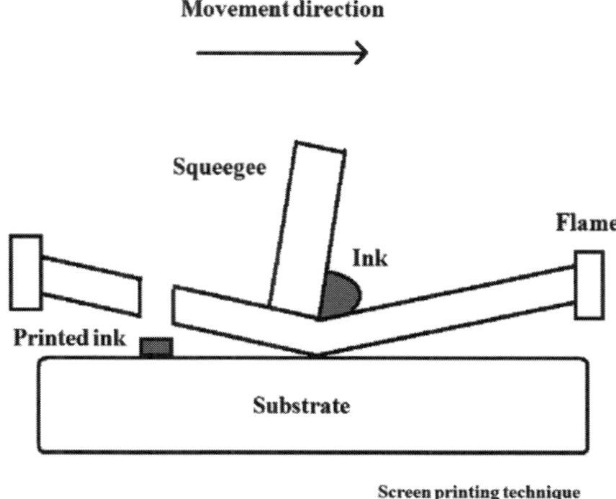

Fig. 16.4. Schematic drawing of screen printing technique.

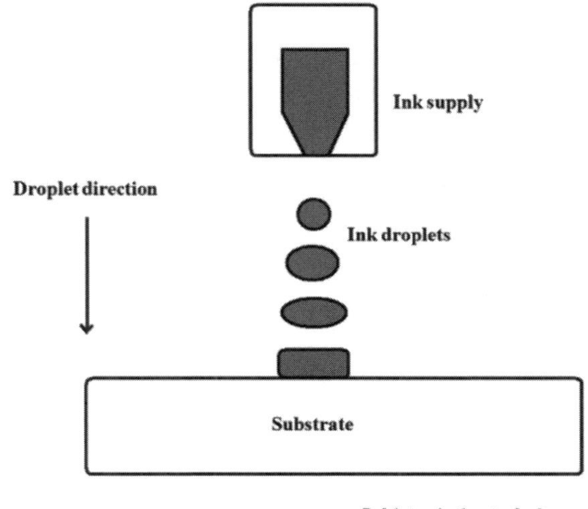

Fig. 16.5. Schematic drawing of inkjet printing technique.

16.3.5 *Inkjet printing*

Inkjet printing is a relatively new process from an industrial point of view for R2R printing technology. The method has been driven forward by the typical low-cost inkjet printer for the home office. The schematic diagram of inkjet printing technique is exhibited in Fig. 16.5. It is well-known that this technique was versatile for several types of ink formulation with many different solvents. The manipulation of this technique can not only be controlled by computer but a digital image can also be created by propelling droplets of ink onto substrate.[21,22] The advantages of this technique are low cost, easily changeable digital print patterns, and low material consumption. Although inkjet printing for industry is R2R compatible, it offers a limited printing speed and unless the ink properties are properly adjusted to a low viscosity (15 mPas) and low evaporation rate, this process commonly has problems with clogging, especially when small nozzles are used. In recent years, there have been major attempts by researcher to develop this technique to print custom-designed display systems, credit cards, and other security tags.

16.4. Applications of printed electronics for electronic components

16.4.1 *Transistors*

Over the last decade, transistors have moved from being a convenient tool for probing the charge transport in high band-gap, low conductivity organic semiconductors to an emerging technology enabling electronic functionalities to be integrated on electronic substrate for a range of applications.[23,24]. From the past, transistor was previously the main key component in most electronic systems.[25] It has been widely used, for example, as matrix controlling elements in displays and as logic gates in digital electronic circuits. A logic inverter can be made by simply connecting a transistor with a resistor. If the inverter has a gain ($dV_{\text{out}}/dV_{\text{in}} > 1$), it can be used as an amplifier or as a building block for more complicated circuits, such as ring, oscillators, flip-flops, ship registers, row decoders, code generator as well as transponder circuits. However, from industry perspective, fabricated system typically requires low power consumption and high yield of product; transistors have been modified as combination of p- and n-channel as exhibited in Fig. 16.6. It consists of at least four layers: (i) source and drain electrode, (ii) semiconductor, (iii) gate dielectric (insulator), and (iv) gate electrode.

Transistor has been continually developed up to the present time. Both inorganic and organic materials have been modified in order to tailor excellent properties of transistor. Commonly, in industry, transistor has been used to fabricate as liquid crystal displays (LCDs) and light-emitting diode (LED). Substrates are typically

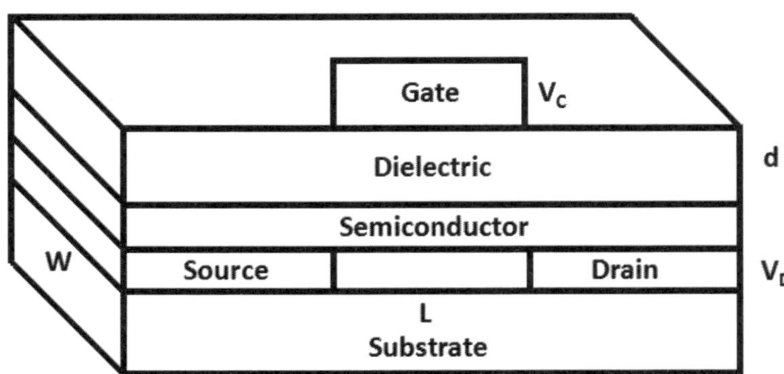

Fig. 16.6. The schematic structure of a field-effect transistor (FET) with a top-gate bottom-contact geometry is shown. The source contact was connected to ground and the applied voltages to the drain and gate contacts are V_D and V_C, respectively. L is the channel length, W is the cannel width, and d is the thickness of the dielectric layer.

engineered polymers and are not flexible. Recent study indicated that nanocellulose substrate can be potential green carrier materials for future commercial transistors. Impressive progress has also been made in realizing logic circuits based on transistor for applications in RFID and chemical or biological sensing. This rapid move toward real-world applications has been enabled by broad, interdisciplinary research efforts to better understand the materials, processing and device architecture requirements for semiconductors to achieve higher level of transistor performance.

16.4.2 *Diode structure*

In electronics, diode is a type of two-terminal electronic component with non-linear resistance and conductance, distinguishing it from components such as two-terminal linear resistors which obey Ohm's law. The most common diode device that is currently used is the semiconductor diode which is composed of a crystalline piece of semiconductor material connected to two electrical terminals. Up to the present time, numerous applications on diode have been developed such as photovoltaic and OLEDs.

16.4.3 *Photovoltaic*

Photovoltaic is a promising future source of sustainable electricity production.[26–29] It has been developed as an alternative energy. However, the main drawback of this technology is concerned with inorganic material with high cost. Scaling up for commercialization needs significant cost reduction. Lately, research and development projects are undertaken to replace inorganic substrates with green natural materials such as cellulose as a carrier substrate. However, a significant work still needs to be done in investigating durability of such devices under stringent climatic conditions.

Based on common use of photovoltaic material, dye solar cell is among emerging solar technologies that has potential for low cost along with relatively high efficiency.[30–33] However, from an industrial point of view, the R2R mass production of print electronic concepts could be groundbreaking.

The basic components of a dye solar cell are the photoelectrode, the counter electrode, and electrolyte as shown in Fig. 16.7. The photoelectrode consists of a

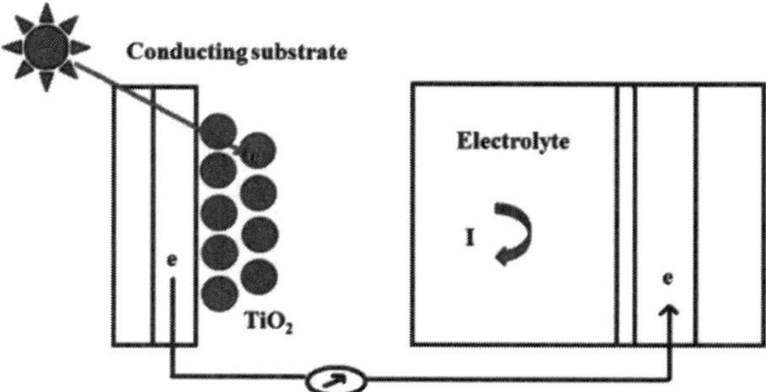

Fig. 16.7. Structure of the dye solar cell.

nanocrystalline TiO_2 layer deposited on substrate with a conducting layer, conventionally a transparent conducting oxide plastic or metal foil.[34] The TiO_2 layer is sensitized with dye forming a monolayer that absorbed the light. Upon the excitation of a dye molecule by absorbing a photon from sunlight which results in an electron injection into the conduction band of the semiconductor (TiO_2), the external circuit depicts this electron and transfers it toward the counter electrode.

16.4.4 *Organic light emitting diodes*

OLEDs[35–37] have attracted considerable interest owing to their promising applications in flat-panel displays by replacing cathode ray tubes (CRTs)[38,39] and LCDs.[40–42]

The basic structure of an OLED consists of a thin film of organic material sandwiched between two electrodes, as depicted in Fig. 16.8. Organic electroluminescent (OEL) materials, based on π-conjugated molecules, were almost insulators, and light was produced by recombination of holes and electrons which has to be injected at the electrodes. The anode is transparent and usually prepared from indium tin oxide (ITO)[43–45], while the cathode is reflective and prepared from metal. The thickness of the organic layer was very thin, between 100 and 150 nm. When a voltage was applied between the electrodes, charges were injected in the organic material, holes from the anode, and electrons from the cathode. Then, the charges move inside the material, generally by hopping processes and then recombine to form excitons. The location of the recombination zone in the diode is a function of the charge mobility of the organic material as well as of the electric field distribution. After diffusion, the exciton recombines and a photon is consequently emitted.

To the best of our knowledge, numerous efforts have been continuously developed for light emitting device. Our research group at University of Toronto has successfully prototyped OLED by using cellulose nanocomposite as flexible substrate.

From the viewpoint of printed electronics, ink solution was prepared as active nanoparticle solution of ceramic or polymer. The combination of ceramic and polymer, called hybrid composite, is sometimes required. Although printed electronics is the excellent fit for R2R process, the optoelectronic quality of OLED is still under

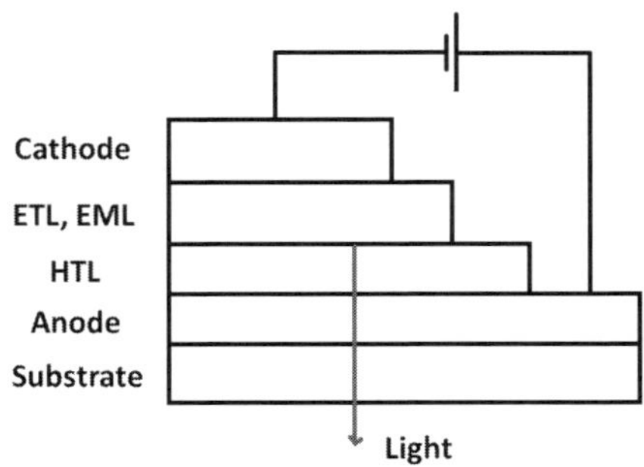

Fig. 16.8. Basic OLED structures.

developed. The push toward developing an electronic printing method with high OLED quality needs to be further investigated.

16.4.5 *Energy storage devices*

Energy storage devices, generally refers to batteries[45] and supercapacitors,[46,47] are made up of two electrodes separated by an electrolyte. For battery, it can be either non-rechargeable (primary) or rechargeable (secondary) form. The electric charge is stored as chemical energy through the electrochemical redox reaction in the electrode materials. However, for supercapacitor, the electrostatic charge is stored through the formation of an electric double layer at the interface between the electrolyte and the electrodes. Battery typically has a larger energy density, while the charging and discharging speeds and the power density are lower. In supercapacitors, on the other hand, the charging does not involve a phase change; consequently, the charging, discharging speed, as well as cyclability are high. The schematic diagram of both battery and supercapacitor is shown in Fig. 16.9.

For transistors and diode, surface roughness of printable substrates could be beneficial for energy storage devices, where a large surface area and electrolyte absorption are preferred. Until now, most research publications reported on studies on the preparation of novel materials and their assembly for energy storage device. However, these techniques are often complex and are high cost. It is inadequate for upscaling to mass production. The push toward the preparation of energy storage device by print electronic method will need further research.

16.4.6 *Electronic memory devices*

Electronic memory devices are theoretically required for storing digital information and are used in applications such as RFID tags. Numerous papers have been reported on the development of electronic memory device up to the present time.[48,49]. A reported non-volatile type of electronic memory device did not require power to maintain the stored information, whereas a volatile memory device lost the information when the power was not applied. Moreover, currently, memory devices

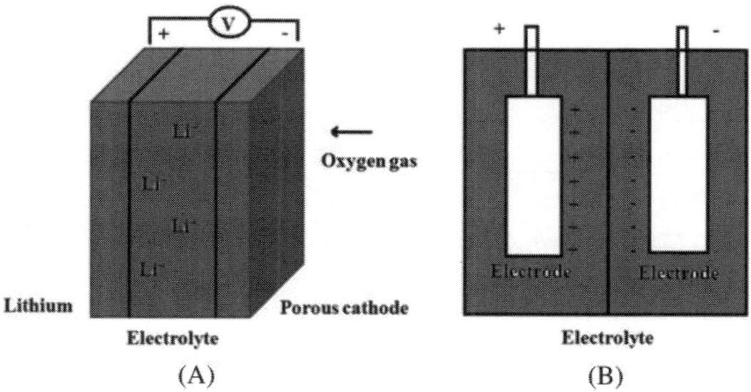

Fig. 16.9. Schematic diagram of energy storage device (A) battery and (B) supercapacitors.

can be categorized into other subcategories, such as write-once read-many times (WORM) memories.[50–53] In order to use electronic memory device with high efficiency, writing/reading speed, retention time, on/off ratio as well as yield of the fabrication process are important parameters that have to be considered to optimize.

Based on RFID system, the fabrication of material has to be focused on the correlation of electronic and magnetic properties. It is used as a wireless non-contact system that uses radiofrequency electromagnetic fields to transfer data from a tag attached to an object for the purposes of automatic identification and tracking. For the tags, it is powered by the electromagnetic fields to read them. On the other hand, the local power source emits radio waves. The tags contain electronically stored information which can be read from up to several meters away. The diagram of RFID system is exhibited in Fig. 16.10. Recently, RFID system was extensively employed in many industries ranging from pharmaceutical science, livestock to any identity card.

From the viewpoint of mass production, printable electronic has been widely employed for RFID system. Kaihovirta *et al.*[54] and Mannerbro *et al.*[55] initially demonstrated the use of electronic printer for RFID fabrication. However, the use of this method is currently still in development stage, the push toward an improved RFID quality is still in progress. In a recent study by the University of Toronto, it was demonstrated that using nanocellulose for RFID application needs a barrier conductive coat before being exposed to high temperature curing conditions of the deposited layers.

16.4.7 *Actuators*

Actuators have been extensively developed in numerous applications, ranging from micro-insect robots, microflying objects to micro-electromechanical systems.[56–58] The theory of actuator is based on piezoelectric that involved the correlation between electronic and mechanic. The charge that accumulated in certain solid materials in response to applied mechanical stress was investigated. The basic diagram of piezoelectric is exhibited in Fig. 16.11. Until today, there is no research data reported on green and flexible actuators.

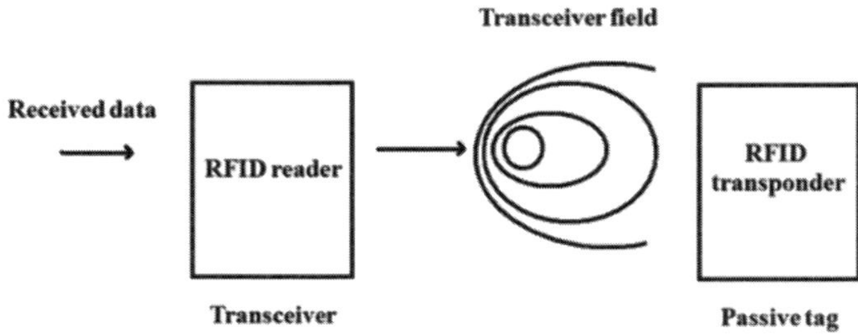

Fig. 16.10. Basic RFID systems.

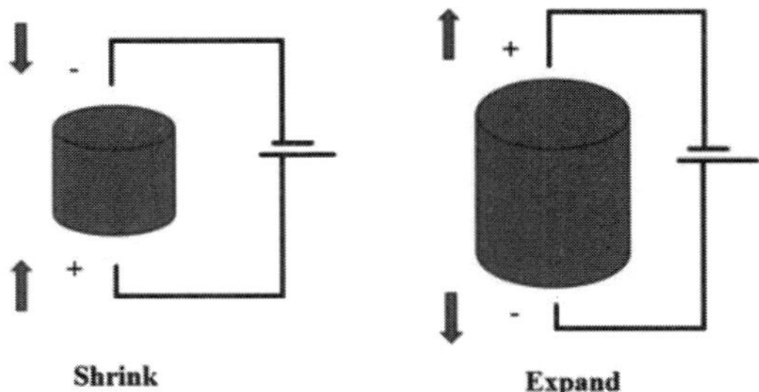

Fig. 16.11. Basic theory of piezoelectric material.

16.5. Conclusion and outlook

Development on printed electronics is becoming a priority of businesses and governments alike. New technology for electronic preparation is becoming a norm of the present day development. Many products are being researched and prototyped. In most cases, it is the efficiency and cost of materials and fabrication preventing them to reach market place.

On the other hand, development of the printed electronics has been reported to make tremendous progress in recent years. It resulted in improving efficiency in many applications. However, further progress is still required in the electronic applications such as stability, performance, and processability. From an industrial viewpoint, printed electronics will be the future for making electronic devices. Green and flexible printed electronics development is now in direct competition with the conventional printed electronics and commercial availability of cheaper and durable substrates will win the initial market.

References

1. S. Ummartyotin, J. Juntaro, C. Wu, M. Sain and H. Manuspiya, Deposition of PEDOT: PSS nanoparticles as a conductive microlayer anode in OLEDs device by desktop inkjet printer, *J. Nanomater.* **1** (2011) 1–7.
2. A. Maattanen, P. Ihalainen, R. Bollstrom, S. Wang, M. Toivakka and J. Peltonen, Enhanced surface wetting of pigment coated paper by UVC irradiation, *Ind. Eng. Industr. Res.* **49** (2010) 11351–11356.
3. H. Kipphan, *Handbook of print media.* Verlag Berlin: Springer (2001).
4. M. Hambsch, K. Reuter, M. Stanel, G. Schmidt, H. Kempa, U. Fugmann, U. Hahn and A.C. Hubler, Uniformity of fully gravure printed organic field-effect transistors, *Mater. Sci. Eng*: B **170** (2010) 93–98.
5. T. Makela, S. Jussila, M. Vilkman, H. Kosonen and R. Korhonen, Roll-to-roll method for producing polyaniline patterns on paper, *Synt. Met.* **135–136** (2003) 41–42.
6. M. Pudas, N. Halonen, P. Granat and J. Vahakangas, Gravure printing of conductive particulate polymer inks on flexible substrate, *Progr. Org. Coat.* **54** (2005) 310–316.
7. H. Sirringhaus, P.J. Brown, R.H. Friend, M.M. Nielsen, K. Bechgaard, B.M.W. Langeveld, A.J.H. Spiering, R.A.J. Janssen, E.W. Meijer, P. Herwig and D.M. de Leeuw, Two-dimensional charge transport in self-organized, high-mobility conjugated polymers, *Nature* **401** (1999) 685–688.

8. D.T. Britton, M. Harting, D. Knoesen, Z. Sigcau, F.P. Nemalili, T.P. Ntsoane, P. Sperr, W. Egger and M. Nippus, Microstructural defect characterization of a-Si:H deposited by low temperature HW-CVD on paper substrates, *Thin Solid Films* **501** (2006) 79–83.
9. M. Jung, J. Kim, J. Noh, N. Lim, C. Lim, G. Lee, J. Kim, H. Kang, K. Jung, A.D. Leonard, J.M. Tour and G. Cho, All-printed and roll-to-roll-printable 13.56 MHz-operated 1-bit RF tag on plastic foils, *IEEE Trans.Elect. Device.* **57** (2010) 571–580.
10. D.Y. Chung, J. Huang, D.D.C. Bradley and A.J. Campbell, High performance, flexible polymer light-emitting diodes (PLEDs) with gravure contact printed hole injection and light emitting layers, *Organic Electron.* **11** (2010) 1088–1095.
11. P. Kopola, T. Aernouts, S. Guillerez, H. Jin, M. Tuomikoski, A. Maaninen and J. Hast, High efficient plastic solar cells fabricated with a high-throughput gravure printing method, *Sol. Energ. Mater. Solar Cells* **94** (2010) 1673–1680.
12. F.C. Krebs, Pad printing as a film forming technique for polymer solar cells, *Sol. Ener. Mater. Solar Cells* **93** (2009) 484–490.
13. J. Noh, D. Yeom, C. Lim, H. Cha, J. Han, J. Kim, Y. Park, V. Subramanian and G. Cho, Scalability of roll-to-roll gravure-printed electrodes on plastic foils, *IEEE Trans. Electron. Packag. Manuf.* **33** (2010) 275–283.
14. M.M. Voigt, A. Guite, D.Y. Chung, R.U.A. Khan, A.J. Campbell, D.D.C. Bradley, F. Meng, J.H.G. Steinke, S. Tierney, I. McCulloch, H. Penxten, L. Lutsen, O. Douheret, J. Manca, U. Brokmann, K. Sonnichsen, H. Dagmar, W. Bock, C. Barron, N. Blanckaert, S. Springer, J. Grupp and A. Mosley, Polymer field effect transistors fabricated by the sequential gravure printing of polythiophene, two insulator layers, and a metal ink gate, *Ad. Func. Mater.* **20** (2010) 239–246.
15. A.C. Hubler, B. Trnovec, T. Zillger, M. Ali, N. Wetzold, M. Mingebach, A. Wagenpfahl, C. Deibel and V. Dyakonov, Printed paper photovoltaic cells, *Ad. En. Mater.* **1** (2011) 1018–1022.
16. A.C. Huebler, F. Doetz, H. Kempa, H.E. Katz, M. Bartzsch, N. Brandt, I. Hennig, U. Fuegmann, S. Vaidyanathan, J. Granstrom, S. Liu, A. Sydorenko, T. Zillger, G. Schmidt, K. Preissler, E. Reichmanis, P. Eckerle, F. Richter, T. Fischer and U. Hahn, Ring oscillator fabricated completely by means of mass-printing technologies, *Org. Electr.* **8** (2007) 480–486.
17. D. Zielke, A.C. Hubler, U. Hahn, N. Brandt, M. Bartzsch, U. Fugmann, T. Fischer, J. Veres and S. Ogier, Polymer-based organic field-effect transistor using offset printed source/drain structures, *Appl. Phys. Lett.* **87** (2005) 123508–123511.
18. M. Harting, J. Zhang, D.R. Gamota and D.T. Britton, Fully printed silicon field effect transistors, *Appl. Phys. Lett* **94** (2009) 193509–193511.
19. J.M. Verilhac, M. Benwadih, A.L. Seiler, S. Jacob, C. Bory, J. Bablet, M. Heitzman, J. Tallal, L. Barbut, P. Frere, G. Sicard, R. Gwoziecki, I. Chartier, R. Coppard and C. Serbutoviez, Step toward robust and reliable amorphous polymer field effect transistors and logic functions made by the use of roll to roll compatible printing processes, *Org. Electron.* **11** (2010) 456–462.
20. S.E. Shaheen, R. Radspinner, N. Peyghambarian and G.E. Jabbour, Fabrication of bulk heterojunction plastic solar cells by screen printing, *Appl. Phys. Lett.* **79** (2001) 2996–2998.
21. S.H. Ko, J. Chung, H. Pan, C.P. Grioropoulos and D. Poulikakos, Fabrication of multilayer passive and active electric components on polymer using inkjet printing and low temperature laser processing, *Sens. Actuat. A: Phys.* **134** (2007) 161–168.
22. S.E. Molesa, S.K. Volkman, D.R. Redinger, A.F. Vornbrock and V. Subramanian, A high performance all-inkjetted organic transistor technology. In *Electron Device Meeting on IEEE International Event* (2004) 1072–1074.

23. Y.M. Sung, J.B. Casady, J.B. Dufrene and A.K. Agarwal, A review of SiC static induction transistor development for high freqneucy power ampliers, *Solid State Electron.* **46** (2002) 605–613.
24. J.S. Park, W.J. Maeng, H.S. Kim and J.S. Park, Review of recent developments in amorphous oxide semiconductor thin-film transistor devices, *Thin Solid Films* **520** (2012) 1679–1693.
25. J.C. Boudenot, From transistor to nanotube, *Comptes Rendus Physique* **9** (2008) 41–52.
26. D. Tobjork and R. Osterbacka, Paper electronics, *Ad. Mater.* **23** (2011) 1935–1961.
27. N. Kaihovirta, T. Makela, X. He, C.J. Wikman, C.E. Wilen and R. Osterbacka, Printed all-polymer electrochemical transistors on patterned ion conducting membranes, *Org. Electron.* **11** (2010) 1207–1211.
28. G. Gelinck, P. Heremans, K. Nomoto and T.D. Anthopoulos, Organic transistors in optical displays and microelectronic applications, *Ad. Mater.* **22** (2010) 3778–3798.
29. H. Sirringhaus, Reliability of organic field-effect transistors, *Ad. Mater.* **21** (2009) 3859–3873.
30. T.P. Brody, The thin film transistor — A late flowering bloom, *IEEE Trans. Electron Device.* **31** (1984) 1614–1628.
31. W.S. Bacon, The transistor's 20th anniversary: how Germanium and a bit of wire changed the world, *Popul. Sci.* **192** (1968) 80–84.
32. D.H. Kim, Y.S. Kim, J. Wu, Z. Liu, J. Song, H.S. Kim, Y.Y. Huang, K.C. Hwang and J.A. Rogers, Ultrathin silicon circuits with strain-isolation layers and mesh layouts for high-performance electronics on Fabric, Vinyl, Leather, and Paper, *Ad. Mater.* **21** (2009) 3703–3707.
33. T.T. Chow, A review on photovoltaic/thermal hybrid solar technology, *Appl. Energy* **87** (2010) 365–379.
34. B. Parida, S. Iniyan and R. Goic, A review of solar photovoltaic technologies, *Renew. Sustain. Energ. Rev.* **15** (2011) 1625–1636.
35. X. Zhang, X. Zhao, S. Smith, J. Xu and X. Yu, Review of R&D progress and practical application of the solar photovoltaic/thermal (PV/T) technologies, *Renew. Sustain. Energ. Rev.* **16** (2012) 599–617.
36. A. Mellit, S.A. Kalogirou, L. Hontoria and S. Shaari, Artificial intelligence techniques for sizing photovoltaic systems: a review, *Renew. Sustain. Energ. Rev.* **13** (2009) 406–419.
37. G. Hashmi, K. Miettunen, T. Peltola, J. Halme, I. Asghar, K. Aitola, M. Toivola and P. Lund, Review of materials and manufacturing options for large area flexible dye solar cell, *Renew. Sustain. Energ. Rev.* **15** (2011) 3717–3732.
38. M. Rita Narayan, Dye sensitized solar cells based on natural photosensitizers, *Renew. Sustain. Energ. Rev.* **16** (2012) 208–215.
39. R. Harikisun and H. Desilvestro, Long term stability of dye solar cells, *Sol. Energy* **85** (2011) 1179–1188.
40. B. Li, L. Wang, B. Kang, P. Wang and Y. Qiu, Review or recent progress in solid state dye-sensitized solar cells, *Sol. Ener. Mater. Solar Cells* **90** (2006) 549–573.
41. M.D.K. Nazeeruddin and M. Gratzel, Conversion and storage of solar energy using dye-sensitized nanocrystalline TiO_2 cells, *Compre. Coord. Chem. II* **9** (2003) 719–758.
42. B. Greffroy, P. Le Roy and C. Prat, Review: organic light emitting diode (OLED) technology: materials, deviced and display technologies, *Pol. Intern.* **55** (2006) 572–582.
43. J. Shinar and R. Shinar, An overview of organic light emitting diodes and their applications, *Compr. Nanosci. Technol.* **1** (2011) 73–107.
44. G. Zhou, W.Y. Wong and S. Suo, Recent progress and current challenges in phosphorescent white organic light emitting diodes (WOLEDs), *J. Photochem. Photobiol. C. Photochem. Rev.* **11** (2010) 133–156.

45. Z.Z. You and J.Y. Dong, Surface modification of ITO electrodes for polymer light emitting devices, *Appl. Surf. Sci.* **253** (2006) 2102–2107.
46. Z. Rao and S. Wang, A review of power battery thermal energy management, *Renew. Sustain. Energy Rev.* **15** (2011) 4554–4571.
47. M. Al Sakka, H. Gualous, J.V. Mierlo and H. Culcu, Thermal modeling and heat management of supercapacitor modules for vehicle applications, *J. Power Sources* **194** (2009) 581–587.
48. G.A. Snook, P. Kao and A.S. Best, Conducting-polymer-based superconductor devices and electrodes, *J. Power Sources* **196** (2011) 1–12.
49. A. Sarac, N. Absi and S. Dauzere-Peres, A literature review on the impact of RFID intechnologies on supply chain management, *Inter. J. Prod. Econ.* **128** (2010) 77–95.
50. R. van der Togt, P.J.M. Bakker and M.W.M. Jaspers, A framework for performance and data quality assessment of radio frequency identification (RFID) systems in health care settings, *J. Biomed. Informat.* **44** (2011) 372–383.
51. M.A. Mamo, E. Segura-Cardenas, A.O. Sustaita, F. Gao, X. Yang and N.C. Greenham, Simple write-once-read-many-times-memory device based on a carbon sphere-poly(vinylphenol) composite, *Org. Electron.* **11** (2011) 1858–1863.
52. J.A. Avila-Nino, E. Segura-Cardenas, A.O. Sustaita, I. Cruz-Cruz, R. Lopez-Sandoval and M. Reyes-Reyes, Nonvolatile write-once-read-many-times memory device with functionalized-nanoshells/PEDOT:PSS nanocomposites, *Mater. Sci. Eng. B* **176** (2011) 462–466.
53. J. Wang, X. Cheng, M. Caironi, F. Gao, X. Yang and N.C. Greenham, Entirely solution-processed write-once-read-many-times memory devices and their operation mechanism, *Org. Electron.* **12** (2011) 1271–1274.
54. N. Kaihovirta, D. Tobjork, T. Makela and R. Osterbacka, Absense of substrate roughness effects on an all-printed organic transistor operating at one volt, *Appl. Phys. Lett.* **93** (2008) 53302–53304.
55. R. Mannerbro, M. Ranlof, N. Robinson and R. Forchheimer, Inkjet printed electrochemical organic electronics, *Synth. Met.* **158** (2008) 556–560.
56. A. de Greef, P. Lambert and A. Delchambre, Towards flexible medical instruments: review of flexible fluidic actuators, *Precision Eng.* **33** (2009) 311–321.
57. B. Watson, J. Friend and L. Yeo, Piezoelectric ultrasonic micro/milli-scale actuators, *Sens. Actuat. A Phys.* **152** (2009) 219–233.
58. C. Li, E.T. Thostenson and T.W. Chou, Sensors and actuators based on carbon nanotubes and their composites: a review, *Compos. Sci. Technol.* **68** (2008) 1227–1249.

Chapter 17

Nanocellulose-modified Wood Adhesives

Wolfgang Gindl-Altmutter and Stefan Veigel
Institute of Wood Technology and Renewable Materials,
Department of Materials Science and Process Engineering,
BOKU-University of Natural Resources and Life Sciences Vienna, Austria

Adhesive bonding is one of the key steps in the production of modern wood composite products. Improvements in wood adhesion are desirable in terms of performance improvement, adhesive savings, or both. The brief overview of the state of the art given in this chapter demonstrates the feasibility of wood adhesive modification through the addition of nanocellulose. Highly significant positive effects of nanocellulose addition on mechanical bond stability in both wet and dry conditions are reported. The effects are consistent for brittle thermosets such as urea–formaldehyde as well as comparably soft thermoplastics such as polyvinylacetate. It is shown that both solid wood adhesion and wood composite production can benefit from the addition of only a few percent of nanocellulose to the adhesive.

17.1. Wood adhesives and their mechanics

Adhesives used in the wood industry may be grouped according to their chemistry, their physical properties, or their intended use in e.g., the construction, furniture, or wood-composite industries.[1–3] When classified according to their chemistry, formaldehyde-based adhesives represent the most important group. Urea, melamine, phenol, and resorcinol are involved as reaction partners, either individually, or in combinations. Furthermore, isocyanate-based adhesives, polyethylene-vinylacetate, and various biobased protein-, tannin-, or lignin-derived adhesives are of relevance. The most important requirements these polymers have to fulfill with regard to the adhesive bonding of wood comprise the wetting of the wood surface in liquid state, the build-up of adhesive forces during cure, and the build-up of sufficient cohesion within the polymer. Apart from these core parameters, a range of criteria have to be considered when an adhesive is formulated for a specific use. Among these criteria, properties as diverse as speed of cure, viscosity, storage life, resistance to hydrolysis, thermal stability, creep characteristics, and many more are of relevance. The ultimate target any wood adhesive has to meet is that it must not fail during service. Due to the complexity of wood adhesive bonds in terms of surface- and adhesive chemistry, surface geometry, and adhesive penetration, it is quite a challenge to develop a comprehensive understanding of the mechanical durability of wood adhesive bonds.[4] In terms of bond mechanics, four specific areas of interest can be identified.[5]

(1) *Wood in the shape of boards, veneer, flakes, strands, particles, or fibers*: The mechanical properties of wood are well characterized in spite of its complex multi-level composite structure,[6] and input data for mechanical considerations can be found in the literature without difficulty.
(2) The interphase, where wood and adhesive interpenetrate. A certain amount of liquid adhesive flows into open microscopic cavities during bonding. Although intensively studied,[7] the contribution of cured adhesive present in wood cavities remains uncertain. In addition to this microscale interphase, many adhesives are also capable of forming a nanoscale interphase, where low molecular weight adhesive fractions diffuse into the wood cell wall and form an interpenetrating network upon cure which often significantly affects cell-wall mechanics.[8–11] It was proposed that this localized cell-wall modification is beneficial with regard to adhesive bond durability.[4]
(3) The interface between the wood cell wall and the adhesive. Very little is known about the actual interface between the wood cell wall and cured adhesive polymers, since conventional test methods for bond strength operate at the mm scale. Due to the complex cellular geometry of the wood surface which becomes even more complicated by uncontrollable crack propagation during testing, it is difficult to achieve insights into the strength of the interface from such tests. Recently, a new indentation-based method was introduced,[12] which could help improve understanding of this important link in the wood-adhesive bond chain.
(4) Bulk cured adhesive in the bond line. Mechanical properties of cured adhesives are available from experiments with cast polymer films,[13,14] and from micromechanical *in situ* measurements in cured adhesive bond lines by means of nanoindentation.[15,16] While it is obvious that the cohesive strength of an adhesive polymer in cured state has to surpass a certain threshold value defined by the strength of the material to be bonded, the stiffness of a cured adhesive also plays an important role in terms of overall bond performance. For a lap-shear assembly, the well-known Volkersen equation nicely describes a relationship between the stiffness of an adhesive and the intensity of stress concentrations in the bonded lap-shear assembly.[17,18] As stress concentration intensity increases with increasing adhesive stiffness, the lap-shear strength measured macroscopically decreases.[19] As shown in Fig. 17.1, the stiffness of adhesives used in wood industry shows extreme variability.

Particularly in the case of adhesives showing either extreme stiffness or pronounced softness, a modification of mechanical properties may be desirable in order to optimize performance. The high stiffness of urea–formaldehyde (UF) is perceived a weakness, which is why attempts at plasticization of this brittle thermoset were made.[21–23] At the other end of the scale, one-component polyurethane wood adhesives were initially characterized by unsatisfactory thermal stability

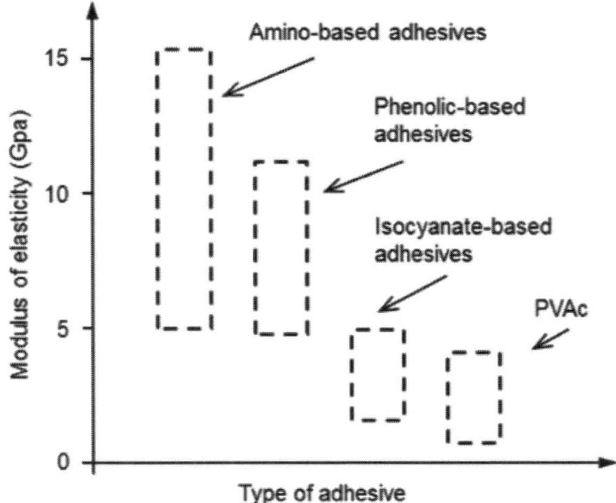

Fig. 17.1. Variability in the stiffness of representative groups of adhesives used in wood industry. (This figure is based on data compiled in Ref. 20.)

and excessive creep, which was substantially improved by varying pre-polymer components.[24,25]

17.2. Modification of wood adhesives by adding particulate or fibrous fillers

The obvious path for the modification of adhesive properties is through the modification of their chemistry. Within a certain range, the mechanics of an adhesive in its cured state can be controlled by fine-tuning its macromolecular structure. Also, e.g., the addition of melamine to UF affects its mechanics in a positive manner.[1] While modifications in the chemistry of adhesives are one path of optimization of mechanical adhesive characteristics, the addition of particulate or fibrous fillers presents a possible alternative route of adhesive modification. By adding fillers, the limitations imposed by polymer chemistry can be overcome. This approach is common in high performance adhesives, which can be reinforced with nanoparticles[26–29] or nanofibers[30–35] from high-strength materials. This approach has also been followed with some wood adhesives. Positive results are reported for nanoclay-filled UF in a number of studies. Here, the addition of nanoclay leads to improved thermal stability,[36] increased water resistance, and internal bonding of panels.[37] No improvement was reported in the manufacture of plywood.[38] Similar to nanoclay, nano-SiO_2-filled UF exhibited significant performance improvements in the bonding of OSB panels.[39] As opposed to UF, the addition of nanoclay does not appear to improve the performance of phenol–formaldehyde and phenol–UF resins.[40] Contrarily, excellent results were achieved by the addition of nanoclay to polyvinylacetate (PVAc) adhesive.[41] It was shown that the shear strength of wood joints increased in dry and wet states by adding nanoclay. Also, the inclusion of nanoclay improved the thermal stability of the adhesive.

Besides the rather modern approach of nanoparticle modification, the classic approach of adding organic flours to UF adhesives should also be mentioned here. This procedure primarily improves the processing characteristics of the adhesive mixture, although it may also contribute to bonding strength by agglutination of starch during hot pressing.[1]

Wood adhesives are designed to bond lignocellulosic parts and particles. Therefore, it is obvious that lignocellulosic particles also present good candidates as fillers of wood adhesives, because good adhesion between polymeric matrix (i.e., adhesive) and the filler can be expected in this case. A very extensive patent application[42] covers virtually all potential combinations of wood adhesives with lignocellulosic and cellulosic fibers and particles from macro to micro to nano. Nonetheless, very little experimental evidence is available with regard to positive effects of lignocellulosic filler in wood adhesives. However, the only example found to the best knowledge of the authors reports a spectacular improvement of up to 20% better strength for phenol–formaldehyde wood adhesive modified by adding steam exploded beech flour.[43] The study suggests that the improved performance results from the physical, rather than the chemical nature of the additive.

As opposed to lignocellulosic fillers, pure cellulosic fillers are better represented in the literature. This is particularly the case for nanocellulose-modified adhesives, which will be treated in this chapter. A differentiation is made into soft adhesives and brittle adhesives, because from the viewpoint of mechanics, the effect of nanocellulose addition is quite different depending on the modulus of the adhesive in question. In soft isocyanate-based adhesives and PVAc, a noticeable increase in stiffness can be expected following a simple rule of mixture approach, due to the comparably high modulus of cellulose. On the other hand, brittle amino-based adhesives are already quite stiff, which is why the stiffening effect of cellulose is obviously much less prominent. In the latter case, toughening mechanisms such as crack-bridging or crack-path deflection may play a more significant role.

17.3. Nanocellulose-modified soft adhesives

Isocyanate-based adhesives and PVAc are remarkably less stiff than phenol- and aminoplastic adhesive polymers (Fig. 17.1). The addition of high-stiffness cellulose nanofibrils (CNFs) or cellulose nanocrystals (CNCs) to such soft polymers obviously results in a significant stiffening of the cured modified adhesive.[44–46] In the case of one-component polyurethane, an isocyanate-based adhesive, CNF cannot be readily dispersed in the adhesive due to their highly polar nature. Chemically modified CNFs were added to one-component polyurethane at weight fractions from 1 to 10% and wood adhesive bonds were successfully produced;[44] however, no significant and consistent improvement of adhesive bond stability in terms of block shear strength was achieved.

In contrast to one-component polyurethane, the addition of CNF and particularly CNC showed significant improvements in wood adhesive bond strength and durability.[45,46] Nanocomposite films of CNF embedded in a PVAc latex adhesive were prepared.[45] Dynamic mechanical analysis (DMA) of the films revealed a strong influence of CNF on the viscoelastic properties of PVAc latex films. Particularly, increasing amounts of CNFs led to increasing reinforcing effects in the glassy state. The performance of solid wood adhesive bonds, however, was affected in a negative manner by the addition of CNFs. A notable exception in this concept was a clearly improved heat resistance and durability in the wet environment.

In a further study with PVAc,[46] CNCs were used for modification instead of CNFs. Very clear and consistent improvements in block shear strength of wood samples bonded with CNC-modified PVAc were observed in the CNC content up to 3%. Modified PVAc showed higher bond strength than the unmodified references in dry and wet conditions, and also at elevated temperature. These improvements go in line with changes observed in CNC-modified PVAc films, where properties such as hardness, modulus of elasticity (MOE) and creep, as well as thermal stability were significantly improved. Structural studies of CNC-modified PVAc revealed that the nanostructure of cured PVAc is different from the unmodified reference in the presence of CNC.

17.4. Nanocellulose-modified brittle adhesives

As described above, cured amino-based adhesives like UF and melamine–UF (MUF) are characterized by a comparably high elastic modulus and therefore a low deformation of the adhesive layer under mechanical loading. As a result, local stress concentrations are generated along the bond line of a wood adhesive joint that reduce the overall bond strength.[47–49] As described in the following section, several attempts have been made to increase the toughness of brittle formaldehyde-based adhesives by the addition of CNFs.

In an initial study,[50] CNFs were prepared by high pressure homogenization of wood pulp and added to a UF-adhesive in a quantity of 0.5–5% in order to improve the mechanical performance of wood adhesive bonds. Bond strength was evaluated using the lap shear test according to the European standard EN 302-1.[51] For comparison, untreated wood pulp was also used as filler for UF and added in the same amount. As shown in Fig. 17.2, specimens prepared with CNF-filled adhesive showed both a significantly higher strength and strain to failure than specimens bonded with pure UF and those filled with pulp fibers. Since failure strain increased considerably, it was concluded that the UF-adhesive was possibly toughened by the addition of CNF. This assumption was supported by the fact that micro-cracks, which can be frequently found in UF bond lines, were not present in CNF-reinforced UF.

Based on these results, the suitability of CNF as filler for wood adhesives was investigated more closely.[52] In this study, the effect of added cellulose filler on the fracture mechanical properties of wood adhesive bonds was analyzed using the double cantilever beam specimen.[53] Three types of CNFs were applied as filler for UF adhesive, namely unmodified CNF prepared from wood pulp, CNF modified by (2,2,6,6-tetramethylpiperidin-1-yl)oxyl (TEMPO) and CNF prepared from bacterial

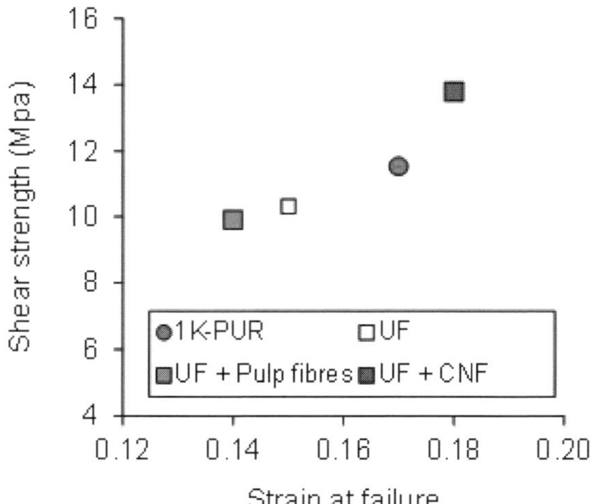

Fig. 17.2. Lap shear test according to EN 302-1: Average shear strength and stain at failure of specimens prepared with unmodified UF adhesive as well as UF containing 1 wt% of CNF and unhomogenised pulp fibers, respectively. One-component poly urethane is shown for reference (own unpublished results).

cellulose. The cellulose content of the adhesive was varied between 0 and 2 wt% based on solid resin.

It was found that even the addition of small amounts of CNF considerably increased adhesive viscosity resulting in a different bond line morphology of cellulose-filled and pure UF. In contrast, the micromechanical characterization of cured bond lines by means of nanoindentation revealed no significant differences, indicating that adhesive cure was not negatively affected by the addition of CNF. Regarding specific fracture energy, CNF-reinforced specimens showed a higher variability than unmodified specimens on average which was attributed to a sub-optimal dispersion of fibrils in the adhesive. When unmodified CNF from wood pulp were used as a filler, the fracture energy of UF-adhesive bonds increased steadily with increasing cellulose content (Fig. 17.3). In this case, the addition of 2 wt% CNF resulted in a 45% higher fracture energy ($353.0 \pm 43.2\,\mathrm{J/m^2}$) as compared to UF-bonds without any cellulose filler ($243.2 \pm 19.5\,\mathrm{J/m^2}$). Although these results confirm the principal feasibility of toughening brittle UF-adhesives by the addition of CNF, the question of the mechanisms that are responsible for this toughening effect could not be answered convincingly.

Apart from bonding of solid wood, CNF-filled adhesives were also successfully applied for the preparation of wood-based panels. UF- and MUF-adhesives were modified with CNF produced by high-pressure homogenization of wood pulp.[54] The cellulose-filled adhesives were used for the preparation of laboratory-scale particle boards (PBs) and oriented strand boards (OSBs) and the rheological behavior of adhesive mixtures as well as mechanical board properties were investigated.

Again, a substantial viscosity increase was determined for CNF-filled adhesive mixtures. Especially in the production of wood-based panels, the adhesive's viscosity must be kept within reasonable limits to ensure a homogeneous distribution of

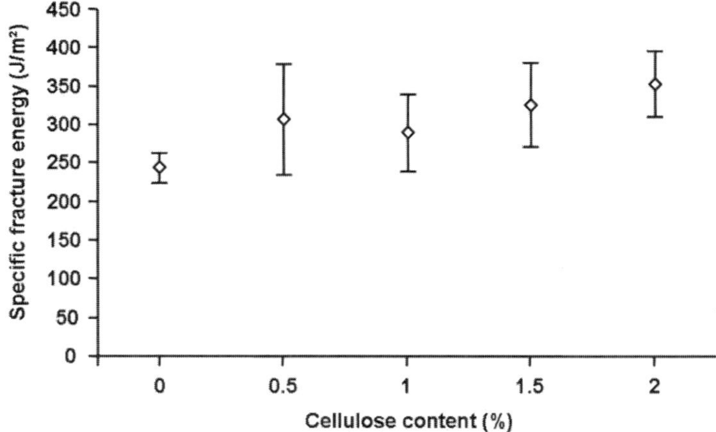

Fig. 17.3. Specific fracture energy of double cantilever beam specimens prepared with UF adhesive containing different amounts of CNFs. (Based on data from Ref. 52.)

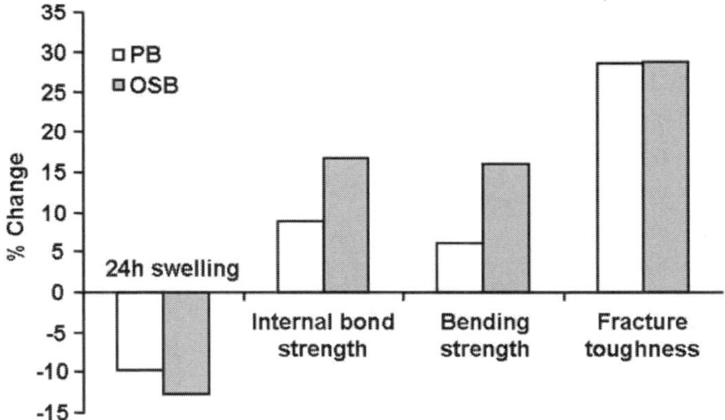

Fig. 17.4. Mechanical properties of laboratory scale PB and OSB prepared with nanocellulose-modified adhesive. The change of properties is given with regard to boards prepared with the same adhesive in unmodified form. (Based on data from Ref. 54.)

adhesive droplets on the surface of wood particles. Upon the addition of 1 wt% of CNFs, adhesive mixtures were still sprayable. Also, from the view of mechanical board performance, this value seems to be close to the optimum. Figure 17.4 shows the relative change in board properties caused by the addition of 1 wt% CNF. It is clearly obvious that for both PB and OSB, the cellulose addition results in a reduced thickness swelling and a higher internal bond and bending strength. Regarding the panel type, CNF-reinforcement worked generally better for OSB than for PB. The most significant increase was found in fracture energy, which was almost 30% higher than for boards produced with the unfilled adhesive. This indicates that wood-based panels can also be substantially toughened by using CNF-filled adhesives.

In the studies mentioned above, cellulose-filled adhesive mixtures were always prepared by adding aqueous fibril suspensions to a commercial adhesive, whereas a completely different approach was chosen in Ref. 55. Here, the possibility of adding the fibril suspension already at the beginning of the adhesive synthesis was studied. Basically, the nanofibrils were properly dispersed in a concentrated formaldehyde solution which was subsequently used for the synthesis of UF. It was expected that stronger interactions between cellulose and adhesive may result from an adhesive synthesis in the presence of CNFs.

Adhesive mixtures with a CNF content of up to 1.6 wt% based on solid resin were produced and used to prepare lap shear specimens.[51] A positive effect on bond strength was only found at an adhesive cellulose content of 0.8% whereas all the other CNF-modified specimens delivered values below that of pure UF-bonds. Accompanying viscosity measurements revealed a value of 0.4 Pas for pure UF, while UF containing 0.8 and 1.6% CNF showed a viscosity of 1.45 and 4.52 Pas, respectively. It was thus concluded that, at least to some extent, the poor performance of specimens containing more than 0.8% CNF can be explained by the rapidly increasing viscosity and a reduced adhesion to the wood surface.

In addition to mechanical testing, the author investigated the thermal and thermomechanical properties of CNF-reinforced UF by means of differential scanning calorimetry (DSC) and DMA. As shown in Fig. 17.5, the DSC curves of cellulose-modified adhesives are characterized by lower exothermic peaks and slightly higher

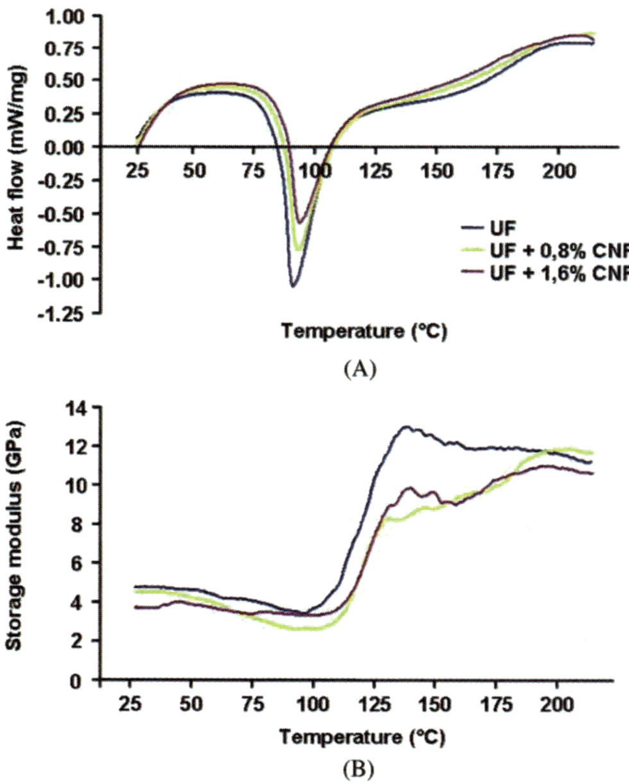

Fig. 17.5. Characterization of the curing behavior of cellulose-modified UF by means of DSC (a) and DMA (b). (Based on data from Ref. 55.)

peak maximum temperatures, indicating a lower enthalpy of reaction and a slightly delayed curing. Similar to DSC, DMA measurements revealed a slower curing of CNF-modified UF indicated by a slower development of storage modulus. However, since the solid resin content of the adhesive mixtures was not constant but decreased considerably for cellulose filled adhesives, this effect may also be attributed to the additional amount of water from the cellulose suspension. Consequently, there was no clear evidence that CNF in amounts of up to 1.6 wt% significantly alter the chemical and mechanical curing behavior of UF.

17.5. Conclusions

At present, the modification of wood adhesives with nanocellulose is a very young subject of research, which is reflected by the relative scarcity of literature. Even so, the work done so far with UF and PVAc demonstrates that significant improvements in adhesive bond performance can be achieved by the addition of nanocellulose. Best results were achieved at 1% up to 3% nanocellulose added, whereby the percentage added is limited by the strong increase in viscosity caused by nanocellulose addition. As the feasibility of nanocellulose modification and its positive effects are being established, future research should be dedicated to two aspects. Firstly, the mechanism of nanocellulose modification in wood adhesives is not clear at present, even though there are indications that nanocellulose somehow interferes with the formation of macromolecular structure during adhesive cure. Secondly, many open questions still have to be addressed before practical, industry-scale application of nanocellulose-modified wood adhesives becomes feasible. Initial results indicate that adhesive cure is not excessively prolonged due to the presence of nanocellulose, which is a very important prerequisite. On the other hand, the severe increase in viscosity caused by the addition of nanocellulose may present a serious obstacle when current processing technologies are used. Thus, both directions of study, i.e., fundamental investigations and application-oriented process development are needed.

Acknowledgment

Funded by the Austrian Science Fund FWF: Project P22516-N22.

References

1. M. Dunky and P. Niemz, *Holzwerkstoffe und leime, technologie und einflussfaktoren*, New York: Springer-Verlag Berlin-Heidelberg (2002).
2. C.R. Frihart, Wood adhesion and adhesives. In *Handbook of wood chemistry and wood composites*, R.M. Rowell (Ed.), Boca Raton, Florida: CRC press (2005).
3. A. Pizzi and K.L. Mittal, *Wood adhesives*, Leiden, the Netherlands: Brill Academic Publishers (2010).
4. C.R. Frihart, Adhesive groups and how they relate to the durability of bonded wood, *J. Adhes. Sci. Technol.* **23** (2009) 601–617.
5. A.A. Marra, *Technology of wood bonding, principles in practice*, New York: Van Nostrand Reinhold (1992).

6. P. Fratzl and R. Weinkamer, Nature's hierarchical materials, *Prog. Mater. Sci.* **52** (2007) 1263–334.
7. F.A. Kamke and J.N. Lee, Adhesive penetration in wood-a review, *Wood Fiber Sci.* **39** (2007) 205–220.
8. W. Gindl, E. Dessipri and R. Wimmer, Using UV-microscopy to study diffusion of melamine-urea-formaldehyde resin in cell walls of spruce wood, *Holzforschung* **56** (2002) 103–107.
9. W. Gindl and H.S. Gupta, Cell-wall hardness and Young's modulus of melamine-modified spruce wood by nano-indentation, *Compos. Part A* **33** (2002) 1141–1145.
10. J. Konnerth, D. Harper, S.H. Lee, T.G. Rials and W. Gindl, Adhesive penetration of wood cell walls investigated by scanning thermal microscopy (SThM), *Holzforschung* **62** (2008) 91–98.
11. F. Stöckel, J. Konnerth, J. Moser, W. Kantner and W. Gindl, Micromechanical properties and wood cell-wall penetration of pMDI, *Wood Sci. Technol.* (2012), DOI: 10.1007/s00226-011-0432-0.
12. M. Obersriebnig, S. Veigel, W. Gindl-Altmutter and J. Konnerth, Determination of adhesive energy at the wood cell-wall/UF interface by nanoindentation (NI), *Holzforschung* (2012), DOI: 10.1515/hf-2011-0205.
13. J.M. Dinwoodie, Causes of deterioration of UF chipboard under cyclic humidity conditions. 1. Performance of UF adhesive films, *Holzforschung* **31** (1977) 50–55.
14. A.J. Bolton and M.A. Irle, Physical aspects of wood adhesive bond formation with formaldehyde based adhesives. 1. The effect of curing conditions on the physical-properties of urea formaldehyde films, *Holzforschung* **41** (1987) 155–158.
15. J. Konnerth, A. Jaeger, J. Eberhardsteiner, U. Müller and W. Gindl, Elastic properties of adhesive polymers. Part II: polymer films and bond lines by means of nanoindentation, *J. Appl. Polym. Sci.* **102** (2006) 1234–1239.
16. F. Stöckel, J. Konnerth, W. Kantner, J. Moser and W. Gindl W, Mechanical characterisation of adhesives in particle boards by means of nanoindentation, *Eur. J. Wood Prod.* **68** (2010) 421–426.
17. O. Volkersen, Die Nietkraftverteilung in zugbeanspruchten Nietverbindungen mit konstanten Laschenquerschnitten, *Luftfahrtforschung* **15** (1938) 41–47.
18. A.D. Crocombe and I.A. Ashcroft, Single lap joint geometry. In *Modeling of adhesively bonded joints*, L.F.M. Da Silva and A. Öchsner (Eds.), Heidelberg: Springer Berlin (2008).
19. W. Gindl-Altmutter, U. Müller and J. Konnerth, The significance of lap-shear testing of wood adhesive bonds by means of Volkersens shear lag model, *Eur. J. Wood Prod.* (2012), DOI: 10.1007/s00107-012-0621-z.
20. F. Stöckel, J. Konnerth and W. Gindl-Altmutter, Mechanical properties of wood adhesives — A review, *Int. J. Adhes. Adhes.* (2012). Submitted.
21. R.O. Ebewele, G.E. Myers, B.H. River and J.A. Koutsky, Polyamine-modified urea-formaldehyde resins. 1. Synthesis, structure, and properties, *J. Appl. Polym. Sci.* **42** (1991a) 2997–3012.
22. R.O. Ebewele, B.H. River, G.E. Myers and J.A. Koutsky, Polyamine-modified urea-formaldehyde resins. 2. Resistance to stress-induced by moisture cycling of solid wood joints and particleboard, *J. Appl. Polym. Sci.* **43** (1991b) 1483–1490.
23. R.O. Ebewele, B.H. River and G.E. Myers, Polyamine-modified urea-formaldehyde-bonded wood joints. 3. Fracture-toughness and cyclic stress and hydrolysis resistance, *J. Appl. Polym. Sci.* **49** (1993) 229–245.
24. S. Clauß, D.J. Dijkstra, J. Gabriel, O. Kläusler, M. Matner, W. Meckel and P. Niemz, Influence of the chemical structure of PUR prepolymers on thermal stability, *Int. J. Adhes. Adhes.* **31** (2011a) 513–523.

25. S. Clauß, J. Gabriel, A. Karbach, M. Matner and P. Niemz, Influence of the adhesive formulation on the mechanical properties and bonding performance of polyurethane prepolymers, *Holzforschung* **65** (2011b) 835–844.
26. Y. Zhang, B. You, H. Huang, S. Zhou, L. Wu and A. Sharma, Preparation of nanosilica reinforced waterborne silylated polyether adhesive with high shear strength, *J. Appl. Polym. Sci.* **109** (2008) 2434–2441.
27. Z. Ahmad, M.P. Ansell and D. Smedley, Epoxy adhesives modified with nano- and micro-particles for in situ timber bonding: effect of microstructure on bond integrity, *Int. J. Mech. Mater. Eng.* **5** (2010) 59–67.
28. M. May, H.M. Wang and R. Akid, Effects of the addition of inorganic nanoparticles on the adhesive strength of a hybrid solgel epoxy system, *Int. J. Adhes. Adhes.* **30** (2010) 505–512.
29. S. Khoee and N. Hassani, Adhesion strength improvement of epoxy resin reinforced with nanoelastomeric copolymer, *Mater. Sci. Eng. A Struct.* **527** (2010) 6562–6567.
30. K.T. Hsiao, J. Alms and S.G. Advani, Use of epoxy/multiwalled carbon nanotubes as adhesives to join graphite fibre reinforced polymer composites. *Nanotechnology* **14** (2003) 791–793.
31. S.M.R. Khalili, A. Shokuhfar, S.D. Hoseini, M. Bidkhori, S. Khalili and R.K. Mittal, Experimental study of the influence of adhesive reinforcement in lap joints for composite structures subjected to mechanical loads, *Int. J. Adhes. Adhes.* **28** (2008) 436–444.
32. S.M.R. Khalili, M.H. Jafarkarimi and M.A. Abdollahi, Creep analysis of fibre reinforced adhesives in single lap joints-Experimental study, *Int. J. Adhes. Adhes.* **29** (2009) 656–661.
33. S.G. Prolongo, M.R. Gude and A. Urena, Synthesis and characterization of epoxy resins reinforced with carbon nanotubes and nanofibers, *J. Nanosci. Nanotechnol.* **9** (2009) 6181–6187.
34. S.G. Prolongo, M.R. Gude and A. Urena, Rheological behavior of nanoreinforced epoxy adhesives of low electrical resistivity for joining carbon fiber/epoxy laminates, *J. Adhes. Sci. Technol.* **24** (2010) 1097–1112.
35. S.H. Yoon, B.C. Kim, K.H. Lee and D.G. Lee, Improvement of the adhesive fracture toughness of bonded aluminum joints using e-glass fibers at cryogenic temperature, *J. Adhes. Sci. Technol.* **24** (2010) 429–444.
36. R. Zahedsheijani, M. Faezipour, A. Tarmian, M. Layeghi and H. Yousefi, The effect of Na+ montmorillonite (NaMMT) nanoclay on thermal properties of medium density fiberboard (MDF), *Eur. J. Wood Prod.* (2012), DOI: 10.1007/s00107-011-0583-6.
37. H. Lei, G. Du, A. Pizzi and A. Celzard, Influence of nanoclay on urea-formaldehyde resins for wood adhesives and its model, *J. Appl. Polym. Sci.* **109** (2008) 2442–2451.
38. K. Doosthoseini and H. Zarea-Hosseinabadi, Using Na+MMT nanoclay as a secondary filler in plywood manufacturing, *J. Indian Acad. Wood Sci.* **7** (2010) 58–64.
39. A. Salari, T. Tabarsa, A. Khazaeian and A. Saraeian, Improving some of applied properties of oriented strand board (OSB) made from underutilized low quality paulownia (Paulownia fortunie) wood employing nano-SiO2, *Ind. Crop. Prod.* **42** (2013) 1–9.
40. H. Lei, G. Du, A. Pizzi, A. Celzard and Q. Fang, Influence of nanoclay on phenol-formaldehyde and phenol-urea-formaldehyde resins for wood adhesives, *J. Adhes. Sci. Technol.* **24** (2010) 1567–1576.
41. A. Kaboorani and B. Riedl, Effects of adding nano-clay on performance of polyvinyl acetate (PVA) as a wood adhesive, *Compos. Part A* **42** (2011) 1031–1039.
42. S. Wang and C. Xing, Wood adhesives containing reinforced additives for structural engineering products, US Patent, No. 20100285295 (2010).

43. M. Takatani, S. Fukumoto, S. Fujita, T. Yamazaki, R. Hamada, T. Kitayama and T. Okamoto, Performance of phenol-formaldehyde based wood adhesives containing steam exploded beech flour, *Holzforschung* **52** (1998) 651–653.
44. K. Richter, N. Bordeanu, F. López-Suevos and T. Zimmermann, Performance of cellulose nanofibrils in wood adhesives. In *Proc. Swiss Bonding 2009*, Switzerland: Rapperswil (2009), pp. 239–246.
45. F. Lopez-Suevos, C. Eyholzer, N. Bordeanu and K. Richter, DMA analysis and wood bonding of PVAc latex reinforced with cellulose nanofibrils, *Cellulose* **17** (2010) 387–398.
46. A. Kaboorani, B. Riedl, B. Planchet, M. Fellin, O. Hosseinaie and S. Wang, Nanocrystalline cellulose (NCC): a renewable nano-material for polyvinyl acetate (PVA) adhesive, *Eur. Polym. J.* (2012). Available at: http://dx.doi.org/10.1016/j.eurpolymj.2012.08.008
47. E. Serrano and P.J. Gustafsson, Influence of bondline brittleness and defects on the strength of timber finger-joints, *Int. J. Adhes. Adhes.* **19**(1) (1999) 9–17.
48. E. Serrano, A numerical study of the shear-strength-predicting capabilities of test specimens for wood–adhesive bonds, *Int. J. Adhes. Adhes.* **24**(1) (2004) 23–35.
49. U. Müller, A. Sretenovic, A. Vincenti and W. Gindl, Direct measurement of strain distribution along a wood bond line — Part I: Shear strain concentration in a lap joint specimen by means of electronic speckle pattern interferometry, *Holzforschung* **59**(3) (2005) 300–306.
50. S.J. Eichhorn, A. Dufresne, M. Aranguren, N.E. Marcovich, J.R. Capadona, S.J. Rowan, C. Weder, W. Thielemans, M. Roman, S. Renneckar, W. Gindl, S. Veigel, J. Keckes, H. Yano, K. Abe, M. Nogi, A.N. Nakagaito, A. Mangalam, J. Simonsen, A.S. Benight, A. Bismarck, L.A. Berglund and T. Peijs, Review: current international research into cellulose nanofibres and nanocomposites, *J. Mater. Sci.* **45**(1) (2010) 1–33.
51. EN 302-1, *Adhesives for load-bearing timber structures — Test methods — Part 1: determination of bond strength in longitudinal tensile shear strength*, Vienna: Austrian Standards Institute (2004).
52. S. Veigel, U. Müller, J. Keckes, M. Obersriebnig and W. Gindl-Altmutter, Cellulose nanofibrils as filler for adhesives: effect on specific fracture energy of solid wood-adhesive bonds, *Cellulose* **18**(5) (2011) 1227–1237.
53. J.M. Gagliano and C.E. Frazier, Improvements in the fracture cleavage testing of adhesively-bonded wood, *Wood Fiber Sci.* **33**(3) (2001) 377–385.
54. S. Veigel, J. Rathke, M. Weigl and W. Gindl-Altmutter, Particle board and oriented strand board prepared with nanocellulose-reinforced adhesive, *J. Nano. Mater.* (2012), DOI: 10.1155/2012/158503.
55. G. Tschurtschenthaler, Die synthese eines cellulose-nanofibrillenverstärkten harnstoff-formaldehyd-leimes, Master thesis, BOKU Vienna (2012).

Index

D

E